Concrete bridge design to
BS 5400

# Concrete bridge design to BS 5400

## L. A. Clark

Construction Press
London and New York

**Construction Press**
*an imprint of*
**Longman Group Limited**
Longman House, Burnt Mill, Harlow,
Essex CM20 2JE, England
*Associated companies throughout the World*

*Published in the United States of America*
*by Longman Inc., New York*

*First published 1983*

**British Library Cataloguing in Publication Data**
Clark, L.A.
  Concrete bridge design to BS 5400.
  1. Bridges, Concrete – Design and construction
  I. Title
  624'.25    TG335

  ISBN 0-86095-893-0

Set in 10/12 Times Roman (VIP)
Printed in Great Britain by The Pitman Press Ltd., Bath

# To Helen

# Acknowledgements

I thank many of my former colleagues at the Cement and Concrete Association for the contributions which they have indirectly made to this book through the discussions which I had with them. I am particularly indebted to George Somerville and Gordon Elliott who, each in his own particular way, encouraged my interest in concrete bridges. In addition, it would not have been possible for me to write this book without the benefit of the numerous discussions which I have had with bridge engineers throughout the United Kingdom – I am grateful to each of them.

My thanks are due to Peter Thorogood and Jim Church who read parts of the manuscript and made many constructive criticisms; also to Julie Hill who, with a small contribution from Christine Cope, carefully and efficiently typed the manuscript.

Finally, prior to writing this book, I had wondered why it is usual for an author to thank his wife and family – now I know! Thus, I wish to thank my wife and daughters for their patience and understanding during the past three years.

L. A. Clark
June, 1981

## Publisher's acknowledgements

Figures 4.5, 4.6, 4.7, 4.8, 8.4 and 10.8 were originally prepared by the author for the Bridge Engineering Standards Division of the Department of Transport under contract. These figures, together with references to the requirements of the Department of Transport's Design Standards, are reproduced with the permission of the Controller of Her Majesty's Stationery Office. Extracts from British Standards are included by permission of the British Standards Institute, 2 Park Street, London W1A 2BS, from whom complete copies can be obtained.

# Contents

# Preface

During the last decade, limit state design has been introduced, both nationally and internationally, into codes of practice for the design of concrete structures. Limit state design in British codes of practice first appeared in 1972 in the building code (CP 110). Since then it has been used in the water retaining structures code (BS 5337) in 1976, the masonry code (BS 5628) in 1978 and, finally, the bridge code (BS 5400) in 1978. The introduction of limit state design to the design of concrete bridges constitutes a radical change in design philosophy because the existing design documents are written, principally, in terms of a working load and permissible stress design philosophy. Thus, the use of BS 5400 may change design procedures, although it is unlikely to change significantly the final section sizes adopted for concrete bridges. This is due to the fact that the loadings and design criteria were chosen so that, in general, bridges designed to BS 5400 would be similar to bridges designed to the then existing design documents.

In view of the different design methods used in BS 5400, a number of bridge engineers have expressed the need for a document which gives guidance in the use of this code of practice. The present book is an attempt to meet this need; its aim is to give the background to the various clauses of BS 5400, which are concerned with concrete bridges, and to compare them with the corresponding clauses in the existing design documents.

After tracing the history of limit state design and explaining its terminology, the analysis, loading and design aspects of BS 5400 are discussed.

BS 5400 permits the use of plastic methods of analysis. However, bridge engineers have complained that there is a lack of guidance in BS 5400 on the use of plastic methods. Therefore, applications of plastic methods are discussed in Chapter 2. In contrast, the reader is assumed to be familiar with current methods of elastic analysis and so these methods are discussed only briefly. However, the evaluation of elastic stiffnesses for various types of bridge deck is discussed in some detail.

The loadings in BS 5400 differ from those in the existing design documents. The two sets of loadings are compared in Chapter 3, where it can be seen that some loadings differ only slightly whereas others differ significantly.

Compared with those of existing documents, the design criteria of BS 5400, and the methods of satisfying them, are very different for reinforced concrete, but very similar for prestressed concrete. These differences are discussed in Chapters 4 to 12.

Worked examples are given at the ends of most chapters. These examples illustrate the applications of various clauses of BS 5400.

Many bridge engineers have expressed the view that BS 5400 does not deal adequately with certain aspects of concrete bridge design. Thus, in addition to giving the background to the BS 5400 clauses and suggesting interpretations of them in ambiguous situations, this book suggests procedures for those aspects of design which are not covered adequately; e.g. shear in composite construction, transverse shear in voided slabs, and the incorporation of temperature loading into the design procedure.

It is hoped that this book will assist practising concrete bridge engineers in interpreting and applying BS 5400. Also it is hoped that it will be of use to undergraduate and postgraduate students taking courses in bridge engineering.

L. A. Clark
June 1981

# Notation

The principal symbols used in this book are as follows. Other symbols are defined in the text.

| | |
|---|---|
| $A_c$ | area of concrete |
| $A_{cf}$ | area of flange of composite beam |
| $A_{ps}$ | area of tendon |
| $A_s$ | area of tension reinforcement |
| $A_s'$ | area of compression reinforcement in beam |
| $A_{sc}$ | area of reinforcement in column |
| $A_{sL}$ | area of longitudinal torsion reinforcement |
| $A_{sv}$ | area of shear reinforcement |
| $A_t$ | area of transverse reinforcement in flange |
| $A_o$ | area within median line of box |
| $a$ | span; acceleration |
| $a'$ | distance measured from compression face of beam |
| $a_b$ | bar spacing |
| $a_{cent}$ | distance between centroids of compressive flange and of composite section |
| $a_{cr}$ | perpendicular distance from crack |
| $a_v$ | shear span |
| $b$ | breadth |
| $b_e$ | width of interface in composite section |
| $C$ | torsional inertia; compressive force; coefficient |
| $C_D$ | drag coefficient |
| $C_L$ | lift coefficient |
| $c$ | cover |
| $c_{min}$ | minimum cover |
| $D$ | internal dissipation of energy |
| $D_c$ | density of concrete |
| $D_x, D_y, D_{xy}, D_1$ | plate bending stiffnesses per unit length |
| $d$ | effective depth; void diameter |
| $d'$ | depth to compression reinforcement in beam |
| $d_c$ | depth of concrete in compression |
| $d_t$ | effective depth in shear |
| $d_o$ | effective depth of half end |
| $E$ | elastic modulus; work done by external loads |
| $E_c$ | elastic modulus of concrete |
| $E_{cf}$ | elastic modulus of flange of composite beam |
| $E_s$ | elastic modulus of steel |
| $e$ | eccentricity |
| $e_i$ | initial column eccentricity |
| $e_{add}$ | additional column eccentricity |
| $F$ | force |

| | |
|---|---|
| $F_{bst}$ | bursting force |
| $F_{bt}$ | tensile force in bar at ultimate limit state |
| $F_c$ | concrete force; centrifugal force |
| $F_s$ | steel force |
| $F_s'$ | force in compression reinforcement |
| $F_t$ | tie force |
| $f$ | stress |
| $f_{ave}$ | average compressive stress in end block |
| $f_b$ | bearing stress |
| $f_{ba}$ | average anchorage bond stress |
| $f_{bs}$ | local bond stress |
| $f_{ci}$ | concrete strength at transfer |
| $f_{cm}$ | average concrete tensile stress between cracks |
| $f_{cp}$ | compressive stress due to prestress |
| $f_{cu}$ | characteristic strength of concrete |
| $f_{cyl}$ | cylinder compressive strength of concrete |
| $f_{ht}$ | hypothetical tensile stress |
| $f_k$ | characteristic strength |
| $f_{pb}$ | tendon stress at failure |
| $f_{pd}$ | design stress of tendon when used as torsion reinforcement |
| $f_{pe}$ | effective prestress |
| $f_{pt}$ | tensile stress due to prestress at an extreme concrete fibre |
| $f_{pu}$ | characteristic strength of tendon |
| $f_r$ | flexural strength (modulus of rupture) of concrete |
| $f_s$ | shear stress |
| $f_{scr}$ | steel stress at a crack at cracking load |
| $f_t$ | design tensile strength of concrete |
| $f_{tm}$ | maximum tensile stress in end block |
| $f_{tp}$ | permissible concrete tensile stress in end block |
| $f_y$ | characteristic strength of reinforcement |
| $f_{yL}$ | characteristic strength of longitudinal torsion reinforcement |
| $f_{yv}$ | characteristic strength of link reinforcement |
| $f_o$ | fundamental natural frequency of unloaded bridge |
| $f_1$ | steel stress ignoring tension stiffening |
| $G$ | shear modulus |
| $H$ | depth of back-fill |
| $h$ | overall depth or thickness |
| $h_b$ | bottom flange thickness |
| $h_e$ | lever arm of cellular slab |
| $h_f$ | T-beam flange thickness |
| $h_{min}, h_{max}$ | minimum and maximum dimensions of rectangle |

| Symbol | Definition |
|---|---|
| $h_t$ | top flange thickness |
| $h_w$ | web thickness |
| $h_{wo}$ | box wall thickness |
| $I$ | second moment of area |
| $I_f$ | second moment of area of flange |
| $I_w$ | second moment of area of web |
| $I_x$ | second moment of area of longitudinal section |
| $I_y$ | second moment of area of transverse section |
| $J_x$ | longitudinal torsional inertia |
| $J_y$ | transverse torsional inertia |
| $K$ | coefficient; factor |
| $K_f$ | flange stiffness |
| $K_w$ | web stiffness |
| $k$ | torsional constant; factor |
| $L$ | span; length |
| $L_t$ | transmission length in box girder |
| $l$ | span; length |
| $l_e, l_{ex}, l_{ey}$ | effective height of column |
| $l_{sb}$ | anchorage length at half joint |
| $l_o$ | clear height of column |
| $l_1$ | length of side span |
| $M$ | moment |
| $M_{add}$ | additional column moment |
| $M_i$ | initial column moment |
| $M_t$ | total column moment; cracking moment |
| $M_x, M_y, M_{xy}$ | plate bending and twisting moments per unit length |
| $M_x^*, M_y^*, M_\alpha^*$ | plate moments of resistance per unit length |
| $M_u$ | moment of resistance of beam or column |
| $M_V$ | Vierendeel bending moment |
| $M_o$ | moment to produce zero stress at level of steel of prestressed beam |
| $M_1, M_2$ | larger and smaller column end moments |
| $m$ | plate yield moment per unit length |
| $m_c$ | corner moment per unit length |
| $m_n$ | normal moment of resistance per unit length of yield line |
| $N$ | axial load; number of cycles of stress to cause fatigue failure |
| $N_c$ | concrete force |
| $N_s$ | steel force; axial load capacity of column at serviceability limit state |
| $N_{uz}$ | axial load capacity of column at ultimate limit state |
| $N_x, N_y, N_{xy}$ | plate in-plane forces per unit length |
| $N_x^*, N_y^*, N_\alpha^*$ | plate resistive forces per unit length |
| $n_w$ | maximum wall load per unit length |
| $P$ | point load; prestressing force |
| $P_k$ | total initial prestressing force; maximum tendon force |
| $P_f$ | effective prestressing force |
| $P_L$ | longitudinal wind load |
| $P_{LL}$ | longitudinal wind load on live load |
| $P_{Ls}$ | longitudinal wind load on superstructure |
| $P_t$ | transverse wind load |
| $P_v$ | vertical wind load |
| $P_x$ | prestressing force at distance $x$ from jack |
| $P_o$ | prestressing force at jack |
| $p$ | uniformly distributed load |
| $Q$ | load |
| $Q^*$ | design load |
| $Q_k$ | characteristic load |
| $Q_x, Q_y$ | plate shear forces per unit length |
| $q$ | uniformly distributed load |
| $R$ | reaction |
| $R^*$ | design resistance |
| $r$ | radius |
| $r_{ps}$ | radius of curvature of duct |
| $S^*$ | design load effect |
| $S_c$ | first moment of area of flange in composite construction |
| $S_x, S_y$ | plate shear stiffnesses per unit length |
| $S_1$ | funnelling factor |
| $S_2$ | gust factor |
| $s$ | spacing |
| $T$ | tension force; torque; time |
| $t$ | thickness; time |
| $t_z$ | temperature at distance $z$ above soffit |
| $u_{crit}$ | length of punching shear critical perimeter |
| $u_s$ | circumference of bar |
| $V$ | shear force |
| $V_c$ | shear force carried by concrete |
| $V_{cr}$ | shear force carried by concrete in flexurally cracked prestressed beam |
| $V_{co}$ | shear force to cause web cracking |
| $V_u$ | maximum allowable shear force |
| $v$ | shear stress; mean hourly wind speed |
| $v_c$ | allowable shear stress; maximum wind gust speed |
| $v_h$ | interface shear stress |
| $v_t$ | torsional shear stress |
| $v_{tmin}$ | value of torsional shear stress above which torsion reinforcement is required |
| $v_{tu}$ | maximum allowable torsional shear stress |
| $W$ | load |
| $w$ | displacement; crack width |
| $w_f, w_w$ | flange and web warping forces per unit length in box girder |
| $w_m$ | mean crack width |
| $x$ | neutral axis depth |
| $x, y, z$ | rectangular co-ordinates |
| $x_1, y_1$ | link dimensions |
| $y$ | parameter defining yield line pattern |
| $y_{po}$ | half length of side of loaded area |
| $y_s$ | static deflection |
| $y_o$ | half length of side of resisting concrete block |
| $z$ | lever arm |
| $\bar{z}$ | distance of section centroid from soffit |
| $\alpha$ | angle; Hillerborg load proportion; parameter defining yield line pattern |
| $\alpha_c$ | coefficient of expansion of concrete |
| $\alpha_n$ | index used in biaxial bending of column |
| $\alpha_s$ | coefficient of expansion of steel |
| $\alpha_z$ | coefficient of expansion at distance $z$ from soffit |
| $\beta$ | percentage redistribution; creep factor; angle |
| $\beta_{cc}$ | creep factor |
| $\gamma_{f1}, \gamma_{f2}, \gamma_{fL}$ | partial safety factors applied to loads |
| $\gamma_{f3}$ | partial safety factor applied to load effects |
| $\gamma_g$ | gap factor |
| $\gamma_{m1}, \gamma_{m2}, \gamma_m$ | partial safety factors applied to material strengths |
| $\gamma_{n1}, \gamma_{n2}, \gamma_n$ | consequence factors |
| $\gamma_x, \gamma_y$ | plate shear strains |
| $\delta$ | deflection; stress transformation factor; logarithmic decrement |
| $\varepsilon$ | strain |
| $\bar{\varepsilon}$ | strain at section centroid |
| $\varepsilon_c$ | creep strain |
| $\varepsilon_{cs}$ | free shrinkage strain |

| | | | |
|---|---|---|---|
| $\varepsilon_{diff}$ | differential shrinkage strain | $\lambda$ | moment reduction factor |
| $\varepsilon_m$ | strain allowing for tension stiffening | $\lambda_w$ | stress block factor for plain concrete walls |
| $\varepsilon_n, \varepsilon_t, \gamma_{nt}$ | direct and shear strains | $\mu$ | coefficient of friction |
| $\varepsilon_s$ | steel design strain; shrinkage strain | $\nu$ | Poisson's ratio |
| $\varepsilon_{ts}$ | tension stiffening strain | $\rho$ | density |
| $\varepsilon_u$ | ultimate concrete strain | $\rho_o$ | shrinkage coefficient |
| $\varepsilon_y$ | yield strain of reinforcement | $\sigma_r$ | stress range |
| $\varepsilon_o$ | soffit strain | $\sigma_H$ | limiting stress range |
| $\varepsilon_1$ | strain ignoring tension stiffening | $\phi$ | bar diameter; creep coefficient; angle |
| $\xi_s$ | depth of slab factor | $\phi_1$ | creep coefficient |
| $\theta$ | rotation; angle | $\psi$ | slope; curvature; dynamic response factor |
| $\theta_n$ | normal rotation in yield line | $\psi_s$ | shrinkage curvature |
| $\theta_t$ | thermal rotation | $\psi_u$ | maximum column curvature at collapse |

*Chapter 1*

# Introduction

## The New Code

### Background

Rules for the design of bridges have been the subject of continuous amendment and development over the years, and a significant development took place in 1967. At that time, a meeting was held to discuss the revision of British Standard BS 153 [1], on which many bridge design documents were based [2]. It was suggested that a unified code of practice should be written in terms of limit state design which would cover steel, concrete and composite steel-concrete bridges of any span. A number of sub-committees were then formed to draft various sections of such a code; the work of these sub-committees has culminated in British Standard 5400 which will, henceforth, be referred to in this book as the Code.

The author understands that the Code Committee did not intend to produce documents which would result in significant changes in design practice but, rather, intended that bridges designed to the Code would be broadly similar to those designed to the then current documents. In addition the sub-committees concerned with the various materials and types of bridges had to produce documents which would be compatible with each other.

In subsequent chapters, the background to the Code is given in detail and suggestions made as to its interpretation in practice. The remainder of this first chapter is concerned with general aspects of the Code.

### Code format

The Code consists of the ten parts listed in Table 1.1 and, at the time of writing, all except Parts 3 and 9 have been published: drafts of these parts are available. Hence, sufficient documents have been published to design concrete bridges. It should be noted that BS 5400 is both a Code of Practice and a Specification. However, not all aspects of the design and construction of bridges are covered; exceptions worthy of mention are the design of

**Table 1.1**  BS 5400 – the ten parts

| Part | Contents |
|------|----------|
| 1 | General statement |
| 2 | Specification for loads |
| 3 | Code of practice for design of steel bridges |
| 4 | Code of practice for design of concrete bridges |
| 5 | Code of practice for design of composite bridges |
| 6 | Specification for materials and workmanship, steel |
| 7 | Specification for materials and workmanship, concrete, reinforcement and prestressing tendons |
| 8 | Recommendations for materials and workmanship, concrete, reinforcement and prestressing tendons |
| 9 | Code of practice for bearings |
| 10 | Code of practice for fatigue |

parapets and such constructional aspects as expansion joints and waterproofing.

The contents of the individual parts are now summarised.

### Part 1

The philosophy of limit state design is presented and the methods of analysis which may be adopted are stated in general terms.

### Part 2

Details are given of the loads to be considered for *all* types of bridges, the partial safety factors to be applied to each load and the load combinations to be adopted.

### Part 3

Design rules for steel bridges are given but reference is not made to Part 3 in this book. At the time of writing it is in draft form.

### Part 4

Design rules for reinforced, prestressed and composite (precast plus in-situ) concrete bridges are given in terms of material properties, design criteria and methods of compliance.

## Part 5

Design rules for steel–concrete composite bridges are given and some of these are referred to in this book.

## Part 6

The specification of materials and workmanship in connection with structural steelwork are given, but reference is not made to Part 6 in this book.

## Part 7

The specification of materials and workmanship in connection with concrete, reinforcement and prestressing tendons is given.

## Part 8

Recommendations are given for the application of Part 7.

## Part 9

The design, testing and specification of bridge bearings are covered. At the time of writing Part 9 is in draft form but some material on bearings is included in Part 2 as an appendix which will eventually be superseded by Part 9.

## Part 10

Loadings for fatigue calculations and methods of assessing fatigue life are given. Part 10 is concerned mainly with steel and steel–concrete composite bridges but some sections are referred to in this book.

## Implementation of Code for concrete bridges

### Highway bridges

From the previous discussion, it can be seen that, if the Code were to be adopted for concrete highway bridges:

1. Part 2 would replace the Department of Transport's Technical Memorandum BE 1/77 [3] and British Standard BS 153 Part 3A [4].
2. Part 4 would replace the Department of Transport's Technical Memoranda BE 1/73 [5] and BE 2/73 [6] and Codes of Practice CP 114 [7], CP 115 [8] and CP 116 [9].
3. Parts 7 and 8 would replace the Department of Transport's Specification for road and bridge works [10].
4. Part 9 would replace the Department of Transport's Memoranda BE 1/76 [11] and IM 11 [12].

At the time of writing, the Department of Transport's views on the implementation of the Code are summarised in their Departmental Standard BD 1/78 [13]. This, essentially, states that the Code will, in due course, be supplemented by Departmental design and specification requirements. In addition, implementation is to be phased over an unstated period of time, with an initial stage of trial applications of the Code to selected schemes. However, it should be noted that several of the Department's Technical Memoranda have been updated to incorporate material from the Code both during the Code's drafting stages and since publication.

### Railway bridges

There should be fewer problems in implementing the Code for railway bridges than for highway bridges, because British Rail have been using limit state design since 1974 [14].

# Development of design standards for concrete structures

Before explaining the philosophy of limit state design it is instructive to consider current design procedures for concrete bridges and to examine the trends that have taken place in the development of codes of practice for concrete structures in general.

Current design procedures for concrete bridges are based primarily on the requirements of a series of Technical Memoranda issued by the Department of Transport (e.g. BE 1/73, BE 2/73 and BE 1/77); these in turn are based on current Codes of Practice for buildings (e.g. CP 114, CP 115 and CP 116), with some important modifications which reflect problems peculiar to bridges.

Essentially, trial structures are analysed elastically to determine maximum values of effects due to specified working loads. Critical sections are then designed on a modular ratio basis to ensure that certain specified stress limitations for both steel and concrete are not exceeded. Thus, the approach is basically one of working loads and permissible stresses, although there are also requirements to check crack widths in reinforced concrete structures and to check the ultimate strength of prestressed concrete structures. This design process has three distinguishing features – it is based on a permissible working stress philosophy, it assumes elastic material properties and it is deterministic. Each of these features will now be discussed with reference to Table 1.2 which summarises the basic requirements of the various structural concrete building codes since 1934.

## Permissible stresses

The permissible working stress design equation is:
stress due to working load ≤ permissible working stress
where,

$$\text{permissible working stress} = \frac{\text{material 'failure' stress}}{\text{safety factor}}$$

Thus stresses are limited at the *working load* essentially to provide an adequate margin of safety against *failure*. Such a design approach was perfectly adequate whilst material strengths were low and the safety factor high because the permissible working stresses were sufficiently low for serviceability considerations (deflections and cracking) not to be critical. In this respect the most important consideration is that of the permissible steel stress and Table 1.2 shows

**Table 1.2** Summary of basic requirements from various Codes of Practice for structural concrete

| Code | Basis of analysis and design | Steel stress limitation (N/mm$^2$) | Required load factor | Additional design requirements |
|---|---|---|---|---|
| DSIR (1934) | Elastic analysis, with variable modular ratio and permissible stresses | 140 for beams 100 for columns $f \not> 0.45 \, fy$ | None for beams 3.0 for columns | None |
| CP 114 (1948) | As above, but m = 15 | 190 in tension 140 in compression $f \not> 0.50 \, fy$ | For columns: 2.0 for steel 2.6 for concrete | Warning against excessive deflections |
| CP 114 (1957) | Either elastic analysis *or* load factor method | 210 in tension 160 in compression $f \not> 0.50 \, fy$ | 2.0 for steel 2.6 for concrete | Span/depth ratios given for beams and slabs. Warning against cracking |
| CP 115 (1959) | Both elastic and ultimate load methods required | Cracking avoided by limiting concrete tension | 1.5D + 2.5L or 2 (D + L) | Warning against excessive deflections |
| CP 114 (as amended 1965) | Either elastic analysis or load factor method | 230 in tension 170 in compression $f \not> 0.55 \, fy$ | 1.8 for steel 2.3 for concrete | More detailed span/depth ratios for deflection. Warning against cracking |
| CP 110 (1972) | Limit state design methods | No direct limit set, except by cracking and deflection requirements | 1.6–1.8 for steel 2.1–2.4 for concrete | Detailed span/depths or calculations for deflection. Specific calculations for crack width required |

that the ratio of permissible steel stress to steel yield strength has gradually increased over the years (i.e. the safety factor has decreased). This, combined with the introduction of high-strength reinforcement, has meant that permissible steel stresses have risen to a level at which the serviceability aspects of design have now to be considered specifically. Table 1.2 shows that, as permissible steel stresses have increased, more attention has been given in building codes to deflection and cracking. In bridge design, the consideration of the serviceability aspects of design was reflected in the introduction of specific crack control requirements in the Department of Transport documents. It can thus be seen that the original simplicity of the permissible working stress design philosophy has been lost by the necessity to carry out further calculations at the working load. Moreover, and of more concern, the working stress designer is now in a position in which he is using a design process in which the purposes of the various criteria are far from self evident.

## Elastic material behaviour

It has long been recognised that steel and concrete exhibit behaviour of a plastic nature at high stresses. Such behaviour exposes undesirable features of working stress design: beams designed on a working stress basis with identical factors of safety applied to the stresses, but with different steel percentages, have different factors of safety against failure, and the capacity of an indeterminate structure to redistribute moments cannot be utilised if its plastic properties are ignored. However, it was not until 1957, with the introduction of the load factor method of design in CP 114, that the plastic properties of materials were recognised, for all structural members, albeit disguised in a working stress format. The concept of considering the elastic response of a structure at its working load and its plastic response at the ultimate load was first codified in the

prestressed concrete code (CP 115) of 1959; and this code may be regarded as the first British limit state design code.

## Deterministic design

The existing design procedure is deterministic in that it is implicitly assumed that it is possible to categorically state that, under a specified loading condition, the stresses in the materials, at certain points of the structure, will be of uniquely calculable values. It is obvious that, due to the inherent variabilities of both loads and material properties, it is not possible to be deterministic and that a probabilistic approach to design is necessary. Statistical methods were introduced into CP 115 in 1959 to deal with the control of concrete quality, but were not directly involved in the design process.

## Limit state design

The implication of the above developments is that it has been necessary:

1. To consider more than one aspect of design (e.g. strength, deflections and cracking).
2. To treat each of these aspects separately.
3. To consider the variable nature of loads and material properties.

The latest building code, CP 110 [15], which introduced limit state design in combination with characteristic values and partial safety factors, was the culmination of these trends and developments. CP 110 made it possible to treat each aspect of design separately and logically, and to recognise the inherent variability of both loads and material properties in a more formal way. Although, with its introduction into British design practice in CP 110 in 1972, limit state design was considered as a revolutionary design approach, it could also be regarded as the formal recognition of trends which have been developing since the first national code was written.

Generally, design standards for concrete bridges have tended to follow, either explicitly or with a slightly conservative approach, the trends in the building codes. This is also the case with BS 5400 Part 4 which, while written in terms of limit state design and based substantially on CP 110, exhibits some modifications introduced to meet the particular requirements of bridge structures.

# Philosophy of limit state design

## What is limit state design?

Limit state design is a design process which aims to ensure that the structure being designed will not become unfit for the use for which it is required during its design life.

The structure may reach a condition at which it becomes unfit for use for one of many reasons (e.g. collapse or excessive cracking) and each of these conditions is referred to as a limit state. In limit state design each limit state is examined separately in order to check that it is not attained. Assessment of whether a limit state is attained could be made on a deterministic or a probabilistic basis. In CP 110 and the Code, a probabilistic basis is adopted and, thus, each limit state is examined in order to check whether there is an *acceptable probability* of it not being achieved. Different 'acceptable probabilities' are associated with the different limit states, but no attempt is made to quantify these in the Code; in fact, the partial safety factors and design criteria, which are discussed later, are chosen to give similar levels of safety and serviceability to those obtained at present. However, typical levels of risk in the design life of a structure are taken to be $10^{-6}$ against collapse and $10^{-2}$ against unserviceability occurring. Thus the chance of collapse occurring is made remote and much less than the chance of the serviceability limit state being reached.

Limit state design principles have been agreed internationally and set out in International Standard ISO 2394 [16]; this document forms the basis of the limit state design philosophy of BS 5400 which is presented in Part 1 of the Code and is now explained.

## Limit states

As implied previously, a limit state is a condition beyond which a structure, or a part of a structure, would become less than completely fit for its intended use. Two limit states are considered in the Code.

### Ultimate limit state

This corresponds to the maximum load-carrying capacity of the structure or a section of the structure, and could be attained by:

1. Loss of equilibrium when a part or the whole of the structure is considered as a rigid body.
2. A section of the structure or the whole of the structure reaching its ultimate strength in terms of post-elastic or post-buckling behaviour.
3. Fatigue failure. However, in Chapter 12, it can be seen that fatigue is considered not under ultimate loads but under a loading similar to that at the serviceability limit state.

### Serviceability limit state

This denotes a condition beyond which a loss of utility or cause for public concern may be expected, and remedial action required. For concrete bridges the serviceability limit state is, essentially, concerned with crack control and stress limitations. In addition, the serviceability limit state is concerned with the vibrations of footbridges; this aspect is discussed in Chapter 12.

## Design life

This is defined in Part 1 of the Code as 120 years. However, the Code emphasises that this does not necessarily mean that a bridge designed in accordance with it will no longer be fit for its purpose after 120 years, nor that it will continue to be serviceable for that length of time, without adequate and regular inspection and maintenance.

## Characteristic and nominal loads

It is usual in limit state design to define loads in terms of their characteristic values, which are defined as those loads with a 5% chance of being exceeded, as illustrated in Fig. 1.1(a). However, for bridges, the statistical data required to derive the characteristic values are not available for all loads; thus, the loads are defined in terms of nominal values. These have been selected on the basis of the existing data and are, in fact, very similar, to the loads in use at the time of writing the Code. For certain bridge loads, such as wind loads, statistical distributions are available; for these a return period of 120 years has been adopted in deriving the nominal loads, since 120 years is the design life specified in the Code.

It is emphasised that the term 'nominal load' is used in the Code for all loads whether they are derived from statistical distributions or based on experience. Values of the nominal loads are assigned the general symbol $Q_k$. They are given in Part 2 of the Code because they are appropriate to all types of bridges.

## Characteristic strengths

The characteristic strength of a material is defined as that strength with a 95% chance of being exceeded (see Fig. 1.1(b)). Since statistical data concerning material properties are generally available, characteristic strengths

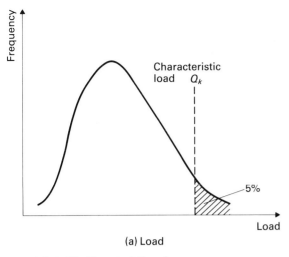

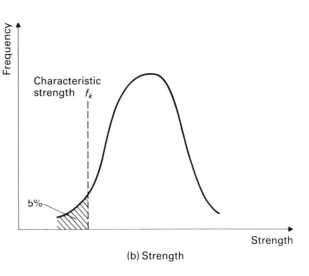

**Fig. 1.1(a),(b)** Characteristic values

can be obtained and this term is thus adopted in the Code. Characteristic strengths are assigned the general symbol $f_k$ and are given, for concrete bridges, in Part 4 of the Code.

## Design loads

At each limit state, a design load is obtained from each nominal load by multiplying the latter by a partial safety factor ($\gamma_{fL}$). The design load ($Q^*$) is thus obtained from

$$Q^* = \gamma_{fL} Q_k \qquad (1.1)$$

The partial safety factor, $\gamma_{fL}$, is a function of two other partial safety factors:

$\gamma_{f1}$, which takes account of the possibility of unfavourable deviation of the loads from their nominal values;

$\gamma_{f2}$, which takes account of the reduced probability that various loadings acting together will all attain their nominal values simultaneously.

It is emphasised that values of $\gamma_{f1}$ and $\gamma_{f2}$ are not given in the Code, but values of $\gamma_{fL}$ are given in Part 2 of the Code. They appear in Part 2 because they are applicable to all bridges; they are discussed in Chapter 3. It should be stated here that the value of $\gamma_{fL}$ is dependent upon a number of factors:

1. Type of loading: it is obviously greater for a highly variable loading such as vehicle loading than for a reasonably well controlled loading such as dead load. This is because in the former case there is a greater chance of an unfavourable deviation from the nominal value.

2. Number of loadings acting together: the value for a particular load decreases as the number of other loads acting with the load under consideration increases. This is because of the reduced probability of all of the loads attaining their nominal values simultaneously.

3. Importance of the limit state: the value for a particular load is greater when considering the ultimate limit state than when considering the serviceability limit state because it is necessary to have a smaller probability of the former being reached.

## Design load effects

The design load effects are the moments, shears, etc., which must be resisted at a particular limit state. They are obtained from the effects of the design loads by multiplying by a partial safety factor $\gamma_{f3}$. The design load effects ($S^*$) are thus obtained from

$$S^* = \gamma_{f3} \text{ (effects of } Q^*\text{)}$$
$$= \gamma_{f3} \text{ (effects of } \gamma_{fL} Q_k\text{)} \qquad (1.2)$$

If linear relationships can be assumed between load and load effects, the design load effects can be determined from

$$S^* = \text{(effects of } \gamma_{f3} \gamma_{fL} Q_k\text{)} \qquad (1.3)$$

It can be seen from Fig. 1.2 that equations (1.2) and (1.3) give the same value of $S^*$ when the relationship between load and load effect is linear, but not when it is non-linear. In the latter case, the point in the design process, at which $\gamma_{f3}$ is introduced, influences the final value of $S^*$.

As is discussed in Chapter 3, elastic analysis will generally continue to be used for concrete bridge design, and thus equation (1.3) will very often be the one used.

The partial safety factor, $\gamma_{f3}$, takes account of any inaccurate assessment of the effects of loading, unforeseen stress distribution in the structure and variations in dimensional accuracy achieved in construction.

Values of $\gamma_{f3}$ are dependent upon the material of the bridge and, for concrete bridges, are given in Part 4 of the Code. The numerical values are discussed in Chapter 4. In addition to the material of the bridge, values of $\gamma_{f3}$ are dependent upon:

1. Type of loading: a lower value is used for an essentially uniformly distributed type of loading (such as dead load) than for a concentrated loading because the effects of the latter can be analysed less accurately.

2. Method of analysis: it is logical to adopt a larger value for an analysis which is known to be inaccurate or unsafe, than for an analysis which is known to be highly accurate or conservative.

3. Importance of the limit state: the consequences of the effects for which $\gamma_{f3}$ is intended to allow are more

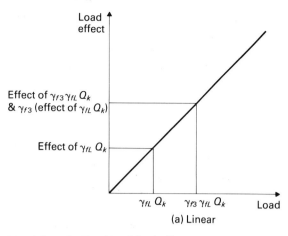

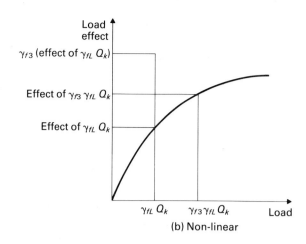

(a) Linear

(b) Non-linear

**Fig. 1.2(a),(b)** Loads and load effects

important at the ultimate than the serviceability limit state and thus a larger value should be adopted for the former.

It should be stated that the concept of using $\gamma_{f3}$ can create problems in design. The use of $\gamma_{f3}$, applied as a general multiplier to load effects to allow for analysis accuracy, has been criticised by Beeby and Taylor [17]. They argue, from considerations of framed structures, that this concept is not defensible on logical grounds, since:

1. For determinate structures there is no inaccuracy.
2. For many indeterminate structures, errors in analysis are adequately covered by the capability of the structure to redistribute moment by virtue of its ductility and hence $\gamma_{f3}$ should be unity.
3. Parts of certain indeterminate structures (e.g. columns in frames) have limited ductility and thus limited scope for redistribution. This means large errors in analysis can arise, and $\gamma_{f3}$ should be much larger than the suggested value of about 1.15 discussed in Chapter 4.
4. There are structures where errors in analysis will lead to moment requirements in an opposite sense to that indicated in analysis. For example, consider the beam of Fig. 1.3: at the support section it is logical to apply $\gamma_{f3}$ to the calculated bending moment, but at section X–X the calculated moment is zero and $\gamma_{f3}$ will have zero effect. In addition, at section Y–Y, where a provision for a *hogging* moment is required, the application of $\gamma_{f3}$ will merely increase the calculated *sagging* moment.

The above points were derived from considerations of framed building structures, but are equally applicable to bridge structures. In bridge design, the problems are further complicated by the fact that, whereas in building design complete spans are loaded, in bridge design positive or negative parts of influence lines are loaded: thus if the influence line is not the 'true' line then the problem discussed in paragraph 4 above is exacerbated, because the designer is not even sure that he has the correct amount of load on the bridge.

In view of these problems it seems sensible, in practice, to look upon $\gamma_{f3}$ merely as a means of raising the global load factor from $\gamma_{fL} \gamma_m$ to an acceptably higher value of

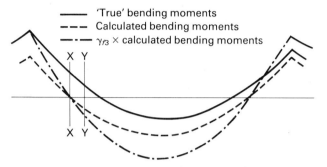

— 'True' bending moments
--- Calculated bending moments
—·— $\gamma_{f3} \times$ calculated bending moments

**Fig. 1.3** Influence of $\gamma_{f3}$ on continuous beams

$\gamma_{f3} \gamma_{fL} \gamma_m$ ($\gamma_m$ is defined in the next section). Indeed, in early drafts of the Code, $\gamma_{f3}$ was called $\gamma_g$ (the gap factor) and Henderson, Burt and Goodearl [18] have stated that the latter was 'not statistical but intended to give a margin of safety for the extreme circumstances where the lowest strength may coincide with the most unlikely severity of loading'. However, it could be argued that $\gamma_g$ was also required for another reason. The $\gamma_{fL}$ values are the same for all bridges, and the $\gamma_m$ values for a particular material, which are discussed in the next section, are the same, irrespective of whether that material is used in a bridge of steel, concrete or composite construction. An additional requirement is that, for each type of construction, designs in accordance with the Code and in accordance with the existing documents should be similar. Hence, it is necessary to introduce an additional partial safety factor ($\gamma_g$ or $\gamma_{f3}$) which is a function of the type of construction (steel, concrete or composite).

Thus $\gamma_{f3}$ has had a rather confusing and debatable history!

## Design strength of a material

At each limit state, design strengths are obtained from the characteristic strengths by dividing by a partial safety factor ($\gamma_m$):

$$\text{design strength} = f_k/\gamma_m \qquad (1.4)$$

The partial safety factor, $\gamma_m$, is a function of two other partial safety factors:

$\gamma_{m1}$, which covers the possible reductions in the strength of the materials in the structure as a whole as compared with the characteristic value deduced from the control test specimens;

$\gamma_{m2}$, which covers possible weaknesses of the structure arising from any cause other than the reduction in the strength of the materials allowed for in $\gamma_{m1}$, including manufacturing tolerances.

It is emphasised that individual values of $\gamma_{m1}$ and $\gamma_{m2}$ are not given in the Code but that values of $\gamma_m$, for concrete bridges, are given in Part 4 of the Code; they are discussed in Chapter 4. The values of $\gamma_m$ are dependent upon:

1.  Material: concrete is a more variable material than steel and thus has a greater $\gamma_m$ value.
2.  Importance of limit state: greater values are used at the ultimate than at the serviceability limit state, because it is necessary to have a smaller probability of the former being reached.

## Design resistance of a structure or a structural element

The design resistance of a structure at a particular limit state is the maximum load that the structure can resist without exceeding the design criteria appropriate to that limit state. For example, the design resistance of a structure could be the load to cause collapse of the structure, or to cause a crack width in excess of the allowable value at a point on the structure.

Similarly, the design resistance of a structural element is the maximum effect that the element can resist without exceeding the design criteria. In the case of a beam, for example, it could be the ultimate moment of resistance, or the moment which causes a stress in excess of that allowed.

The design resistance ($R^*$) is obviously a function of the characteristic strengths ($f_k$) of the materials and of the partial safety factors ($\gamma_m$):

$$R^* = \text{function } (f_k/\gamma_m) \qquad (1.5)$$

As an example, when considering the ultimate moment of resistance ($M_u$) of a beam

$$R^* = M_u \qquad (1.6)$$

and (see Chapter 5)

$$\text{function } (f_k/\gamma_m) = (f_y/\gamma_{ms})A_s \left( d - \frac{\alpha(f_y/\gamma_{ms})A_s}{(f_{cu}/\gamma_{mc})b} \right) \qquad (1.7)$$

where $f_y, f_{cu}$ = $f_k$ of steel and concrete, respectively
$\gamma_{ms}, \gamma_{mc}$ = $\gamma_m$ of steel and concrete, respectively
$A_s$ = steel area
$b$ = beam breadth
$d$ = beam effective depth
$\alpha$ = concrete stress block parameter

However, in some situations the design resistance is calculated from

$$R^* = [\text{function } (f_k)]/\gamma_m \qquad (1.8)$$

where $\gamma_m$ is now a partial safety factor applied to the resistance (e.g. shear strength) appropriate to characteristic

strengths. In such situations either values of $R^*$ or values of the function of $f_k$ are given in the Code. An example is the treatment of shear, which is discussed fully in Chapter 6: $R^*$ values for various values of $f_k$ are tabulated as allowable shear stresses.

## Consequence factor

In addition to the partial safety factors $\gamma_{fL}$, $\gamma_{f3}$ and $\gamma_m$, which are applied to the loads, load effects and material properties, there is another partial safety factor ($\gamma_n$) which is mentioned in Part 1 of the Code.

$\gamma_n$ is a function of two other partial safety factors:
$\gamma_{n1}$, which allows for the nature of the structure and its behaviour;
$\gamma_{n2}$, which allows for the social and economic consequences of failure.

Logically, $\gamma_{n1}$ should be greater when failure occurs suddenly, such as by shear or by buckling, than when it occurs gradually, such as in a ductile flexural failure. However, it is not necessary for a designer to consider $\gamma_{n1}$ when using the Code because, when necessary, it has been included in the derivation of the $\gamma_m$ values or of the functions of $f_k$ used to obtain the design resistances $R^*$.

Regarding $\gamma_{n2}$, the consequence of failure of one large bridge would be greater than that of one small bridge and hence $\gamma_{n2}$ should be larger for the former. However, the Code does not require a designer to apply $\gamma_{n2}$ values: it argues that the total consequences of failure are the same whether the bridge is large or small, because a greater number of smaller bridges are constructed. Thus, it is assumed that, for the sum of the consequences, the risks are broadly the same.

Hence, to summarise, neither $\gamma_{n1}$ nor $\gamma_{n2}$ need be considered when using the Code.

## Verification of structural adequacy

For a satisfactory design it is necessary to check that the design resistance exceeds the design load effects:

$$R^* \geqslant S^* \qquad (1.9)$$
$$\text{or function } (f_k, \gamma_m) \geqslant \text{function } (Q_k, \gamma_{fL}, \gamma_{f3}) \qquad (1.10)$$

This inequality simply means that adequate load-carrying capacity must be ensured at the ultimate limit state and that the various design criteria at the serviceability limit state must be satisfied.

## Summary

The main difference in the approach to concrete bridge design in the Code and in the current design documents is the concept of the partial safety factors applied to the loads, load effects and material properties. In addition, as is shown in Chapter 4, some of the design criteria are different. However, concrete bridges designed to the Code should be very similar in proportions to those designed in recent years because the design criteria and partial safety factors have been chosen to ensure that 'on average' this will occur.

There is thus no short-term advantage to be gained from using the Code and, indeed, initially there will be the disadvantage of an increase in design time due to unfamiliarity. Hopefully, the design time will decrease as designers become familiar with the Code and can recognise the critical limit state for a particular design situation.

The advantage of the limit state format, as presented in the Code, is that it does make it easier to incorporate new data on loads, materials, methods of analysis and structural behaviour as they become available. It is thus eminently suitable for future development based on the results of experience and research.

## Chapter 2

# Analysis

## General requirements

The general requirements concerning methods of analysis are set out in Part 1 of the Code, and more specific requirements for concrete bridges are given in Part 4.

## Serviceability limit state

Part 1 permits the use of linear elastic methods or non-linear methods with appropriate allowances for loss of stiffness due to cracking, creep, etc. The latter methods of analysis must be used where geometric changes significantly modify the load effects; but such behaviour is unlikely to occur at the serviceability limit state in a concrete bridge.

Although non-linear methods of analysis are available for concrete bridge structures [19], they are more suited to checking an existing structure, rather than to direct design; this is because prior knowledge of the reinforcement at each section is required in order to determine the stiffnesses. Thus the most likely application of such analyses is that of checking a structure at the serviceability limit state, when it has already been designed by another method at the ultimate limit state. Hence, it is anticipated that analysis at the serviceability limit state, in accordance with the Code, will be identical to current working load linear elastic analysis.

Part 4 of the Code gives the following guidance on the stiffnesses to be used in the analysis at the serviceability limit state.

The flexural stiffness may be based upon:

1. The concrete section ignoring the presence of reinforcement.
2. The gross section including the reinforcement on a modular ratio basis.
3. The transformed section consisting of the concrete in compression combined with the reinforcement on a modular ratio basis.

However, whichever option is chosen, it should be used consistently throughout the structure.

Axial, torsional and shearing stiffnesses may be based upon the concrete section ignoring the presence of the reinforcement. The reinforcement can be ignored because it is difficult to allow for it in a simple manner, and it is considered to be unlikely that severe cracking will occur due to these effects at the serviceability limit state.

Strictly, the moduli of elasticity and shear moduli to be used in determining any of the stiffnesses should be those appropriate to the mean strengths of the materials, because when analysing a structure it is the overall response which is of interest. If there is a linear relationship between loads and their effects, the values of the latter are determined by the relative and not the absolute values of the stiffnesses. Consequently, the same effects are calculated whether the material properties are appropriate to the mean or characteristic strengths of materials. Since the latter are used throughout the Code, and not the mean strengths, the Code permits them to be used for analysis. Values for the short term elastic modulus of normal weight concrete are given in a table in Part 4 of the Code, and Appendix A of Part 4 of the Code states that half of these values should be adopted for analysis purposes at the serviceability limit state. The tabulated values have been shown [20] to give good agreement with experimental data. Poisson's ratio for concrete is given as 0.2. The elastic modulus for reinforcement and prestressing steel is given as 200 kN/mm², except for alloy bars to BS 4486 [21] and 19-wire strand to BS 4757 section 3 [22], in which case it is 175 kN/mm².

It is also stated in Part 4 that shear lag effects may be of importance in box sections and beam and slab decks having large flange width-to-length ratios. In such cases the designer is referred to the specialist literature, such as Roik and Sedlacek [23], or to Part 5 of the Code, which deals with steel–concrete composite bridges. Part 5 treats the shear lag problem in terms of effective breadths, and gives tables of an effective breadth parameter as a function of the breadth-to-length ratio of the flange, the longitudinal location of the section of interest, the type of loading (distributed or concentrated) and the support conditions. The tables were based [24] on a parametric study of shear lag in steel box girder bridges [25]. However, they are considered to be applicable to concrete flanges of composite bridges [26] and, within the limitations of the effective breadth concept, they should also be applicable to concrete bridges.

## Ultimate limit state

At the ultimate limit state, Part 1 of the Code permits the adoption of either elastic or plastic methods of analysis. Plastic methods are inferred to be those based upon considerations of collapse mechanisms, or upon non-elastic distributions of stresses or of stress resultants. Although such methods exist for certain types of concrete bridge structure (e.g. yield line theory and the Hillerborg strip method for slabs), it is envisaged that the vast majority of structures will continue to be analysed elastically at the ultimate limit state. However, a simple plastic method could be used for checking a structure at the ultimate limit state when it has already been designed at the serviceability limit state. Such an approach would be most appropriate to prestressed concrete structures.

A design approach which is permitted in Part 4 of the Code, and which is new to bridge design, although it is well established in building codes, is redistribution of elastic moments. This method is discussed later in this chapter.

The stiffnesses to be adopted at the ultimate limit state may be based upon nominal dimensions of the cross-sections, and on the elastic moduli; or the stiffnesses may be modified to allow for shear lag and cracking. As for the serviceability limit state, whichever alternative is selected, it should be used consistently throughout the structure.

Part 4 of the Code also permits the designer to modify elastic methods of analysis where experiment and experience have indicated that simplifications in the simulation of the structure are possible. An example of such a simplification would be an elastic analysis of a deck in which the torsional stiffnesses are put equal to zero, although they would be known to have definite values. Such a simplification would result in a safe lower bound design, as explained later in this chapter, and would avoid the common problems of interpreting and designing against the torques and twisting moments output by the analysis. However, the author is not aware of any experimental data which, at present, justify such simplifications.

## Elastic analysis at the ultimate limit state

The validity of basing a design against collapse upon an elastic analysis has been questioned by a number of designers, it being thought that this constitutes an anomaly. In particular, for concrete structures, it is claimed that such an approach cannot be correct because the elastic analysis would generally be based upon stiffnesses calculated from the uncracked section, whereas it is known that, at collapse, the structure would be cracked. Although it is anticipated that uncracked stiffnesses will usually be adopted for analysis, it is emphasised that the use of cracked transformed section stiffnesses are permitted.

In spite of the doubts that have been expressed, one should note that it is perfectly acceptable to use an elastic analysis at the ultimate limit state and an anomaly does not arise, even if uncracked stiffnesses are used. The basic reason for this is that an elastic solution to a problem satisfies equilibrium everywhere and, if a structure is designed in accordance with a set of stresses (or stress resultants) which are in equilibrium and the yield stresses (or stress resultants) are not exceeded anywhere, then a safe lower bound design results. Clark has given a detailed explanation of this elsewhere [27].

It is emphasised that the elastic solution is merely one of an infinity of possible equilibrium solutions. Reasons for adopting the elastic solution based upon uncracked stiffnesses, rather than an inelastic solution, are:

1. Elastic solutions are readily available for most structures.
2. Prior knowledge of the reinforcement is not required.
3. Problems associated with the limited ductility of structural concrete are mitigated by the fact that all critical sections tend to reach yield simultaneously; thus stress redistribution, which is dependent upon ductility, is minimised.
4. Reasonable service load behaviour is assured.

## Local effects

When designing a bridge deck of box beam or beam and slab construction, it is necessary to consider, in addition to overall global effects, the local effects induced in the top slab by wheel loads. Part 4 states that the local effects may be calculated elastically, with due account taken of any fixity existing between the slabs and webs. This conforms with the current practice of assuming full fixity at the slab and web junctions and using either Pucher's influence surfaces [28] or Westergaard's equations [29].

As an alternative to an elastic method at the ultimate limit state, yield line theory, which is explained later, or another plastic analysis may be used. The reference to another plastic analysis was intended by the drafters to permit the use of the Hillerborg strip method, which is also explained later. However, this method is not readily applicable to modern practice, which tends to omit transverse diaphragms, except at supports, with the result that top slabs are, effectively, infinitely wide and supported on two sides only.

In order to reduce the number of load positions to be considered when combining global and local effects, it is permitted to assume that the worst loading case for this particular aspect of design occurs in the regions of sagging moments of the structure as a whole. When making this suggestion, the drafters had transverse sagging effects primarily in mind because these are the dominant structural effects in design terms. However, the worst loading case for transverse hogging would occur in regions of global and local hogging, such as over webs or beams in regions of global transverse hogging; whereas the worse loading case for longitudinal effects could be in regions of either global compression or tension in the flange, in combination with the local longitudinal bending.

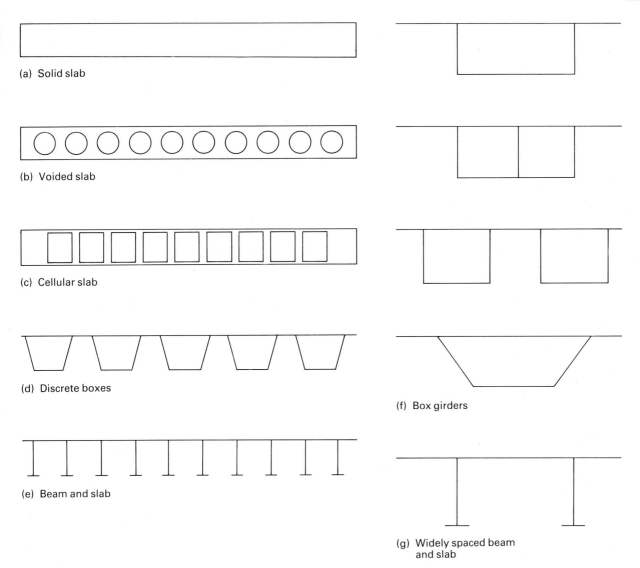

(a) Solid slab

(b) Voided slab

(c) Cellular slab

(d) Discrete boxes

(e) Beam and slab

(f) Box girders

(g) Widely spaced beam and slab

**Fig. 2.1(a)–(g)** Bridge deck types

# Types of bridge deck

## General

Prior to discussing the available methods of analysis for bridge decks, it is useful to consider the various types of deck used in current practice, and to examine how these can be divided for analysis purposes.

In Fig. 2.1, the various cross-sections are shown diagrammatically. Those in Fig. 2.1 (a–e) are generally analysed as two-dimensional infinitesimally thin structures, and the effects of down-stand beams or webs are ignored; whereas those in Fig. 2.1 (f and g) are generally analysed as three-dimensional structures, and the behaviour of the actual individual plates which make up the cross-section considered.

The choice of a type of deck for a particular situation obviously depends upon a great number of considerations, such as span, site conditions, site location and availability of standard sections, materials and labour. These points are referred to by a number of authors [30–33] and only brief discussions of the various types of deck follow.

## Solid slabs

Solid slab bridges can be either cast in-situ, of reinforced or prestressed construction, or can be of composite construction, as shown in Fig. 2.2. In the latter case precast prestressed beams, with bottom flanges, are placed adjacent to each other and in-situ concrete placed between and over the webs of the precast beams to form a composite slab. The precast beams are often of a standardised form [34,35].

Solid slabs are frequently the most economic form of construction for spans up to about 12 m, for reinforced concrete in-situ construction, and up to about 15 m, for composite slabs using prestressed precast beams. The latter are available for span ranges of 7–16 m [34] and 4–14 m [35].

It is obviously valid to analyse either in-situ or composite slabs as thin plates.

## Voided slabs

For spans in excess of about 15 m the self weight effects of solid slabs become prohibitive, and voids are introduced

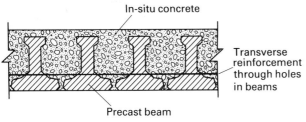

Fig. 2.2 Composite solid slab

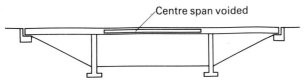

Fig. 2.3 Continuous slab bridge

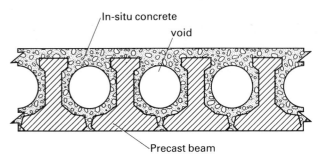

Fig. 2.4 Composite voided slab

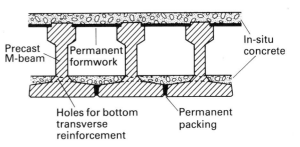

Fig. 2.5 Composite cellular slab using M-beams

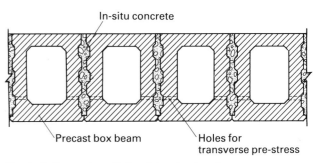

Fig. 2.6 Composite cellular slab using box beams

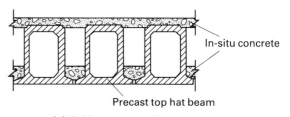

(a) Bridge cross section

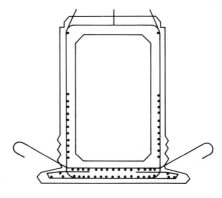

(b) Beam detail

Fig. 2.7(a),(b) Composite cellular slab using top hat beams [38]

to reduce these effects. It is often necessary in continuous slab bridges to make the centre span voided, as shown in Fig. 2.3, in order to prevent uplift at the end supports under certain conditions of loading. Voided slabs are generally used for spans of up to about 18 m and 25 m for reinforced and post-tensioned construction respectively.

It should be mentioned that the cost of forming the voids by means of polystyrene, heavy cardboard, thin wood or thin metal generally exceeds the cost of the concrete replaced. Hence, economies arise only from the reduction in the self weight effects, and, in the case of prestressed construction, from the reduced area of concrete to be stressed.

Voided slabs can be either cast in-situ, of reinforced or prestressed construction, or can be of composite construction as shown in Fig. 2.4. The latter are constructed in a similar manner to solid composite slabs, but void formers are placed between the webs of the precast beams prior to placing the in-situ concrete.

The presence of voids in a slab reduces the shear stiffness of the slab in a direction perpendicular to the voids. The implication of this is that it is not necessarily valid to analyse such slabs by means of a conventional thin plate analysis which ignores shearing deformations. Analyses which include the effects of shearing deformations are available and are discussed later in this Chapter. However, for the majority of practical voided slab cross-sections these effects can be ignored.

## Cellular slabs

The introduction of rectangular, as opposed to circular, voids in a slab obviously further reduces the self weight effects; but causes greater shear flexibility of the slab, and can result in construction problems for in-situ cellular slabs due to the difficulty of placing the concrete beneath the voids. The in-situ construction problems can be overcome by using precast prestressed beams in combination with in-situ concrete to form a composite cellular slab as shown in Figs. 2.5 to 2.7.

The M-beam form of construction shown in Fig. 2.5 can be used for spans in the range 15 to 29 m [36], but has not proved to be popular because of the two stages of in-situ

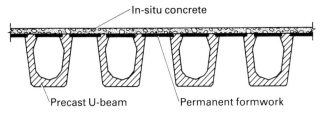

**Fig. 2.8** U-beam deck

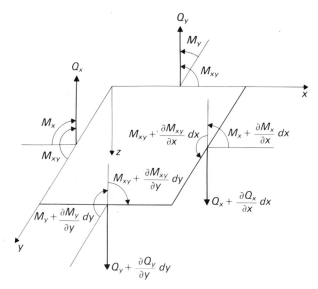

**Fig. 2.9** Stress resultants acting on a plate element

concreting required on site and the necessity to thread transverse reinforcement through holes at the bottom of the webs of the precast beams [37].

There are a variety of standard box beam sections [34,35] (see Fig. 2.6) which can be used for spans in the range 12 to 36 m, but the need for transverse prestressing tendons through the deck creates site problems.

The precast top hat beam (see Fig. 2.7) which was developed by G. Maunsell and Partner [38] has the advantage that no transverse reinforcement or prestressing tendons have to be threaded through the beams.

In all forms of cellular slab construction a considerable proportion of the cross-section is voided. It is generally necessary to adopt either a thin plate analysis which allows for the effects of shearing deformations, or to apply an appropriate modification to an analysis which ignores them, as mentioned later in this chapter.

## Discrete box beams

Discrete box beam decks can be constructed by casting an in-situ top slab on precast prestressed U-beams [35,39] as shown in Fig. 2.8. The advantage of such a form of construction is that it is not necessary to thread transverse reinforcement or prestressing tendons through the beams. In addition, some benefit is gained from the torsional stiffness of a closed box section, although this effect is not as beneficial as it would be if the beams were connected through the bottom, in addition to the top, flanges.

The structural behaviour of such decks is extremely complex due to the fact that the cross-section consists of alternate flexible top slabs and stiff boxes. Strictly, such decks should be analysed by methods which consider the behaviour of the individual plates which make up the cross-section but, in practice, they are often analysed by means of a grillage representation.

## Beam and slab

Beam and slab type of construction, consisting of precast prestressed beams in combination with an in-situ top slab, is frequently used for all spans; and precast beams are available which can be used for span ranges of 12 to 36 m, in the case of I-beams [34], and 15 to 29 m, in the case of M-beams [36].

Beam and slab bridges are generally analysed as plane grillages. This is not strictly correct because the neutral surface is a curved, rather than a plane, surface but, in practice, it is reasonable to consider it as a plane.

## Box girders

There is a wide range of box girder cross-sections and methods of construction. The latter include precast or in-situ, reinforced or prestressed and the use of segmental construction. Box girders are generally adopted for spans in excess of about 30 m and a useful review of the various structural forms has been carried out by Swann [40].

The structural behaviour of box girders and methods for their analysis have been discussed by Maisel and Roll [41].

# Elastic methods of analysis

## General

It is assumed that the reader is familiar with the elastic methods of analysis currently used in practice and only a brief review of the various methods follows.

## Orthotropic plate theory

An orthotropic plate is one which has different stiffnesses in two orthogonal directions. Thus a voided slab is orthotropic, and a beam and slab deck, when analysed by means of a plate analogy, is also orthotropic. It is emphasised that bridge decks are generally orthotropic due to geometric rather than material differences in two orthogonal directions.

If a plate element subjected to an intensity of loading of $q$ is considered in rectangular co-ordinates $x,y$ which co-incide with the directions of principal orthotropy, then the bending moments per unit length $(M_x, M_y)$, twisting moment per unit length $(M_{xy})$ and shear forces per unit length $(Q_x, Q_y)$ which act on the element are shown in Fig. 2.9.

For equilibrium of the element it is required that [42]

$$\frac{\partial^2 M_x}{\partial x^2} - 2\frac{\partial^2 M_{xy}}{\partial x\,\partial y} + \frac{\partial^2 M_y}{\partial y^2} = -q \qquad (2.1)$$

One should note that the equilibrium equation (2.1) applies to any plate and is independent of the plate stiffnesses.

The constitutive relationships in terms of the plate displacement ($w$) in the $z$ direction and the shear strains ($\gamma_x$, $\gamma_y$) in the $x$ and $y$ directions respectively are [43]

$$M_x = -D_x\,\frac{\partial}{\partial x}\left(\frac{\partial w}{\partial x} - \gamma_x\right) - D_1\,\frac{\partial}{\partial_y}\left(\frac{\partial w}{\partial y} - \gamma_y\right) \qquad (2.2)$$

$$M_y = -D_y\,\frac{\partial}{\partial y}\left(\frac{\partial w}{\partial y} - \gamma_y\right) - D_1\,\frac{\partial}{\partial x}\left(\frac{\partial w}{\partial x} - \gamma_x\right) \qquad (2.3)$$

$$M_{xy} = -D_{xy}\left[\frac{\partial}{\partial y}\left(\frac{\partial w}{\partial x} - \gamma_x\right) + \frac{\partial}{\partial x}\left(\frac{\partial w}{\partial y} - \gamma_y\right)\right] \qquad (2.4)$$

$$Q_x = S_x\gamma_x \qquad (2.5)$$

$$Q_y = S_y\gamma_y \qquad (2.6)$$

where $D_x$, $D_y$ are the flexural stiffnesses per unit length, $D_1$ is the cross-flexural stiffness per unit length, $D_{xy}$ is the torsional stiffness per unit length and $S_x$, $S_y$ are the shear stiffnesses per unit length.

In conventional thin plate theory, it is assumed that $S_x = S_y = \infty$ or $\gamma_x = \gamma_y = 0$, i.e. that the plate is stiff in shear: it is then possible to combine equations (2.2) to (2.4) to give the following fourth order governing differential equation for a shear stiff plate [42]

$$D_x\,\frac{\partial^4 w}{\partial x^4} + 2(D_1 + 2D_{xy})\,\frac{\partial^4 w}{\partial x^2\partial y^2} + D_y\,\frac{\partial^4 w}{\partial y^4} = q \qquad (2.7)$$

If the plate has finite values of the shear stiffnesses then Libove and Batdorf [44] have shown how it is possible to obtain a sixth order governing differential equation. However, as discussed later, it is reasonable for many bridge decks to assume that one of the shear stiffnesses is infinite and the other finite: it is then possible to obtain fairly simple solutions to the governing equations.

## Series solutions

Many bridge decks are essentially prismatic rectangular plates which are simply supported along two edges, and, in such situations, it is possible to solve equation (2.7) by making use of Fourier sine series for the displacement ($w$) and the load ($q$) as follows (see Fig. 2.10)

$$w = \sum_{m=1}^{\infty} Y_m(y)\sin\frac{m\pi x}{L} \qquad (2.8)$$

$$q = \sum_{m=1}^{\infty} q_m(y)\sin\frac{m\pi x}{L} \qquad (2.9)$$

These expressions are chosen because they satisfy the simply supported boundary conditions at $x = 0$ and $L$. Any

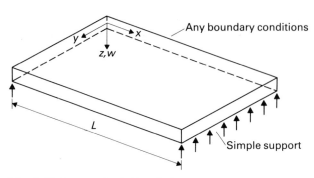

**Fig. 2.10** Rectangular bridge deck

boundary conditions on the two other edges can be dealt with in the analysis.

The series solution for bridge decks was originally due to Guyon [45] and Massonnet [46], and was then developed by Morice and Little [47] who, together with Rowe, produced design charts which enable the calculations to be carried out by hand [48]. Cusens and Pama [49] have published a more general treatment of the method which extends its range of application and have also presented design charts for calculations by hand.

In addition to the above charts, computer programs exist for performing series solutions such as the Department of Transport's program ORTHOP [50].

It is emphasised that simple series solutions cannot be obtained for non-prismatic decks in which the cross-section varies longitudinally; nor for skew decks, because it is not possible to satisfy the skew boundary conditions. Although the series solutions are for single span simply supported decks, it is also possible to apply them to decks which are continuous over discrete supports by using a flexibility approach in which the discrete supports are considered to be redundancies and zero displacement imposed at each [49]. This approach is used in the ORTHOP program referred to above.

## Series solutions for shear deformable plates

If the bridge deck shown in Fig. 2.10 is considered to be a voided or cellular slab, with the voids running in the span direction $x$, then it is reasonable to consider the deck to be shear stiff longitudinally ($S_x = \infty$) but to be shear deformable transversely. In such a case it is possible to combine equations (2.1) to (2.6) so that a series solution can be obtained. This has been done by Morley [51] by representing the transverse shear, force by the following Fourier sine series, in addition to using equations (2.8) and (2.9)

$$Q_y = \sum_{m=1}^{\infty} Q_{ym}\sin\frac{m\pi x}{L} \qquad (2.10)$$

This representation requires that, at the supports, $Q_y = 0$, and the method is only applicable if there are rigid end diaphragms at the supports. Morley [51] presents design charts which enable solutions to be obtained by hand, and

Elliott [52] has published a computer program which solves the same problem. An alternative approach has been presented by Cusens and Pama [49].

## Folded plate method

The cellular slab, the discrete boxes or the box girders shown in Fig. 2.1 can be considered to be composed of a number of individual plates which span from abutment to abutment and are joined along their edges to adjacent plates. Such an assemblage of plates can be solved by the folded plate method which was originally due to Goldberg and Leve [53], and was subsequently developed by De Fries-Skene and Scordelis [54] into the form in which it is incorporated into the Department of Transport's computer program MUPDI [55].

The folded plate method considers both in-plane and bending effects in each plate and can thus deal with local bending and distortional effects. It is a powerful method, but the bridge must be right and have simple supports at which there are diaphragms which can be considered to be rigid in their own planes but flexible out of plane; in addition, the section must be prismatic because the solution is obtained in terms of Fourier series. Continuous bridges can be considered in the same way as that discussed previously for the series solution of plates.

## Finite elements

In the finite element approach, a structure is considered to be divided into a number of elements which are connected at specified nodal points. The method is the most versatile of the available methods and, in principle, can solve almost any problem of elastic bridge deck analysis. The reader is referred to one of the standard texts on finite element analysis for a full description of the method.

There are a great number of finite element programs available which can handle a variety of structural forms. In addition, there is a great variety of element shapes and types: the latter include one-dimensional beam elements, two-dimensional plane stress and plate bending elements, and three-dimensional shell elements. The following Department of Transport programs are readily available:

1.  STRAND 2 [56] is for the analysis of reinforced and prestressed concrete slabs and uses a triangular plate bending element in combination with a triangular plane stress element: in addition, beam elements, which are assumed to have the same neutral axis as that of the plate, can be used.
2.  QUEST [57] is intended for the analysis of box girders and uses quadrilateral thin shell elements which consider both bending and membrane stress resultants.
3.  CASKET [58] is a general purpose finite element program with facilities for plane stress, plate bending, beam, plane truss, plane frame, space truss and space frame elements, which are all compatible with each other.

## Finite strips

The finite strip method is a particular type of finite element analysis in which the elements consist of strips which run the length of the structure and are connected along the strip edges. The method is thus particularly suited to the analysis of box girders and cellular slabs since they can be naturally divided into strips.

The in-plane and out-of-plane displacements within a strip are considered separately, and are represented by Fourier series longitudinally and polynomials transversely. Since Fourier series are used longitudinally, the method is only applicable to right prismatic structures with simply supported ends. However, intermediate supports can be considered in the same way as that discussed previously for the series solution of plates.

The finite strip method was originally developed by Cheung [59] who adopted a third order polynomial for the out-of-plane displacement function, and specified two degrees of freedom (vertical displacement and rotation) along the edges of each strip.

The in-plane displacement function is a first order polynomial, which implies a linear distribution of in-plane displacement across a strip, and there are two degrees of freedom (longitudinal and transverse displacements) along the edges of each strip.

The use of a third order polynomial for the out-of-plane displacement function results in discontinuities of transverse moments at the strip edges, because only compatibility of deflection and slope is ensured. Hence, a large number of strips is required in order to obtain an accurate prediction of transverse moments. However, this can be overcome by introducing a fifth order polynomial, which ensures compatibility of curvature in addition to deflection and slope. However, two additional 'degrees of freedom' have to be introduced in order to determine the constants of the polynomial; these 'degrees of freedom' are the curvatures at the strip edges [60]. An alternative formulation, which also uses a fifth order polynomial, involves the introduction of an auxiliary nodal line in each strip and only has the two degrees of freedom of deflection and slope [49].

The auxiliary nodal line technique can also be adopted for the in-plane effects, and a second order polynomial is then used for the in-plane displacement function.

## Grillage analysis

In a grillage analysis, the structure is idealised as a grillage of interconnected beams. The beams are assigned flexural and torsional stiffnesses appropriate to the part of the structure which they represent. A generalised slope-deflection procedure, or a matrix stiffness method, is then used to calculate the vertical displacements and the rotations about two horizontal axes at the joints. Hence the bending moments, torques and shear forces of the grillage beams at the joints can be determined.

Since the grillage method represents the structure by means of beams, and cannot thus simulate the Poisson's ratio effects of continua, it should, strictly, be used only for grillage structures. Nevertheless, the grillage method is a very popular method of analysis among bridge engineers, and it has been applied to the complete range of concrete bridge structures [61]. When applied to voided or cellular slabs or to box girders, a shear deformable grillage is frequently used [61] in which shear stiffnesses, as well as flexural and torsional stiffnesses, are assigned to the grillage members; and the slope-deflection equations, or stiffness matrices, modified accordingly.

Guidance on the simulation of various types of bridge deck by a grillage is given by Hambly [61] and West [62].

## Influence surfaces

A number of sets of influence surfaces have been produced in tabulated and graphical form for the analysis of isotropic plates, and these are extremely useful in the preliminary design stage of orthotropic, as well as isotropic, plates. The influence surfaces have, generally, been derived experimentally or by means of finite difference techniques.

Influence surface values are available for rectangular isotropic slabs [28], skew simply supported isotropic slabs [63, 64, 65], skew simply supported torsionless slabs [66], and skew continuous isotropic slabs [67, 68, 69].

# Elastic stiffnesses

## Plate analysis

When idealising a bridge deck as an orthotropic plate and using a series solution, finite plate elements or finite strips, the following stiffnesses are suggested for the various types of deck. Whenever a composite section is being considered, due allowance should be made for any difference between the elastic moduli of the two concretes by means of the usual modular ratio approach. It is assumed throughout that the longitudinal shear stiffness ($S_x$) is infinite.

### Solid slab

$$D_x = D_y = \frac{Eh^3}{12(1 - v^2)} \tag{2.11}$$

$$D_{xy} = \frac{Gh^3}{12} \tag{2.12}$$

$$D_1 = vD_y \tag{2.13}$$

$$S_y = \infty \tag{2.14}$$

where $h$ is the slab thickness, $E$ and $v$ are the elastic modulus and Poisson's ratio respectively of concrete, and $G$ is the shear modulus which is given by

$$G = \frac{E}{2(1 + v)} \tag{2.15}$$

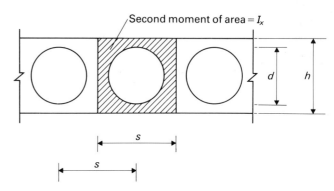

**Fig. 2.11** Voided slab geometry

### Voided slab

Elliott and Clark [70] have reported the results of finite element analyses which were carried out to determine the flexural and torsional stiffnesses of voided slabs, and which were checked experimentally. It was found that, for a Poisson's ratio of 0.2, which is a reasonable value to adopt for concrete, the following equations for the stiffnesses gave values which agreed closely with those of the analyses and the experiments.

With the notation of Fig. 2.11,

$$D_x = \frac{EI_x}{s(1 - v^2)} \tag{2.16}$$

$$D_y = \frac{Eh^3}{12} \left[ 1 - 0.95 \left( \frac{d}{h} \right)^4 \right] \tag{2.17}$$

$$D_{xy} = \frac{Gh^3}{12} \left[ 1 - 0.84 \left( \frac{d}{h} \right)^4 \right] \tag{2.18}$$

$$D_1 = vD_y \tag{2.19}$$

The author is not aware of any reliable published data on the transverse shear stiffness of voided slabs. However, an idea of the significance of transverse shear flexibility can be obtained from the results of a test on a model voided slab bridge, with a depth of void to slab depth ratio of 0.786, reported by Elliott, Clark and Symmons [71]. The experimental deflections and strains, under simulated highway loading, were compared with those predicted by shear-stiff orthotropic plate analysis. It was found that the observed load distribution was slightly inferior to the theoretical distribution in the uncracked state, but was very similar after the bridge had been extensively cracked due to longitudinal bending. In design terms, it thus seems reasonable to ignore the effects of transverse shear flexibility and to take the transverse shear stiffness as infinity. Since the model slab had a void ratio which is very close to the practical upper limit in prototype slabs, it is suggested that the transverse shear stiffness can be taken as infinity for all practical voided slabs. However, if it is desired to specify a finite value of the transverse shear stiffness, then approximate calculations carried out by Elliott [72] suggest that, for practical void ratios, it is about 15% of the value of a solid slab of the same overall depth ($h$). The solid slab value is given by [73]

$$S_y = \frac{5}{6} Gh \tag{2.20}$$

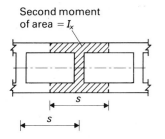

Second moment of area $= I_x$

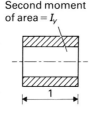

Second moment of area $= I_y$

**Fig. 2.12** Cellular slab geometry

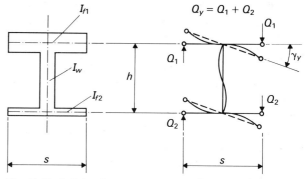

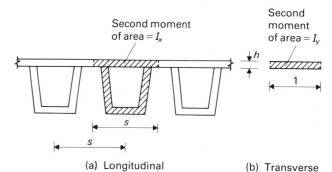

**Fig. 2.13** Cellular distortion

Hence, for a voided concrete slab

$$S_y \simeq 0.15 \times \frac{5}{6} \ Gh = \frac{Gh}{8} \tag{2.21}$$

### Cellular slab

Elliott [52] has suggested the following values for relatively thin-walled cellular slabs having an area of void not less than about one-third of the gross area. The notation is illustrated in Fig. 2.12.

$$D_x = \frac{EI_x}{s(1-v^2)} \tag{2.22}$$

$$D_y = \frac{EI_y}{1-v^2} \tag{2.23}$$

$$D_{xy} = GI_y \tag{2.24}$$

$$D_1 = v\,D_y \tag{2.25}$$

The transverse shear stiffness can be obtained by considering the distortion of a cell and assuming that points of contraflexure occur in the flanges midway between the webs as shown in Fig. 2.13 (as originally suggested by Holmberg [74]). The shear stiffness can then be shown to be

$$S_y = \frac{24\,K_w}{s} \ \frac{12 + K_1 + K_2}{12 + 4(K_1 + K_2) + K_1 K_2} \tag{2.26}$$

where

$$K_1 = \frac{K_w}{K_{f1}} \tag{2.27}$$

$$K_2 = \frac{K_w}{K_{f2}} \tag{2.28}$$

$$K_w = \frac{EI_w}{h(1-v^2)} \tag{2.29}$$

$$K_{f1} = \frac{EI_{f1}}{s(1-v^2)} \tag{2.30}$$

$$K_{f2} = \frac{EI_{f2}}{s(1-v^2)} \tag{2.31}$$

### Discrete boxes

With the notation of Fig. 2.14

$$D_x = \frac{EI_x}{s} \tag{2.32}$$

$$D_y = EI_y = \frac{Eh^3}{12} \tag{2.33}$$

$$D_1 = 0 \tag{2.34}$$

Second moment of area $= I_x$

Second moment of area $= I_y$

(a) Longitudinal      (b) Transverse

**Fig. 2.14(a),(b)** Discrete box geometry

It is suggested that the torsional stiffness be calculated by assuming that the 'longitudinal torsional stiffness' is equal to that of the shaded section shown in Fig. 2.14(a), and the 'transverse torsional stiffness' is zero. If the sections of top slab which do not form part of the box section are ignored, the 'longitudinal torsional stiffness' is given by the usual thin-walled box formula.

$$GJ_x = G\left(\frac{4A_0^2}{\oint \frac{dl}{t}}\right) \tag{2.35}$$

where $A_0$ is the area within the median line of the box, $t$ is the box thickness at distance $l$ from an origin, and the integration path is the median line. Thus the 'longitudinal torsional stiffness per unit width' is given by

$$\frac{GJ_x}{s} = \frac{G}{s}\left(\frac{4A_0^2}{\oint \frac{dl}{t}}\right) \tag{2.36}$$

Since the 'transverse torsional stiffness' is taken to be zero, the total torsional stiffness per unit width is also given by equation (2.36). Clark [75] has demonstrated that $D_{xy}$ is one-quarter of the total torsional stiffness per unit width calculated as above; hence

$$D_{xy} = \frac{G}{4s}\left(\frac{4A_0^2}{\oint \frac{ds}{t}}\right) \tag{2.37}$$

Rather than calculate a specific value for the transverse shear stiffness ($S_y$), it is suggested that the torsional stiffness ($D_{xy}$) be modified by applying the relevant reduction factors given by Cusens and Pama [49].

### Beam and slab

With the notation of Fig. 2.15

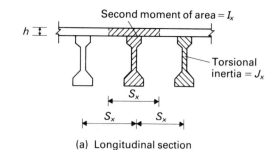

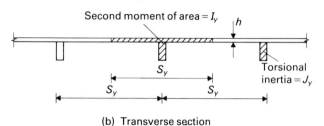

(a) Longitudinal section

(b) Transverse section

**Fig. 2.15(a),(b)** Beam and slab geometry

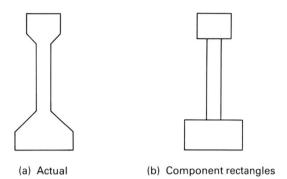

(a) Actual          (b) Component rectangles

**Fig. 2.16(a),(b)** Torsional inertia of beam

The latter calculation is an approximation and, in fact, underestimates the true torsional inertia because the junction effects, where adjacent rectangles join, are ignored. Jackson [77] has suggested a modification to allow for the junction effects, but it is not generally necessary to carry out the modification for practical sections [62].

$$D_x = \frac{EI_x}{s_x} \tag{2.38}$$

$$D_y = \frac{EI_y}{s_y} \tag{2.39}$$

$$D_{xy} = \frac{Gh^3}{12} + \frac{G}{4}\left(\frac{J_x}{s_x} + \frac{J_y}{s_y}\right) \tag{2.40}$$

$$D_1 = 0 \tag{2.41}$$

$$S_y = \infty \tag{2.42}$$

The torsional inertias ($J_x$ and $J_y$) of the individual longitudinal and, if present, transverse beams can be calculated by dividing the actual beams into a number ($n$) of component rectangles as shown in Fig. 2.16. The torsional inertia of the $i$th component rectangle of size $b_i$ by $h_i$ is given by

$$\left.\begin{array}{ll} J_i = k\,b_i^3 h_i & b_i \leqslant h_i \\ J_i = k\,b_i h_i^3 & b_i \geqslant h_i \end{array}\right\} \tag{2.43}$$

The coefficient ($k$) is dependent upon the aspect ratio of the rectangle, where the aspect ratio is always greater than or equal to unity and is defined as $b_i/h_i$ or $h_i/b_i$ as appropriate. Values of $k$ are given in Table 2.1 [76]. The total torsional inertia of a beam is then obtained by summing those of the individual rectangles

$$J_x \text{ or } J_y = \sum_{i=1}^{n} J_i \tag{2.44}$$

**Table 2.1** Torsional inertia constant for rectangles

| Aspect ratio | $k$ | Aspect ratio | $k$ |
|---|---|---|---|
| 1.0 | 0.141 | 2.5 | 0.249 |
| 1.1 | 0.154 | 2.8 | 0.258 |
| 1.2 | 0.166 | 3.0 | 0.263 |
| 1.3 | 0.175 | 4.0 | 0.281 |
| 1.4 | 0.186 | 5.0 | 0.291 |
| 1.5 | 0.196 | 6.0 | 0.298 |
| 1.8 | 0.216 | 7.5 | 0.305 |
| 2.0 | 0.229 | 10.0 | 0.312 |
| 2.3 | 0.242 | $\infty$ | 0.333 |

## Grillage analysis

### General

Guidance on the evaluation of elastic stiffnesses for various types of deck, for use with grillage analysis, has been given by Hambly [61] and West [62]. In general, Hambly's recommendations are modifications of those given previously in this chapter for plate analysis, whereas West's differ quite considerably when calculating torsional stiffnesses.

It should be noted that an orthotropic plate has a single torsional stiffness ($D_{xy}$), whereas an orthogonal grillage can have different torsional stiffnesses ($GC_x$, $GC_y$) in the directions of its two sets of beams. Furthermore, since a grillage cannot simulate the Poisson effect of a plate, Poisson's ratio does not appear in the expressions for the flexural stiffnesses, and a grillage stiffness equivalent to $D_1$ does not occur. It is again assumed that the longitudinal shear stiffness is infinite.

The following general recommendations are a combination of those of Hambly [61] and West [62] and of the author's personal views. Reference should be made to Hambly [61] for more detailed information concerning edge beams and other special cases. The recommendations are given in terms of inertias rather than stiffnesses, because this is the form in which the input to a grillage analysis program is generally required.

As for orthotropic plate stiffnesses, discussed earlier, differences between the elastic moduli of the concretes in a composite section should be taken into account by the modular ratio approach.

### Solid slab

The inertias of an individual grillage beam should be obtained from the following inertias per unit width by multiplying them by the breadth of slab represented by the grillage beam.

$$\text{Longitudinal flexural} = \text{transverse flexural} = \frac{h^3}{12} \tag{2.45}$$

Longitudinal torsional = transverse torsional = $\dfrac{h^3}{6}$ (2.46)

Transverse shear area = $\infty$ (2.47)

### Voided slab

The inertias of an individual grillage beam should be obtained from the following inertias per unit width by multiplying by the breadth of slab represented by the grillage beam: the notation is given in Fig. 2.11.

Longitudinal flexural $= \dfrac{I_x}{s}$ (2.48)

Transverse flexural $= \dfrac{h^3}{12}\left[1 - 0.95\left(\dfrac{d}{h}\right)^4\right]$ (2.49)

Longitudinal torsional = transverse torsional =

$$\dfrac{h^3}{6}\left[1 - 0.84\left(\dfrac{d}{h}\right)^4\right]$$ (2.50)

Transverse shear area $= \infty$ or $\dfrac{h}{8}$ (2.51)

### Cellular slab

The inertias of an individual grillage beam should be obtained from the following inertias per unit width by multiplying them by the breadth of slab represented by the grillage beam: the notation is given in Figs. 2.12 and 2.13.

Longitudinal flexural $= \dfrac{I_x}{s}$ (2.52)

Transverse flexural $= I_y$ (2.53)

Longitudinal torsional = transverse torsional = $2I_y$ (2.54)

Transverse shear area =

$$\dfrac{24\,K_w}{s}\,\dfrac{12 + K_1 + K_2}{12 + 4\,(K_1 + K_2) + K_1 K_2}$$ (2.55)

where

$K_1 = \dfrac{K_w}{K_{f1}}$ (2.56)

$K_2 = \dfrac{K_w}{K_{f2}}$ (2.57)

$K_w = \dfrac{I_w}{h}$ (2.58)

$K_{f1} = \dfrac{I_{f1}}{s}$ (2.59)

$K_{f2} = \dfrac{I_{f2}}{s}$ (2.60)

### Discrete boxes

Various writers [39, 61, 62] have proposed different methods of simulating a deck of discrete boxes by means of a grillage. It is not clear which is the most appropriate simulation to adopt in a particular situation, but whichever simulation is chosen the author would recommend calculating the stiffnesses as suggested by the proponent of the chosen simulation.

### Beam and slab

A longitudinal grillage beam can represent part of the top slab plus either a single physical beam, or a number of physical beams. If a longitudinal grillage beam represents $n$ physical beams, where $n$ is not necessarily an integer, then its inertias are given, with the notation of Fig. 2.15, by

Flexural $= nI_x$ (2.61)

Torsional $= n\left(J_x + \dfrac{s_x h^3}{6}\right)$ (2.62)

A transverse grillage beam can represent either part of the top slab, or part of the top slab with a transverse physical diaphragm. The shear area it taken to be infinity and the flexural and torsional inertias to be, with the notation of Fig. 2.15, and $s_g$ being the spacing of the tranverse grillage beams,

Flexural $= \dfrac{s_g}{s_y}\,I_y$ (2.63)

Torsional $= s_g\left(\dfrac{J_y}{S_y} + \dfrac{h^3}{6}\right)$ (2.64)

# Plastic methods of analysis

## Introduction

In this section, examples of plastic methods of analysis which could be used in bridge design are given. However, as mentioned previously, it is unlikely that, with the possible exception of yield line theory for slabs, such methods will be incorporated into design procedures in the near future. Before discussing the various methods, it is necessary to introduce some concepts used in the theory of plasticity and limit analysis.

## Limit analysis

It is useful at this stage to distinguish between the terms limit analysis and limit state design. Limit analysis is a means of assessing the ultimate collapse load of a structure, whereas limit state design is a design procedure which aims to achieve both acceptable service load behaviour and sufficient strength. Thus limit analysis can be used for calculations at the ultimate limit state in a limit state design procedure.

A concept within limit analysis is that it is often not possible to calculate a unique value for the collapse load of a structure: this is alien to one's experience with the theory of elasticity, where a single value of the load, required to produce a specific stress, at a particular point in a structure, can be calculated. In limit analysis, all that it is generally possible to state is that the collapse load is between two values, known as upper and lower bounds to the collapse load. For certain structures coincidental upper and lower bounds can be obtained, and thus the unique value of the collapse load can be determined. However, this is not the general case and, for a vast number of commonly occurring structures, coincidental upper and lower bounds have not been determined. It is thus necessary to consider two distinct types of analysis within limit analysis; namely, upper and lower bound methods.

An upper bound method is unsafe in that it provides a value of the collapse load which is either greater than or equal to the true collapse load. The procedure for calculating an upper bound to the collapse load can be thought of in terms of proposing a valid collapse mechanism and equating the internal plastic work to the work done by the external loads.

A lower bound method is safe in that it provides a value of the collapse load which is either less than or equal to the true collapse load. A lower bound to the collapse load is the load corresponding to any statically admissible stress (or stress resultant) field which nowhere violates the yield criterion. A statically admissible stress field is one which everywhere satisfies the equilibrium equations for the structure. The expression 'nowhere violates the yield criterion' essentially means that the section strength at each point of the structure should not be exceeded. It is important to note that neither deformations in any form nor stiffnesses are mentioned when considering lower bound methods.

## Lower bound methods

As has been indicated earlier, any elastic method of analysis is a lower bound method, in terms of limit analysis, because it satisfies equilibrium. There are other lower bound methods available which employ inelastic stress distributions; but these have been developed generally with buildings, rather than bridges, in mind. The inelastic lower bound design of bridges is complicated by the more complex boundary conditions, and the fact that bridges are designed for different types of moving concentrated loads. In the following, lower bound methods which could be adopted for bridges are described.

### Hillerborg strip method

One inelastic lower bound method, which is mentioned specifically in the Code, is the Hillerborg strip method [78] for slabs, in which the two dimensional slab problem is reduced to one dimensional beam design in two directions. This is achieved by the designer *choosing* to make $M_{xy} = 0$ throughout the slab. Thus the slab equilibrium equation (2.1) reduces to

$$\frac{\partial^2 M_x}{\partial x^2} + \frac{\partial^2 M_y}{\partial y^2} = -q \qquad (2.65)$$

It is further assumed that, at any point, the load intensity $q$ can be split into components $\alpha q$ in the $x$ direction and $(1 - \alpha) q$ in the $y$ direction, so that equation (2.65) can be split into two equilibrium equations

$$\left. \begin{aligned} \frac{\partial^2 M_x}{\partial x^2} &= -\alpha q \\[2ex] \frac{\partial^2 M_y}{\partial y^2} &= -(1 - \alpha)q \end{aligned} \right\} \qquad (2.66)$$

Equations (2.66) are the equilibrium equations for beams running in the $x$ and $y$ directions. The designer is free to choose any value of $\alpha$ that he wishes, but zero or unity is frequently chosen.

The above simplicity of the strip method breaks down when concentrated loads are considered, because it is necessary to introduce one of the following: complex moment fields including twisting moments [78]; strong bands of reinforcement [79]; or one of the approaches suggested by Kemp [80, 81].

A further problem in applying the strip method to bridge design is that difficulties arise with slabs supported only on two opposite edges, as occurs with slab bridges and top slabs in modern construction, where the tendency is to omit transverse diaphragms.

In view of the above comments it is considered unlikely that the method will be used in bridge deck design, but it is possible that it could be used for abutment design as discussed in Chapter 9.

### Lower bound design of box girders

A lower bound approach to the design of box girders has been suggested by Spence and Morley [82]. The basis of the method is that, instead of designing to resist elastic distortional and warping stresses, which have a peaky longitudinal distribution, it is assumed that, in the vicinity of an eccentric concentrated load, there is a transmission zone, having a length a little greater than that of the load, which is considered to act as a diaphragm and to transfer the load into pure torsion, bending and shear in the remainder of the beam. Thus, outside the transmission zone, the beam walls are subjected to only in-plane stress resultants.

The design procedure is to statically replace an eccentric load by equivalent pairs of symmetric and anti-symmetric loads over the webs. The anti-symmetric loads, $W$ in Fig. 2.17, result in a distortional couple. The length ($L_t$) of the transmission zone is then *chosen* by the designer and equilibrium of the zone under the warping forces ($w_f$ and $w_w$) and corner moments ($m_c$) considered. It follows [82] that these can be calculated from

$$m_c = \frac{Wb}{8L_t} \qquad (2.67)$$

$$2m_c = w_f b = w_w c \qquad (2.68)$$

The beam, as a whole, is then analysed, for bending and torsion, as if a rigid diaphragm were positioned at the load: this results in a set of in-plane forces in the walls of the box. These in-plane forces are superimposed on the corner moments and warping moments, and reinforcement designed as described in Chapter 5.

## Moment redistribution

A lower bound method of analysis, which has been permitted for building design for some time, is the redistribution of elastic moments in indeterminate structures. The Code permits this method to be used for prismatic beams.

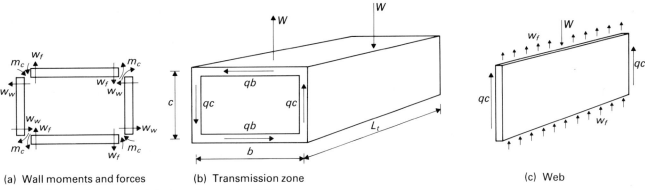

(a) Wall moments and forces     (b) Transmission zone        (c) Web

**Fig. 2.17(a)–(c)** Lower bound design of box girder [82]

Although the Code does not state the fact, the restriction to prismatic beams was intended to preclude redistribution of moments in all structures and structural elements except 'small' beams, such as are used for the longitudinal beams of beam and slab construction. This was because it was thought that there was a lack of knowledge of redistribution in deep members such as box girders.

The concept of moment redistribution can be illustrated by considering an encastré beam of span $L$ carrying an ultimate uniformly distributed load of $W$. The elastic bending moment diagram, assuming zero self weight, is shown in Fig. 2.18(b). If the beam is designed to resist a hogging moment at the supports of $\lambda WL/12$, instead of $WL/12$, then the beam will yield at the supports at a load of $\lambda W$ and, in order to carry the design load of $W$, the mid-span section must be designed to resist a sagging moment of $[(WL/24) + (WL/12) - (\lambda WL/12)]$, as shown in Fig. 2.18(c).

A similar argument would obtain if the mid-span section were designed for a moment less than $WL/24$.

If the beam could be considered to exhibit true plastic behaviour, with unlimited ductility, then any value of $\lambda$ could be chosen by the designer. However, concrete has limited ductility in terms of ultimate compressive strain, and this limits the ductility of a beam in terms of rotation capacity. As $\lambda$ decreases, the amount of rotation, after initial yield, is increased. Thus, $\lambda$ should not be so small that the rotation capacity is exhausted. It should also be noted that the rotation capacity required at collapse is a function of the difference between the elastic moment and the reduced design moment. Thus, it is convenient to think in terms of this difference, and to definite the amount of redistribution as $\beta = 1 - \lambda$. An upper limit to $\beta$ has to be imposed, *when there is a reduction in moment*, because of the limited ductility discussed above.

The amount of redistribution permitted also has to be limited for another reason: although a beam designed for a certain amount of redistribution will develop adequate strength, it could exhibit unsatisfactory serviceability limit state behaviour, since, at this stage, the beam would behave essentially elastically. As the difference between the ultimate elastic moment and the reduced design moment increases, the behaviour at the serviceability limit state, in terms of stiffness (and, thus, cracking) deteriorates. Hence, the amount of redistribution must be restricted.

The Code states four conditions which must be fulfilled when redistributing moments, and these are now discussed.

Total = $W$

(a) Loading

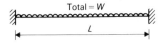

$-WL/12$

$WL/24$

(b) Elastic

Elastic
Redistributed

$-WL/12$
$-\lambda WL/12$    $WL/12 - \lambda WL/12$

$WL/24 + WL/12 - \lambda WL/12$

(c) Redistribution

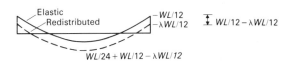

Elastic   'Free' BM. $WL/8$

$WL/8$

Redistributed

(d) Overall equilibrium

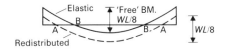

Elastic ultimate
Elastic service

$-0.7WL/12$    $0.7WL/24$    Redistribution $= \beta WL/12$

Redistributed

(e) Serviceability conditions

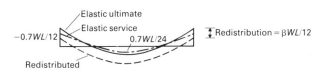

$-300\,\text{kNm}$        $-200\,\text{kNm}$

$280\,\text{kNm}$

(f) Elastic ultimate moment envelope

**Fig. 2.18(a)–(f)** Moment redistribution

First, overall equilibrium must be maintained by keeping the range of the bending moment diagram equal to the 'free' bending moment (see Fig. 2.18(d)).

Second, if the beam is designed to resist the redistributed moments shown in Fig. 2.18(d), then, in the regions A–B, sagging reinforcement would be provided. However, these regions would be subjected to hogging moments at the serviceability limit state, where elastic conditions obtain. The Code thus requires every section to be designed to resist a moment of not less than 70%, for reinforced concrete, nor 80%, for prestressed concrete, of the moment, at that section, obtained from an elastic moment envelope covering all combinations of ultimate loads. For the single load case considered in Fig. 2.18(a), this implies that the resistance moment at any section should be not less than (for reinforced concrete) that appropriate to the chain dotted line of Fig. 2.18(e). The values of 70% and 80% originated in CP 110, where the ratio of $(\gamma_{fL} \gamma_{f3})$ at the serviceability limit state to that at the ultimate limit state is in the range 0.63 to 0.71. By providing reinforcement, or prestress, to resist 70 or 80% respectively of the maximum elastic moments, it is ensured that elastic behaviour obtains up to about 70 or 80% of the ultimate load, i.e. at the service load. In the Code, the service load is in the range 0.58 to 0.76 of the ultimate load and, hence, the limits of 70 and 80% are reasonable.

Third, the Code requires that the moment at a *particular* section may not be *reduced* by more than 30%, for reinforced concrete, nor 20%, for prestressed concrete, of the maximum moment at *any* section. Thus, in Fig. 2.18(f), the moment at any section may not be reduced by more than 90 kNm for reinforced concrete. This seems illogical, since it is the reduction in moment, expressed as a percentage of the moment at the *section under consideration*, which is important when considering limited ductility. It is unclear why the CP 110 committee used the moment at any section. However, this condition is always over-ruled by the second condition which implies that the moment at a *particular* section may not be reduced by more than 30 or 20% of the maximum moment at *that* section. There is no limit to the amount that the moment at a section can be *increased*, because this does not increase the rotation requirement at that section, and the third condition is intended to restrict the rotation which would occur at collapse. Beeby [83] has suggested that the Code limits were not derived from any particular test data, but were thought to be reasonable. However, they can be shown to be conservative by examining test data.

The fourth, and final, condition imposed by the Code is that the neutral axis depth must not exceed 0.3 of the effective depth, if the full allowable *reduction* in moment has been made. As the neutral axis depth is increased, the amount of permissible redistribution reduces linearly to zero when the neutral axis depth is 0.6 of the effective depth, for reinforced concrete, and 0.5, for prestressed concrete. The reason for these limits is that the rotation at failure, when crushing of the concrete occurs, is inversely proportional to the neutral axis depth. Thus, if the concrete crushing strain is independent of the strain gradient across the section, the rotation capacity and, hence, the permissible redistribution increases as the neutral axis depth decreases. Beeby [83] has demonstrated that the Code limits on neutral axis depth are conservative.

A general point regarding moment redistribution is that, when this is carried out, the shear forces are also redistributed: the author would suggest designing against the greater of the non-redistributed and redistributed shear forces.

Moment redistribution has been described in some detail because it is a new concept in bridge design documents, but it must be stated that it is difficult to conceive how it can be simply applied in practice to bridges. This is because, in order to maintain equilibrium by satisfying equation (2.1), any redistribution of longitudinal moments should be accompanied by redistribution of transverse and twisting moments. It would appear that redistribution in a deck can only be achieved by applying an imaginary 'loading' which causes redistribution. For example, longitudinal support moments could be reduced, and span moments increased, by applying an imaginary 'loading' consisting of a displacement of the support; the moments due to this imaginary 'loading' would then be added to those of the conventional loadings.

## Upper bound methods

Upper bound methods are more suitable for analysis (i.e. calculating the ultimate strength of an existing structure) than for design; however, as will be seen, it is possible to use an upper bound method (yield line theory) to design slab bridges and top slabs.

### Introduction to yield line theory

The reader is referred to one of the specialist texts, e.g. Jones and Wood [84], for a detailed explanation of yield line theory, since only sufficient background to apply the method is given in the following.

The first stage in the yield line analysis of a slab is to propose a valid collapse mechanism consisting of lines, along which yield of the reinforcement takes place, and rigid regions between the yield lines. A possible mechanism for a slab bridge subjected to a point load is shown in Fig. 2.19; it can be seen that the geometry of the yield line pattern can be specified in terms of the parameter $\alpha$. In this pattern, only a single parameter is required to define the geometry, but, in a more general case, there could be a number of parameters; thus, in general, if there are $n$ parameters, each will be designated $\alpha_i$ where $i$ takes the values of one to $n$.

A point on the slab is then given a unit deflection; the deflection at any other point, and the rotations in the yield lines, can be calculated in terms of $\alpha_i$. In the slab bridge example of Fig. 2.19, it would be most convenient to consider the point of application of the load $P$ to have unit deflection.

Once the deflections at every point are known, the work done by the external loads, in moving through their deflections, can be calculated from

$$E = \Sigma[\int\int p \, \delta \, dx \, dy] \qquad (2.69)$$
$$\text{each rigid region}$$

where $p$ is the load per unit area on an element of slab of side $dx$, $dy$ and $\delta$, which is a function of $\alpha_i$, is the deflection of the element. The summation in equation (2.69) is

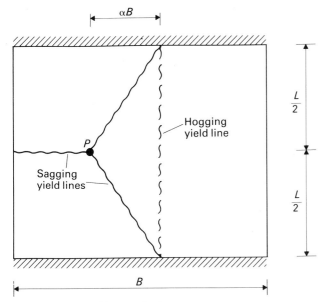

Fig. 2.19 Slab bridge mechanism

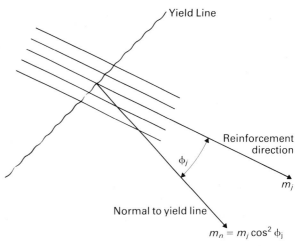

Fig. 2.20 Normal moment in yield line

carried out for each rigid region between yield lines, and the integration for each rigid region is carried out over its area. In practice, one is generally concerned with point loads, line loads and uniformly distributed loads. For a point load equation (2.69) reduces to the load multiplied by its deflection, and for a line load, or a uniformly distributed load, it reduces to the total load multiplied by the deflection of its centroid. These calculations are illustrated in the examples at the end of this chapter.

Similarly, once the rotations in the yield lines are known, the internal dissipation of energy in the yield lines can be calculated from

$$D = \Sigma \left[ \int m_n \, \theta_n \, dl \right] \qquad (2.70)$$
$$\text{each line}$$

where $\theta_n$, which is a function of $\alpha_i$, is the normal rotation in a particular yield line, $m_n$ is the normal moment of resistance per unit length of the yield line and $l$ is the distance along the yield line. The summation in equation (2.70) is carried out for each yield line, and the integration for each yield line is carried out over its length.

In general, a yield line will be crossed by a number of sets of reinforcement, each at an angle $\phi_j$ to the normal to the yield line, as shown in Fig. 2.20. If the moment of resistance per unit length of the $j$th set of reinforcement is $m_j$, in the direction of the reinforcement, then its contribution to the moment of resistance normal to the yield line is $m_j \cos^2 \phi_j$. Hence, the total normal moment of resistance is given by

$$m_n = \Sigma \, m_j \cos^2 \phi_j \qquad (2.71)$$

Since $\phi_j$ are each functions of $\alpha_i$, so also is $m_n$.

In practice, it is often easier to calculate the dissipation of energy in a yield line by considering the rotations ($\theta_j$) of the yield line in the direction of each set of reinforcement, and by considering the projections ($l_j$) of the yield line in the directions normal to each set of reinforcement. The dissipation of energy is then given by

$$D = \Sigma \left[ \Sigma \int m_j \, \theta_j \, dl_j \right] \qquad (2.72)$$
$$\text{each line}$$

The use of this equation is illustrated in the examples at the end of this chapter.

The next stage in the analysis is to equate the external work done to the internal dissipation of energy, and to arrange the resulting equation as an expression for the applied load ($P$ in the example of Fig. 2.19) as a function of $\alpha_i$. The minimum value of $P$ for the proposed collapse mechanism can then be found by differentiating the expression for $P$ with respect to each of $\alpha_i$, and equating to zero. The resulting set of $n$ simultaneous equations can be solved to give $\alpha_i$ and hence $P$.

It is emphasised that, although the resulting value of $P$ is the lowest value for the chosen mechanism, it is not necessarily the lowest value that could be obtained for the slab. This is because there could be an alternative mechanism which would give a lower minimum value of $P$. It is thus necessary to propose a number of different collapse mechanisms and to carry out the above calculations and minimisation for each mechanism.

A major drawback of yield line theory is that the engineer cannot be sure, even after he has examined a number of mechanisms, that there is not another mechanism which would give a lower value of the collapse load. The engineer is thus dependent on his experience, or that of others, when proposing collapse mechanisms: fortunately, the critical mechanisms have been documented for a number of practical situations.

### Yield line analysis of slab bridges

Yield line theory can be used for calculating the ultimate strength of a slab bridge which has been designed by another method. The various possible critical collapse mechanisms, and their equations, have been documented by Jones [85] for the general case of a simply supported skew slab subjected to either a uniformly distributed load or a single point load. Granholm and Rowe [86] give guidance on choosing the critical mechanism for a simply supported skew slab bridge subjected to a uniformly distributed load plus a group of point loads (i.e. a vehicle), and they also give the equations for such loadings. Clark [87] has extended these solutions to allow for different uniform load intensities on various areas of the bridge and for the application of a knife edge load. Although the above

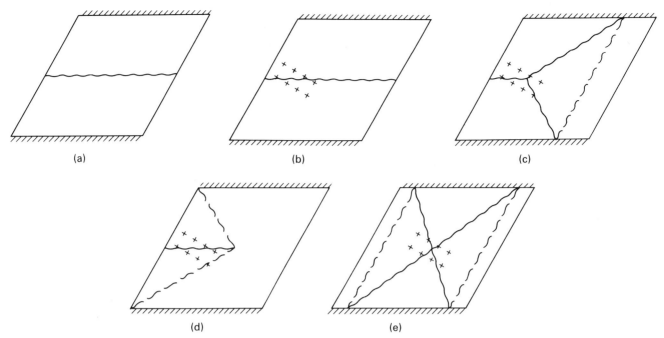

**Fig. 2.21(a)–(e)** Skew slab bridge mechanisms

authors give general equations for the various mechanisms and loadings, it is very often simpler, in practice, to work from first principles, as shown in the examples at the end of this chapter.

The possible critical mechanisms which should be examined when assessing the ultimate strength of a simply supported skew slab bridge are those shown in Fig. 2.21.

In the case of a continuous slab, similar mechanisms to those of Fig. 2.21 would form but, in each case, there would also be a hogging yield line at the interior support as shown in Fig. 2.22(a) and (b). Alternatively, a local mechanism could develop around the interior piers as shown in Fig. 2.22(c). Although these mechanisms are not considered explicitly in the literature, similar mechanisms are analysed in [84] and [85] and thus guidance is available in these texts.

## Yield line design of slab bridges

Although yield line theory is more suitable for analysis, it can be used for the design of slab bridges. When used for design, as opposed to analysis, the calculation procedure is very similar. A collapse mechanism is proposed; the external work done calculated from equation (2.69), in terms of the *known* loads $p$; the internal dissipation of energy calculated from equation (2.72), in terms of the *unknown* moments of resistance $m_j$; and the external work done equated to the internal dissipation of energy. However, instead of *minimising* the load with respect to the parameters $\alpha_i$ which define the yield line pattern, as is done when analysing a slab, the moments of resistance $m_j$ are *maximised* with respect to the parameters $\alpha_i$. Hence, in general, there are an infinite number of possible design solutions which result in different relative values of the moments of resistance. In practice, the designer *chooses* ratios for the moments of resistance, and it is usual to choose ratios which do not depart too much from the ratios of the equivalent elastic moments. Sometimes this is not necessary, because

the equations for one possible collapse mechanism involves only one set of reinforcement. It is thus possible to simplify the design procedure because it is possible to calculate $m_j$ for that set of reinforcement directly. The values of $m_j$ for the other sets of reinforcement can then be calculated by considering alternative mechanisms. As an example, if the transverse reinforcement is parallel to the abutments, mechanisms (a) and (b) of Fig. 2.21 involve only the longitudinal reinforcement. Hence, a suggested design procedure for simply supported skew slab bridges is to first provide sufficient longitudinal reinforcement to prevent mechanisms (a) and (b) forming under their appropriate design loads. The amount of transverse reinforcement required to prevent mechanisms (c) and (e) forming can then be calculated by setting up equations (2.69) and (2.72), equating them, and maximising the unknown transverse moment of resistance with respect to the parameters $\alpha_i$ defining the collapse mechanism. This procedure is illustrated in Example 2.2 at the end of this Chapter.

Clark [87] has designed model skew slab bridges by this method and then tested them to failure: it was found that the ratios of experimental to calculated ultimate load were 1.07 to 1.25 with a mean value of 1.16 for six tests. Granholm and Rowe [86] also tested model skew slab bridges and obtained values of the above ratio of 0.95 to 1.12 with a mean value of 1.08 for eleven tests. Thus, although yield line theory is theoretically an upper bound method, it can be seen, generally, to result in a safe estimate of the ultimate load and, thus, to a safe design if used as a design method.

It should be emphasised that, because of the values of the partial safety factors that have been adopted in the Code, it will often be found that the calculated amount of transverse reinforcement is less than the Code minima of 0.15% and 0.25% for high yield and mild steel respectively. When this occurs, the latter values should obviously be provided.

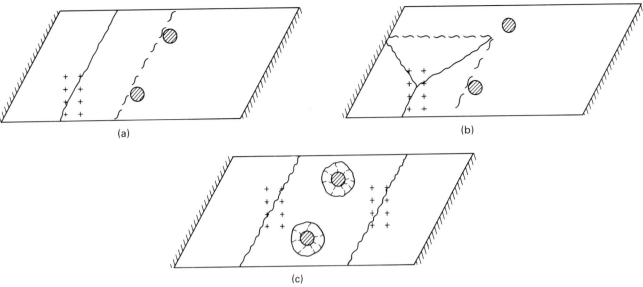

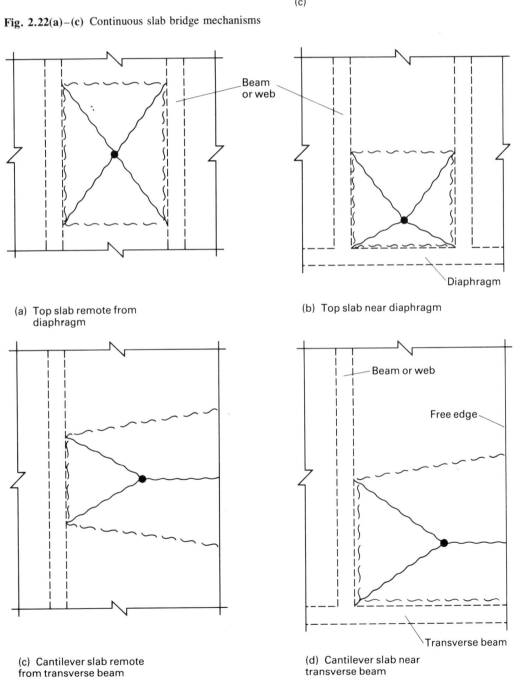

**Fig. 2.22(a)–(c)** Continuous slab bridge mechanisms

Beam or web

(a) Top slab remote from diaphragm

(b) Top slab near diaphragm

Diaphragm

(c) Cantilever slab remote from transverse beam

(d) Cantilever slab near transverse beam

Beam or web

Free edge

Transverse beam

**Fig. 2.23(a)–(d)** Top slab and cantilever slab mechanisms

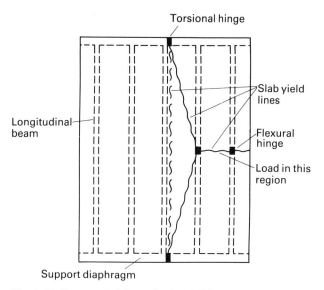

**Fig. 2.24** Beam and slab mechanism [88]

For a continuous slab, the design procedure is similar, except that the ratio of longitudinal hogging to sagging moment of resistance must be chosen by the designer. It is suggested that this ratio should be chosen to be similar to the ratio of the elastic support and span moments. The longitudinal reinforcement can then be determine from mechanism (a) of Fig. 2.22, and the transverse reinforcement from mecahnisms (b) and (c).

### Application to top slabs

Yield line theory can be used for the analysis and design of top slabs in beam and slab, cellular slab and box girder construction, and also for cantilever slabs. One should note that the application to top slabs of the simple yield line theory outlined earlier is conservative because it considers only flexural action, and the beneficial effects of membrane action are ignored; as, indeed, they are also for elastic design.

Possible collapse mechanisms for top slabs and cantilever slabs subjected to a uniformly distributed load plus a point load are shown in Fig. 2.23.

A possible design procedure is to calculate the amounts of longitudinal and transverse reinforcement required to resist the *global* bending effects, and then to determine the additional amounts of reinforcement required to prevent the mechanisms of Fig. 2.23 forming under the local effect loadings. It will be necessary for the designer to choose ratios of the transverse bottom to transverse top to longitudinal bottom to longitudinal top moments of resistance, or to predetermine some of the values. This procedure is illustrated by Example 2.3 at the end of this chapter.

### Upper bound methods for beam and slab bridges

An upper bound method has been proposed for beam and slab bridges by Nagaraja and Lash [88]. The type of mechanism considered is shown in Fig. 2.24, and consists of yield lines in the slabs, plus flexural hinges in the longitudinal beams and torsional hinges in the support diaphragms. The method of solution is similar to that described above for a slab bridges, and consists of equating the work

done by the loads to the energy dissipated in the yield lines and the flexural and torsional hinges. The rotations in the yield lines can be calculated from the geometry of the mechanism, as can the rotations in the flexural and torsional hinges. The ultimate moment of resistance of the beams, and the ultimate torque that can be resisted by the diaphragms, can be calculated by the methods described in Chapters 5 and 6. Nagaraja and Lash [88] give the equations for various mechanisms, but it is again suggested that, in practice, the calculations are carried out from first principles.

Nagaraja and Lash [88] compared ultimate loads calculated by such an approach with those recorded from tests on one-tenth scale models, and obtained ratios of experimental to calculated ultimate live loads of 0.90 to 1.12 with a mean value of 1.03 for twelve tests.

### Upper bound methods for box girders

Methods are available for carrying out upper bound analyses of box girders by considering overall collapse mechanisms, some of which involve distortion of the cross-section. Spence and Morley [82] and Morley and Spence [89] have considered the combined flexural plus distortional collapse mechanisms of Fig. 2.25 for simply supported single cell box girders with no internal diaphragms. The method of analysis is similar to that described previously for slabs; i.e., the work done by the applied loads is equated to the dissipation of energy in the mechanism. For mechanisms (a), (b) and (c), energy is dissipated in yielding the longitudinal reinforcement at midspan, in twisting the box walls and end diaphragms, and in rotation of the corner hinges. For mechanism (d), the webs do not twist, but energy is dissipated in the shear discontinuities along the corners near the loaded web. Spence and Morley [82] tested thirteen model girders without side cantilevers, and found that eight of these failed in pure bending and five exhibited failures involving distortion. The failure loads of those failing in pure bending were 0.88 to 0.97 of the theoretical pure bending failure load, and the failure loads of those failing in the distortional mode were 0.85 to 1.00 of the theoretical failure load calculated from consideration of mechanism (a). The tests indicated that the displacements at collapse were not sufficient for the webs, flanges and diaphragms to yield in torsion, and it was thus suggested that the appropriate dissipation terms be omitted from the equations.

Morley and Spence [89] have indicated how continuous single cell girders may be analysed. The loaded span fails by the formation of a mechanism similar to one of those in Fig. 2.25: the non-failing spans move outwards in order that the longitudinal stresses in the failing span satisfy the condition of zero axial force near to the supports. The dissipation of energy due to longitudinal extension and contraction in the support regions can then be calculated.

Cookson [90] has extended the work of Spence and Morley to multi-cell girders. Possible mechanisms are shown in Fig. 2.26: the critical mechanisms are generally those involving local failure of a single cell or a pair of cells. These mechanisms involve shear discontinuities and their analysis is quite complex. In addition, it appears that

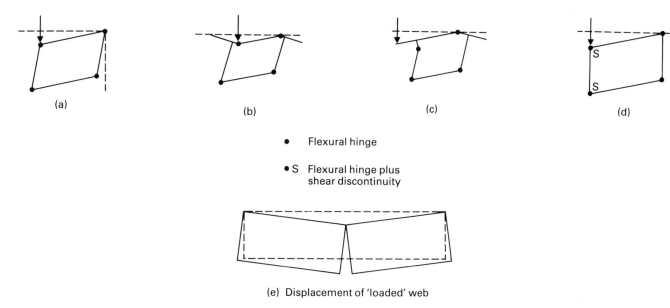

(a)   (b)   (c)   (d)

●    Flexural hinge

●S   Flexural hinge plus
shear discontinuity

(e) Displacement of 'loaded' web

**Fig. 2.25(a)–(e)** Single cell box girder mechanisms [82, 89]

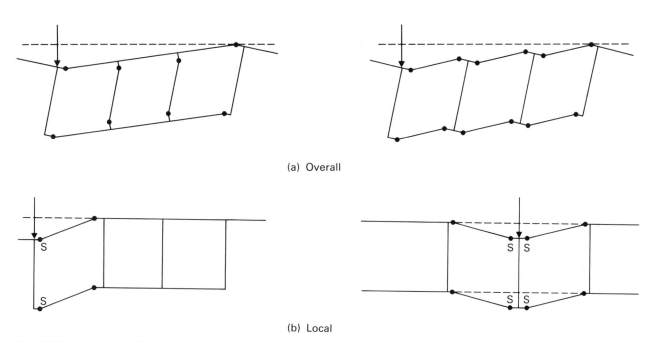

(a) Overall

(b) Local

**Fig. 2.26(a),(b)** Multi-cell box girder mechanisms [89, 90]

the amount of energy dissipated in the shear discontinuities is less than that predicted theoretically. Further work thus needs to be carried out before the method can be applied in practice.

# Model analysis and testing

The Code specifically permits the use of model analysis and testing to determine directly the load effects in a structure, or to justify a particular theoretical analysis.

Testing may also be used to determine the ultimate resistance of cross-sections which are not specifically covered by the Code.

# Examples

## 2.1 Yield line analysis of composite slab bridge

The validity of applying yield line theory to composite slab bridges has been demonstrated experimentally by Best and Rowe [91].

The method will be used to calculate the number of units of HB loading which would cause failure of a right simply supported composite slab bridge of 10 m span and 11.3 m breadth. The longitudinal and transverse sagging moments of resistance per unit width have been calculated by the methods of Chapter 5 to be 856 and 70 kN m/m respectively; and the hogging moments of resistance will

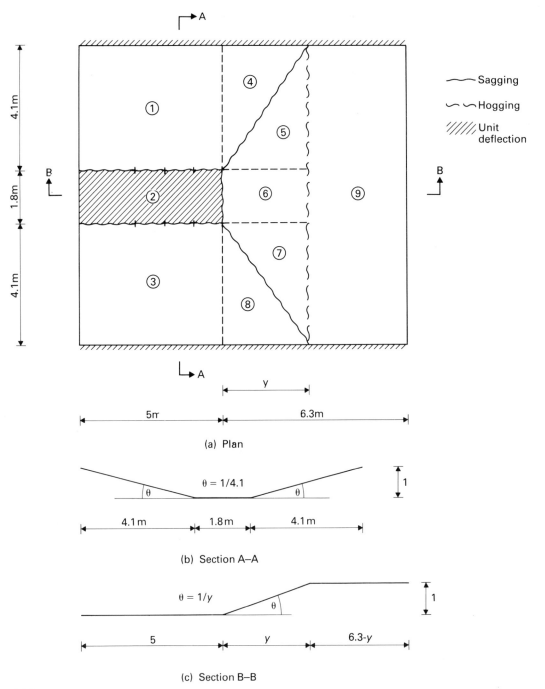

(a) Plan

(b) Section A–A

(c) Section B–B

**2.27(a)–(c)** Example 2.1

be assumed to be zero. The centre line of the external wheels of the HB vehicle cannot be less than 2 m from the free edge of the slab. The self weight of the slab can be taken as 12 kN/m², the superimposed dead load as 2.5 kN/m², and the parapets to apply line loadings of 3.5 kN/m along the free edges.

In view of the span, only one bogie of the HB vehicle will be considered, and the critical mechanism is that shown in Fig. 2.27(a); it is sufficiently accurate to assume that the HB bogie is positioned symmetrically about mid-span. In practice, each of the mechanisms of Fig. 2.21 would be considered to determine which would be critical.

If the shaded area of Fig. 2.27(a) deflects unity, then, from Figs. 2.27(b) and (c), the rotations in the longitudinal and transverse steel directions are 1/4.1 and 1/$y$ respectively, where $y$ is the unknown parameter which defines

the geometry of the mechanism. The projected lengths of the yield lines in the longitudinal and transverse steel directions are $(5 + y)$ and 10 m respectively. The dissipation of internal energy is thus given by (see equation (2.72))

Longitudinal steel $= 2(856)\,(1/4.1)\,(5 + y)$
$\qquad\qquad\qquad = 2088 + 418y$
Transverse steel $= (70)\,(1/y)\,(10) = 700/y$
$\therefore$ Total dissipation $= D = 2088 + 418y + (700/y)$

The HB vehicle deflects unity and, thus, if the HB load is $P$, the external work done is

$$E(\text{HB}) = P \times 1 = P$$

The work done by the uniformly distributed load is best calculated by dividing the slab into the nine regions shown in Fig. 2.27(a); the work done in each region is then the load intensity multiplied by the area of the region and by the deflection of the centroid of the area. Hence, and noting that region 9 does not deflect,

$$E(\text{u.d.1}) = (12 + 2.5) [2(5 \times 4.1) (1/2) + (5 \times 1.8) (1)$$
$$+ 4 (1/2 \times 4.1y) (1/3) + (1.8y) (1/2)]$$
$$= 428 + 53y$$

Similarly, the external work done by the parapet line loading is

$$E(\text{k.e.1}) = 3.5 [2(4.1) (1/2) + (1.8) (1)] = 21$$

$\therefore$ Total external work done $= E = P + 449 + 53y$

Now $E = D$

$\therefore P + 449 + 53y = 2088 + 418y + (700/y)$

or $P = 1639 + 365y + (700/y)$

In order to find the value of $y$ for a minimum $P$,

$$\frac{dP}{dy} = 365 - (700/y^2) = 0$$

$\therefore y = 1.385$ m

This value is less than 6.3 m, thus the proposed mechanism would be able to form.

$\therefore P = 2650$ kN   or   132 units of HB

## 2.2 Yield line design of skew slab bridge

A solid slab highway bridge has a right span of 10 m, a structural depth of 580 mm, a skew of 30° and the cross-section shown in Fig. 2.28(a). The specified highway loading is HA and 45 units of HB. The nominal superimposed dead load is equivalent to a uniformly distributed load of 2.5 kN/m², and the nominal parapet loading is 3.5 kN/m along each free edge. It is required to calculate, by yield line theory, the moments of resistance to be provided by bottom reinforcement, placed parallel to the slab edges, for load combination 1 (see Chapter 3). It will be assumed that no top steel is to be provided for strength.

In accordance with Chapter 3, the carriageway width is 9.3 m, and consists of three 3.1 m notional lanes.

It is explained in Chapter 1 that the design load effects (the moments of resistance in this example) are given by $\gamma_{f3}$ multiplied by the effects of $\gamma_{fL} Q_k$. However, when using yield line theory, it is necessary (see Chapter 4) to calculate the design load effects from a 'design' load of $\gamma_{f3} \gamma_{fL} Q_k$. The relevant nominal loads and partial safety factors discussed in Chapter 3 are given in Table 2.2

**Table 2.2** Example 2.2 'design' loads

| Load | $\gamma_{f_3}$ | $\gamma_{fL}$ | Nominal | 'Design' |
|------|------|------|---------|----------|
| Dead | 1.15 | 1.2 | 13.92 kN/m² | 19.2 kN/m² |
| Surfacing | 1.15 | 1.75 | 2.5 kN/m² | 5.03 kN/m² |
| Parapet | 1.15 | 1.75 | 3.5 kN/m | 7.04 kN/m |
| HA (alone) | 1.1 | 1.5 | 9.68 kN/m² plus 33.5 kN/m | 16.0 kN/m² plus 55.3 kN/m |
| HA (with HB) | 1.1 | 1.3 | ditto | 13.8 kN/m² plus 47.9 kN/m |
| HB | 1.1 | 1.3 | 450 kN/axle | 644 kN/axle |
| Footway | 1.1 | 1.5 | 4.0 kN/m² | 6.6 kN/m² |

The longitudinal reinforcement is designed by considering mechanism (b) of Fig. 2.21 which is found to be more critical than mechanism (a) under HA loading. The loads and mechanism are shown in Fig. 2.28(b).

For unit deflection of the yield line, the total rotation in the yield line $= 2(1/5 \sec 30) = 0.346$

If the moment of resistance per unit length of the longitudinal steel is $m_1$, the dissipation is given by

$$D = m_1(0.346) (13.3) = 4.6 m_1$$

The external work components are

$E(\text{HB}) = 2(644/4) [(3.47/5) + (3.97/5) + (4.47/5)$
$\qquad\qquad\qquad + (4.97/5)] = 1087$

$E(\text{k.e.1.}) = (47.9 + 47.9/3) (3.1 \sec 30) (1) = 229$

$E(\text{parapet}) = 4(7.04) (5 \sec 30) (1/2) = 81$

$E(\text{u.d.1.1}) = 4(24.23) (5 \times 0.5 \sec 30) (1/2) = 140$

$E(\text{u.d.1.2}) = 4(30.83) (5 \times 1.5 \sec 30) (1/2) = 534$

$E(\text{u.d.1.3}) = 2(24.23) (5 \times 3.1 \sec 30) (1/2) = 434$

$E(\text{u.d.1.4}) = 2(38.03) (5 \times 3.1 \sec 30) (1/2) = 681$

$E(\text{u.d.1.5}) = 2(28.83) (5 \times 3.1 \sec 30) (1/2) = 516$

$\therefore$ Total work done $= E = 3702$

Now $D = E$

$\therefore 4.6 m_1 = 3702$

$\therefore m_1 = 805$ kN m/m

The transverse reinforcement is designed by considering mechanism (c) of Fig. 2.21 which can be idealised to that shown in Fig. 2.28(c) [86, 87].

For unit deflection of the shaded area, the rotation in the direction of the longitudinal steel is $1/4.874 = 0.205$

Projected length normal to longitudinal steel $=$
$\qquad 5.1 + y \sec 30 = 5.1 + 1.155y$

Longitudinal steel dissipation $=$
$2(805) (0.205) (5.1 + 1.155y) = 1683 + 382y$

Rotation in direction of transverse steel $= 1/y$

Projected length normal to transverse steel $= 10$

If the moment of resistance per unit length of the transverse steel is $m_2$, the dissipation is given by $(m_2) (1/y) (10) = 10m_2/y$

$\therefore$ Total dissipation $= D = 1683 + 382y + 10m_2/y$

The external work components are (noting that it is sufficiently accurate to assume that each wheel of the HB vehicle displaces unity)

$E(\text{HB}) = 2(644) (1) = 1288$

$E(\text{parapet}) = (7.04) [(1.8) (1) + 2(4.874) (1/2)] = 47$

$E(\text{k.e.1.}) = (47.9) (y) (1/2) = 24y$

$E(\text{u.d.1.1}) = 2(24.23) (4.874 \times 0.5) (1/2) = 59$

$E(\text{u.d.1.2}) = 2(30.83) (4.874 \times 1.5) (1/2) = 225$

$E(\text{u.d.1.3}) = 2(24.23) (4.874 \times 3.1) (1/2) = 366$

$E(\text{u.d.1.4}) = 4(38.03) (1/2 \times 4.221y) (1/3) = 107y$

$E(\text{u.d.1.5}) = (38.03) (1.559y) (1/2) = 30y$

$E(\text{u.d.1.6}) = (24.23) (1.8 \times 3.1) (1) = 135$

$E(\text{u.d.1.7}) = (30.83) (1.8 \times 1.5) (1) = 83$

$E(\text{u.d.1.8}) = (24.23) (1.8 \times 0.5) (1) = 22$

$\therefore$ Total work done $= 2225 + 161y$

Now $D = E$

$\therefore 1683 + 382y + 10m_2/y = 2225 + 161y$

$\therefore m_2 = 54.2y - 22.1y^2$

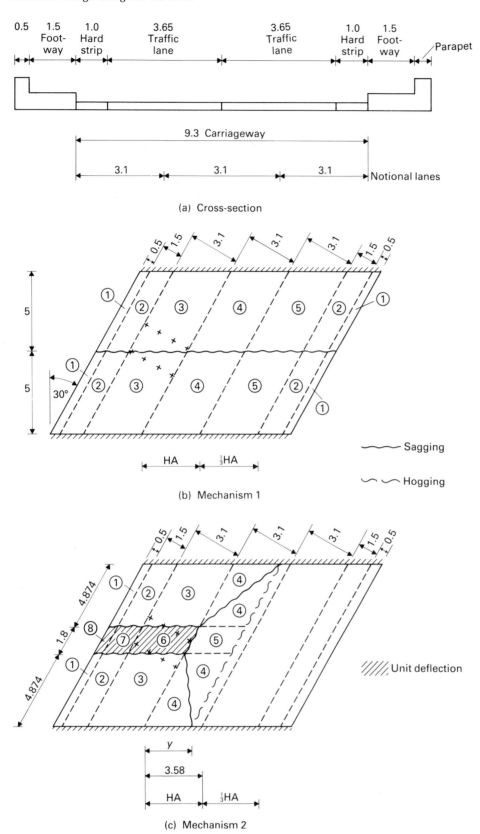

0.5  1.5  1.0  3.65  3.65  1.0  1.5
Foot-  Hard  Traffic  Traffic  Hard  Foot-  Parapet
way  strip  lane  lane  strip  way

9.3 Carriageway

3.1  3.1  3.1  Notional lanes

(a) Cross-section

5

5

30°

HA  ⅓HA

———— Sagging

～～ ～ Hogging

(b) Mechanism 1

4.874

1.8

4.874

///// Unit deflection

y

3.58

HA  ⅓HA

(c) Mechanism 2

**Fig. 2.28(a)–(c)** Example 2.2

In order to find the value of $y$ for a maximum $m_2$,

$$\frac{dm_2}{dy} = 54.2 - 44.2y = 0$$

$$\therefore y = 1.226 \text{ m}$$

This value is less than 3.58 m, thus the mechanism would form as shown in Fig. 2.28(c).

$$\therefore m_2 = 33.2 \text{ kN m/m}$$

The amount of reinforcement, that would develop this moment, would be less than the Code minima discussed in Chapter 10 and, thus, the latter value should be provided. This will generally be the case when designing in accordance with yield line theory.

Finally, although no top steel is required to develop adequate strength, some will be required in the obtuse corners to control cracking.

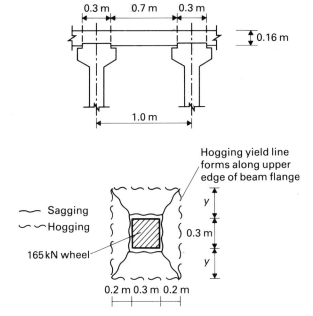

**Fig. 2.29** Example 2.3

## 2.3 Yield line design of top slab

A bridge deck consists of M-beams at 1.0 m centres with a 160 mm thick top slab. It is required to design the top slab reinforcement, in accordance with yield line theory, to withstand the HA wheel load.

It is necessary to pre-determine some of the reinforcement areas, and it will thus be assumed that the Code minimum area of 0.15% of high yield steel is provided in both the top and bottom of the slab in a direction parallel to the M-beams. Such reinforcement would provide sagging and hogging moments of resistance per unit length of 5.35 kN m/m. Let the sagging and hogging moments of resistance per unit length normal to the beams be $m_1$ and $m_2$ respectively.

The 'design' loads are (see Chapter 3 for details of nominal loads and partial safety factors)

HA wheel load $= (1.1)\,(1.5)\,(100) = 165$ kN
Self weight $= (1.15)\,(1.2)\,(3.84) = 5.30$ kN/m$^2$
Surfacing (say) $= (1.15)\,(1.75)\,(2.5) = 5.03$ kN/m$^2$
Total u.d.l. $= 10.33$ kN/m$^2$

The contact area of the HA wheel load is a square of side 300 mm (see Chapter 3). Dispersal through the surfacing will be conservatively ignored.

The collapse mechanism will be as shown in Fig. 2.29 in which the parameter $y$ defines the geometry of the mechanism. If the wheel deflects unity, the rotations parallel and perpendicular to the beams are $(1/y)$ and $(1/0.2)$ respectively.

The internal dissipation of energy is

$$D = 2(m_1 + m_2)\,(1/0.2)\,(0.3 + 2y) +$$
$$2(5.35 + 5.35)\,(1/y)\,(0.7)$$
$$= (m_1 + m_2)\,(3 + 20y) + 14.98/y$$

The external work done is

$$E = (165)\,(1) + 10.33\,[2(0.3)\,(y)\,(1/2)$$
$$+ 2(0.3)\,(0.2)\,(1/2) + 8(1/2)\,(0.2)\,(y)\,(1/3)]$$
$$= 165.6 + 5.857y$$

Now $D = E$

$$\therefore (m_1 + m_2)\,(3 + 20y) + 14.98/y = 165.6 + 5.857y$$

$$\therefore (m_1 + m_2) = \frac{165.6 + 5.857y - 14.98/y}{3 + 20y}$$

In order to find the value of $y$ for a maximum $(m_1 + m_2)$.

$$\frac{d(m_1 + m_2)}{dy} = 0$$

From which, $y = 0.239$ m
$\therefore (m_1 + m_2) = 13.4$ kNm/m

Any values of $m_1$ and $m_2$ may be chosen, provided that they sum to at least 13.4 kNm/m and that they are not less than the required *global* transverse sagging and hogging moment of resistance respectively.

## Chapter 3

# Loadings

## General

As explained in Chapter 1, nominal loads are specified in Part 2 of the Code, together with values of the partial safety factor $\gamma_{fL}$, which are applied to the nominal loads to obtain the design loads. It should be noted that, in the Code, the term 'load' includes both applied forces and imposed deformations, such as those caused by restraint of movements due to changes in temperature.

The nominal loads are very similar to those which appear as working loads in the present design documents BS 153 and BE 1/77.

## Loads to be considered

The Code divides the nominal loads into two groups: namely, permanent and transient.

### Permanent loads

Permanent loads are defined as dead loads, superimposed dead loads, loads due to filling materials, differential settlement and loads derived from the nature of the structural material. In the case of concrete bridges, the latter refers to shrinkage and creep of the concrete.

### Transient loads

All loads other than the permanent loads referred to above are transient loads: these consist of wind loads, temperature loads, exceptional loads, erection loads, the primary and secondary highway loadings, footway and cycle track loadings, and the primary and secondary railway loadings.

Primary highway and railway loadings are vertical live loads, whereas the secondary loadings are the live loads due to changes in speed or direction. Hence the secondary highway loadings include centrifugal, braking, skidding and collision loads; and the secondary railway loadings include lurching, nosing, centrifugal, traction and braking loads.

## Load combinations

There are three principal (combinations 1 to 3) and two secondary (combinations 4 and 5) combinations of load.

### Combination 1

The loads to be considered are the permanent loads plus the appropriate primary live loads for highway and footway or cycle track bridges; or the permanent loads plus the appropriate primary and secondary live loads for railway bridges.

### Combination 2

The loads to be considered are those of combination 1 plus wind loading plus erection loads when appropriate.

### Combination 3

The loads to be considered are those of combination 1 plus those arising from restraint of movements, due to temperature range and differential temperature distributions, plus erection loads when appropriate.

### Combination 4

This combination only applies to highway and footway or cycle track bridges.

The loads to be considered for highway bridges are the

**Table 3.1**  $\gamma_{fL}$

| Load | Limit state | $\gamma_{fL}$ for combination | | | | |
|---|---|---|---|---|---|---|
| | | 1 | 2 | 3 | 4 | 5 |
| Dead – steel | U | 1.05 | 1.05 | 1.05 | 1.05 | 1.05 |
| | S | 1.00 | 1.00 | 1.00 | 1.00 | 1.00 |
| – concrete | U | 1.15 | 1.15 | 1.15 | 1.15 | 1.15 |
| | S | 1.00 | 1.00 | 1.00 | 1.00 | 1.00 |
| Superimposed dead | U | 1.75 | 1.75 | 1.75 | 1.75 | 1.75 |
| | S | 1.20 | 1.20 | 1,20 | 1.20 | 1.20 |
| Reduced value for dead and superimposed dead if a more severe effect results | U | 1.00 | 1.00 | 1.00 | 1.00 | 1.00 |
| Wind – during erection | U | | 1.10 | | | |
| | S | | 1.00 | | | |
| – plus dead plus superimposed dead | U | | 1.40 | | | |
| | S | | 1.00 | | | |
| – plus all other combination 2 loads | U | | 1.10 | | | |
| | S | | 1.00 | | | |
| – relieving effect | U | | 1.00 | | | |
| | S | | 1.00 | | | |
| Temperature – range | U | | | 1.30 | | |
| | S | | | 1.00 | | |
| – friction at bearings | U | | | | | 1.30 |
| | S | | | | | 1.00 |
| – difference | U | | | 1.00 | | |
| | S | | | 0.80 | | |
| Differential settlement | U | not specified | | | | |
| | S | | | | | |
| Earth pressure – fill or surcharge | U | 1.50 | 1.50 | 1.50 | 1.50 | 1.50 |
| | S | 1.00 | 1.00 | 1.00 | 1.00 | 1.00 |
| – relieving effect | U | 1.00 | 1.00 | 1.00 | 1.00 | 1.00 |
| Erection | U | | 1.15 | 1.15 | | |
| Highway primary – HA alone | U | 1.50 | 1.25 | 1.25 | | |
| | S | 1.20 | 1.00 | 1.00 | | |
| – HA with HB or HB alone | U | 1.30 | 1.10 | 1.10 | | |
| | S | 1.10 | 1.00 | 1.00 | | |
| Highway secondary – centrifugal | U | | | | 1.50 | |
| | S | | | | 1.00 | |
| – longitudinal HA | U | | | | 1.25 | |
| | S | | | | 1.00 | |
| – longitudinal HB | U | | | | 1.10 | |
| | S | | | | 1.00 | |
| – skidding | U | | | | 1.25 | |
| | S | | | | 1.00 | |
| – parapet collision | U | | | | 1.25 | |
| | S | | | | 1.00 | |
| – support collision | U | | | | 1.25 | |
| | S | | | | 1.00 | |
| Foot/cycle track | U | 1.50 | 1.25 | 1.25 | 1.25 | |
| | S | 1.00 | 1.00 | 1.00 | 1.00 | |
| Railway | U | 1.40 | 1.20 | 1.20 | | |
| | S | 1.10 | 1.00 | 1.00 | | |

permanent loads plus a secondary live load with its associated primary live load: each of the secondary live loads discussed later in this chapter are considered individually.

The loads to be considered for footway or cycle track bridges are the permanent loads plus the secondary live load of a vehicle colliding with a support.

## Combination 5

The loads to be considered are the permanent loads plus the loads due to friction at the bearings.

## Partial safety factors

The values of the partial safety factor $\gamma_{fL}$ to be applied at the ultimate and serviceability limit states for the various load combinations are given in Table 3.1. The individual values are not discussed at this juncture, but the following general points should be noted:

1. Larger values are specified for the ultimate than for the serviceability limit state.
2. The values are less for reasonably well defined loads, such as dead load, than for more variable loads, such

as live or superimposed dead load. Hence the greater uncertainty associated with the latter loads is reflected in the values of the partial safety factors.

3. The value for a live load, such as HA load, is less when the load is combined with other loads, such as wind load in combination 2 or temperature loading in combination 3, than when it acts alone, as in combination 1. This is because of the reduced probability that a number of loads acting together will all attain their nominal values simultaneously. This fact is allowed for by the partial safety factor $\gamma_{f2}$, which was discussed in Chapter 1 and which is a component of $\gamma_{fL}$.

4. A value of unity is specified for certain loads (e.g. superimposed dead load) when this would result in a more severe effect. This concept is discussed later.

5. The values for dead and superimposed dead load at the ultimate limit state can be different to the tabulated values, as is discussed later when these loads are considered in more detail.

# Application of loads

## General

The general philosophy governing the application of the loads is that the worst effects of the loads should be sought. In practice, this implies that the arrangement of the loads on the bridge is dependent upon the load effect being considered, and the critical section being considered. In addition, the Code requires that, when the most severe effect on a structural element can be diminished by the presence of a load on a certain portion of the structure, then the load is considered to act with its least possible magnitude. In the case of dead load this entails applying a $\gamma_{fL}$ value of 1.0: it is emphasised that this value is applied to *all* parts of the dead load and not solely to those parts which diminish the load effect. In the case of superimposed dead load and live load, these loads should not be applied to those portions of the structure where their presence would diminish the load effect under consideration.

Influence lines are frequently used in bridge design and, in view of the above, it can be seen that superimposed dead load and live load should be applied to the *adverse* parts of an influence line and not to *relieving* parts. It is not intended that parts of parts of influence lines should be loaded. For example, the loading shown in Fig. 3.1(a) should not be considered.

## Highway carriageways and lanes

### Carriageway

The carriageway is defined as the traffic lanes plus hard shoulders plus hard strips plus marker strips. If raised kerbs are present, the carriageway width is the distance between the raised kerbs. In the absence of raised kerbs,

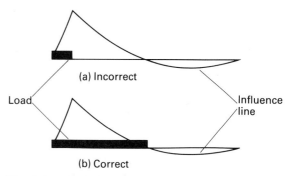

**Fig. 3.1(a),(b)** Influence line loading

the carriageway width is the distance between the safety fences less the set-back for the fences: the set-back must be in the range 0.6 to 1.0 m.

### Traffic lane

The lanes marked on the running surface of the bridge are referred to as traffic lanes. Hence traffic lanes in the Code are equivalent to working lanes in BE 1/77. However, the traffic lanes in the Code have no significance in deciding how live load is applied to the bridge.

### Notional lanes

These are notional parts of the carriageway which are used solely for applying the highway loading. They are equivalent to the traffic lanes in BE 1/77 and they are determined in a similar manner, although the actual numerical values are a little different for some carriageway widths.

## Overturning of structure

The stability of a structure against overturning is calculated at the ultimate limit state. The criterion is that the least restoring moment due to unfactored *nominal* loads should be greater than the greatest overturning moment due to *design* loads.

## Foundations

The soil mechanics aspects of foundations should be assessed in accordance with CP 2004 [92], which is not written in terms of limit state design. Hence these aspects should be considered under *nominal* loads. However, when carrying out the structural design of a foundation, the reaction from the soil should be calculated for the appropriate *design* loads.

# Permanent loads

## Dead load

The nominal dead loads will generally be calculated from the usual assumed values for the specific weights of the materials (e.g. 24 kN/m³ for concrete). When such assumed values are used it is necessary, at the ultimate limit state, to adopt $\gamma_{fL}$ values of 1.1 for steel and 1.2 for concrete rather than the values of 1.05 and 1.15 respectively given in Table 3.1. The latter values are only used when the nominal dead loads have been accurately assessed from the final structure. Such an assessment would require the material densities to be confirmed and the weight of, for example, reinforcement to be ascertained. It is thus envisaged that in general the larger $\gamma_{fL}$ values will be adopted for design purposes. The emphasis placed on checking dead loads in the Code is because the dead load factor of safety is less than that previously implied by BE 1/77.

As indicated earlier, when discussing the $\gamma_{fL}$ values, it is necessary to consider the fact that a more severe effect, due to dead load at a particular point of a structure, could result from applying a $\gamma_{fL}$ value of 1.0 to the entire dead load rather than a value of, for concrete, 1.2.

## Superimposed dead load

The partial safety factor given in Table 3.1 for superimposed dead load appears to be rather large. The reason for this is to allow for the fact that bridge decks are often resurfaced, with the result that the actual superimposed dead load can be much greater than that assumed at the design stage [18]. However, by agreement with the appropriate authority, the values may be reduced from 1.75 at the ultimate limit state, and 1.2 at the serviceability limit state, to not less than 1.2 and 1.0 respectively. It is then the responsibility of the appropriate authority to ensure that the superimposed dead load assumed for design is not exceeded in reality.

As for dead load, the possibility of a more severe effect, due to applying a $\gamma_{fL}$ value of 1.0 to the entire superimposed dead load, should be considered. In addition, the removal of superimposed dead load from parts of the structure, where they would have a relieving effect, should be considered.

## Filling material

The nominal loads due to fill should be calculated by conventional principles of soil mechanics. The partial safety factor of 1.5 at the ultimate limit state seems to be high, particularly when compared with that of 1.2 for concrete. However, the reason for the large value is that the pressures on abutments, etc., due to fill, are considered to be calculable only with a high degree of uncertainty, particularly for the conditions after construction [18].

It seems reasonable to apply a factor of 1.5 when considering the lateral pressures due to the fill; but, when the vertical effects of the fill are considered, it seems more logical to treat the fill as a superimposed dead load and to argue that a $\gamma_{fL}$ value of 1.2 should be adopted because the volume of fill will be known reasonably accurately.

## Shrinkage and creep

Shrinkage and creep only have to be taken into account when they are considered to be important. Obvious situations are where deflections are important and in the design of the articulation for a bridge.

In terms of section design, procedures exist in Part 4 of the Code to allow for the effects of shrinkage and creep on the loss of prestress and in certain forms of composite construction. These aspects are discussed in Chapters 7 and 8.

## Differential settlement

In the Code, as in BE 1/77, the onus is placed upon the designer in deciding whether differential settlement should be considered in detail.

## Transient loads

### Wind

The clauses on wind loading are based upon model tests carried out at the National Physical Laboratory and which have been reported by Hay [93]. The tests were carried out in a constant airstream of 25 m/s, and the model cross-sections were very much more appropriate to steel than to concrete bridges.

The clauses are very similar to those in BE 1/77, and the calculation procedure is as lengthy. However, it is emphasised that, according to the Code, it is not necessary to consider wind loading in combination with temperature loading. In addition, as is also the case in BE 1/77, wind loading does not have to be applied to the superstructure of a beam and slab or slab bridge having a span less than 20 m and a width greater than 10 m. However, a number of overbridges have widths less than 10 m, and the exclusion clause is not applicable to these.

In general the calculation procedure is as follows.

The mean hourly wind speed is first obtained for the location under consideration from isotachs plotted on a map of the British Isles.

The maximum wind gust speed and the minimum wind gust speed are then calculated for the cases of live load both on and off the bridge. The minimum gust speed is appropriate to those areas of the bridge where the wind has a relieving effect, and is used with the reduced $\gamma_{fL}$ values of Table 3.1. The gust speeds are obtained by multiplying the mean hourly wind speed by a number of factors, which depend upon:

1. The return period: the isotachs are for a return period of 120 years (the design life of the bridge), but the Code permits a return period of 50 years to be adopted for foot or cycle track bridges, and of 10 years for erection purposes.
2. Funnelling: special consideration needs to be given to bridges in valleys, etc.

3. Gusting: a gust factor, which is dependent upon the height above ground level and the horizontal loaded length, is applied. The gust factor may be reduced for foot or cycle track bridges according to the height above ground level.

In addition the minimum wind gust speed is dependent upon an hourly speed factor which is a function of the height above ground level.

The next stage of the calculation is to determine the transverse and longitudinal wind loads (which are dependent on the gust speed, an exposed area and a drag coefficient), and the vertical wind load (which is dependent on the gust speed, the plan area and a lift coefficient).

This part of the calculation is probably the most complex and requires a certain amount of engineering judgement to be made. This is because the Code gives cross-sections for which drag coefficients may be obtained from the Code, and also gives cross-sections for which drag coefficients may not be derived and for which wind tunnel tests should be carried out.

It should also be stated that an 'overall depth' is required to determine the exposed area and the drag coefficient, but, in general, different values of the 'overall depth' are used for the two calculations.

Finally the transverse, longitudinal and vertical wind loads are considered in the following four combinations:

1. Transverse alone
2. Transverse ± vertical
3. Longitudinal alone
4. 50% transverse + longitudinal ± 50% vertical

There are thus less combinations than exist in BE 1/77.

## Temperature

The clauses on temperature loading are based upon studies carried out by the Transport and Road Research Laboratory and the background to the clauses has been described by Emerson [94, 95]. The clauses are essentially identical to those in BE 1/77 except that temperature loading does not have to be considered with wind loading.

There are, effectively, two aspects of temperature loading to be considered; namely, the restraint to the overall bridge movement due to temperature range, and the effects of temperature differences through the depth of the bridge.

### Temperature range

The temperature range for a particular bridge is obtained by first determining the maximum and minimum shade air temperature, for the location of the bridge, from isotherms plotted on maps of the British Isles. The isotherms were derived from Meteorological Office data and are for a return period of 120 years (the design life of the bridge). The shade air temperatures may be adjusted to those appropriate to a return period of 50 years for foot or cycle track bridges, for the design of joints and during erection. An adjustment should also be carried out for the height above mean sea level.

The minimum and maximum effective bridge temperatures can then be obtained from tables which relate shade air temperature to effective bridge temperature. The latter is best thought of as that temperature which controls the overall longitudinal expansion or contraction of the bridge. The tables referred to above were based upon data obtained from actual bridges [94]. The effective bridge temperatures are dependent upon the depth of surfacing, and a correction has to be made if this differs from the 100 mm assumed for concrete bridges in the tables. Emerson [95] has suggested that such an adjustment should also take account of the shape of the cross-section of the bridge.

The effective bridge temperatures are used for two purposes.

First, when designing expansion joints, the movement to be accommodated is calculated in terms of the expansion from a datum effective bridge temperature, at the time the joint is installed, up to the maximum effective bridge temperature and down to the minimum effective bridge temperature. The resulting movements are taken as the nominal values, which have to be multiplied by the $\gamma_{fL}$ values of Table 3.1. The coefficient of thermal expansion for concrete is given as $12 \times 10^{-6}/°C$, except for limestone aggregate concrete when it is $7 \times 10^{-6}/°C$. The author would suggest also considering the latter value for lightweight concrete [96].

Second, if the movement calculated as above is restrained, stress resultants are developed in the structure. These stress resultants are taken to be nominal loads, but this contradicts the definition of a load which, according to the Code, includes 'imposed deformation such as those caused by restraint of movement due to changes in temperature'. The author would thus argue that the movement is the load, and that any stress resultants arising are load effects. This difference of approach is important when designing a structure to resist the effects of temperature and is elaborated in Chapter 13.

The Code indicates how to calculate the nominal loads when the restraint to temperature movement is accompanied by flexure of piers or shearing of elastomeric bearings.

Coefficients of friction are given for roller and sliding bearings: these are used in conjunction with nominal dead and superimposed dead load to calculate the nominal load due to frictional bearing restraint. The values are the same as those in BE 1/77 and are based partly upon [97].

It should be noted that, as in BS 153, the effects of frictional bearing restraint are considered in combination with dead and superimposed dead load only. This is because the resistance to movement of roller or sliding bearings is least when the vertical load is a minimum. Hence movement could take place under dead load conditions and, having moved, the restraining force is relaxed [18].

### Temperature differences

Due to the diurnal variations in solar radiation and the relatively small thermal conductivity of concrete, severe non-linear vertical temperature differences occur through the depth of a bridge. Two distributions of such differences

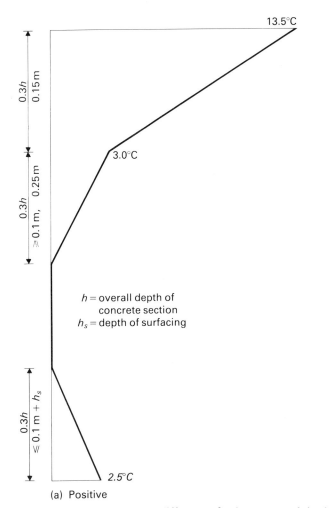

(a) Positive

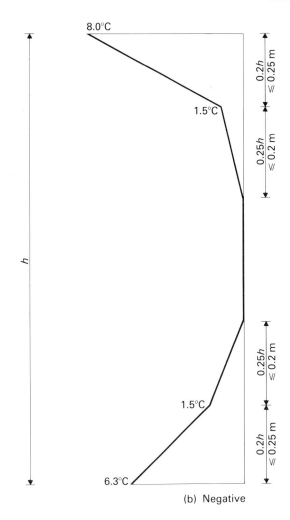

(b) Negative

**Fig. 3.2(a),(b)** Temperature differences for 1 m structural depth

are given in the Code (see Fig. 3.2). The temperature differences depend on the depth of concrete in the bridge: those shown in Fig. 3.2 are for a depth of 1 m. One of the distributions is for positive temperature differences, and is appropriate when there is a heat gain through the top surface; and the other is for negative temperature differences, and is appropriate when there is a heat loss from the top surface. The temperature distributions are composed of four or five straight lines which approximate the non-linear distributions, which have been calculated theoretically and measured on actual bridges by Emerson [95, 98]. The approximation has been shown to be adequate for design purposes by Blythe and Lunniss [99].

The Code distributions have been chosen to give the greatest temperature differences that are likely to occur in practice. It is not possible to think of them in terms of a return period, but they are likely to occur more than once a year [95].

The Code states that the effects of the temperature differences in Fig. 3.2 should be regarded as nominal values and that these effects, multiplied by $\gamma_{fL}$, should be regarded as design effects. This again appears to be an inconsistent use of terminology. The author would suggest that the nominal loads are the imposed deformations due to either internal or external restraint of the free movements implied by the temperature difference; and that the $\gamma_{fL}$ values should be applied to these to give design loads in the form of design imposed deformations. Any stresses or

stress resultants which are developed in response to the imposed deformations would then be the load effects. Such arguments regarding definitions may seem pedantic but they are important when designing a structure to resist the effects of temperature differences and are elaborated in Chapter 13.

Regarding the $\gamma_{fL}$ values, it should be noted from Table 3.1 that the serviceability limit state value is 0.8. This means that the final effects are only 80% of those calculated in accordance with BE 1/77, which adopts the same temperature difference diagrams. The reason for adopting a $\gamma_{fL}$ value of 0.8 at the serviceability limit state is not clear but, in drafts of the Code prior to May 1977, values of 1.0 and 1.2 for the serviceability and ultimate limit states, respectively, were specified. It thus appears that the Part 2 Committee thought it reasonable to reduce each of these by 0.2 in view of the reduced probability of a severe temperature difference occurring at the same time as a bridge is heavily loaded with live load.

The temperature differences given in the Code were calculated for solid slabs, but it is considered that the inaccuracy involved in applying them to other cross-sections is outweighed by certain assumptions made in the calculation. Measurements on box girders [100] and beam and slab [101] construction have shown that the temperature differences are very similar to those predicted for a solid slab of the same depth.

In addition, the temperature differences given in the

37

main body of the Code are for a surfacing depth of 100 mm. An appendix to Part 2 of the Code gives temperature differences for other depths of surfacing which are based upon the work of Emerson [95].

### Combination of temperature range and difference

A severe positive temperature difference can occur at any time between May and August, and measurements have shown that the lowest effective bridge temperature likely to co-exist with the maximum positive temperature difference is 15°C [95].

A severe negative temperature difference can occur at any time of the day, night or year. However, it is considered unlikely that a severe difference would occur between about ten o'clock in the morning and midnight on, or after, a hot sunny day. Thus, it is considered that a severe negative difference is unlikely to occur at an effective bridge temperature within 2°C of the maximum effective bridge temperature [95].

The above co-existing effective bridge temperatures have been adopted in the Code.

## Exceptional loads

These include the loads due to otherwise unaccounted effects such as earthquakes, stream flows, ice packs, etc. The designer is expected to calculate nominal values of such loads in accordance with the probability approach given in Part 1 of the Code and discussed in Chapter 1.

Snow loads should generally be ignored except for certain circumstances, such as when dead load stability could be critical.

## Erection loads

At the serviceability limit state, it is required that nothing should be done during erection which would cause damage to the permanent structure, or which would alter its response in service from that considered in design.

At the ultimate limit state, the Code considers the loads as either temporary or permanent and draws attention to the possible relieving effects of the former.

The importance of the method of erection, and the possibility of impact or shock loadings, are emphasised.

As already mentioned, wind and temperature effects during erection should generally be assessed for 10- and 50-year return periods respectively. For snow and ice loading, a distributed load of 500 N/m² will generally be adequate; this loading does not have to be considered in combination with wind loading.

## Primary highway loading

### General

The primary effects of highway loading are the vertical loads due to the mass of the traffic, and are considered as static loads.

The standard highway loading consists of normal (HA) loading and abnormal (HB) loading. The original basis of these loadings has been described by Henderson [102]. Both of these loadings are deemed to include an allowance for impact.

It should be noted from Table 3.1 that, at the serviceability limit state in load combination 1, the $\gamma_{fL}$ value for HA loading is 1.2. This value was chosen [18] because it was considered to reflect the difference between the uncertainties of predicting HA loading and dead load, which has a value of 1.0. Presumably, the HB value of 1.1 was chosen to be between the HA and dead load values.

### HA loading

HA loading is a formula loading which is intended to represent normal actual vehicle loading. The HA loading consists of *either* a uniformly distributed load plus a knife edge load *or* a single wheel load. The validity of representing actual vehicle loading by the formula loading has been demonstrated by Henderson [102], for elastic conditions, and by Flint and Edwards [103], for collapse conditions.

*Uniformly distributed load* The uniformly distributed component of HA loading is 30 kN per linear metre of notional lane for loaded lengths (*L*) up to 30 m and is given by 151 $(1/L)^{0.475}$ kN per linear metre of notional lane for longer lengths, but not less than 9 kN per linear metre. The loading is compared with the BE 1/77 loading in Fig. 3.3.

It can be seen that the two loadings are very similar; however, the upper cut-off is now 30 kN/m at 30 m instead of 31.5 kN/m at 23 m, and there is now a lower cut-off of 9 kN/m at 379 m. The latter has been introduced because of the lack of dependable traffic statistics for long loaded lengths [18]. Figure 3.3 gives the loading per linear metre of notional lane and the load intensity is always obtained by dividing by the lane width irrespective of the lane width. This is different to BE 1/77 and means that, for a loaded length less than 30 m, the load intensity can be in the range 7.9 to 13.0 kN/m² compared with the BE 1/77 range of 8.5 to 10.5 kN/m².

*Loaded length* The loaded length referred to above is the length of the base of the positive or negative portion of the influence line for a particular effect at the design point under consideration. Thus for a single span member, the loaded length for the span moment is the span. However, for a two span continuous member, having equal spans of 20 m, as shown in Fig. 3.4, the loaded length for calculating the support moment would be 40 m and hence a loading of 26.2 kN/m would be applied; and the loaded length for calculating the span moment would be 20 m and a loading of 30 kN/m would be applied.

For multispan members, each case will have to be considered separately. Thus the moment at support B of the four span member of Fig. 3.5 would be calculated by considering spans AB and BC loaded with loading appropriate to a loaded length of 2*L*, or spans AB, BC and DE loaded with loading appropriate to a loaded length of 3*L*: the former is likely to be more severe for most situations. The

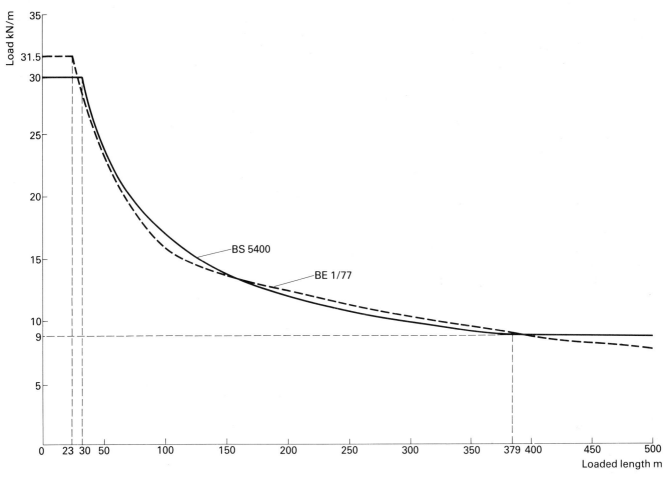

**Fig. 3.3** HA uniformly distributed load

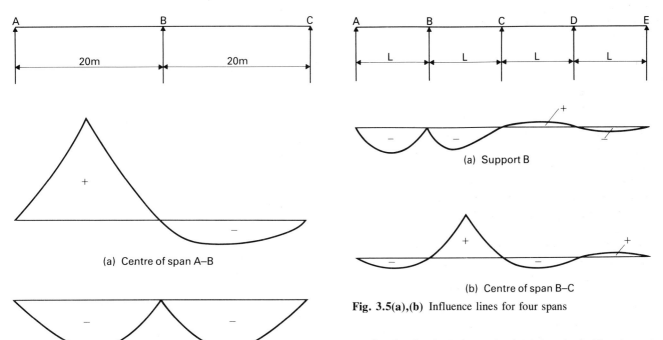

(a) Centre of span A–B

(b) Support B

**Fig. 3.4(a),(b)** Influence lines for two spans

(a) Support B

(b) Centre of span B–C

**Fig. 3.5(a),(b)** Influence lines for four spans

moment in span BC would be calculated by considering span BC loaded with loading appropriate to a loaded length of $L$, or spans BC and DE loaded with loading appropriate to a loaded length of $2L$.

*Knife edge load* It is emphasised that the knife edge part of HA loading is not intended to represent a heavy axle, but is merely a device to enable the same uniformly distributed loading to be used to simulate the shearing and bending effects of actual vehicle loading [102]. The Code value of the load is the same as that in BE 1/77, and is 120 kN per notional lane.

The load per metre is always obtained by dividing by the notional lane width and is thus in the range 31.6 to

52.2 kN/m, which should be compared with the BE 1/77 range of 32.4 to 52.2 kN/m.

The knife edge load is generally positioned perpendicular to the notional lane except when considering supporting members, in which case it is positioned in line with the bearings, and when considering skew decks, in which case it can be positioned parallel to the supporting members or perpendicular to the free edges. This clause is thus more precise than its equivalent in BE 1/77, which requires, for skew slabs, the knife edge load to be placed in a direction which produces the worst effect. It is understood that the intention of the Code drafters was that the intensity of loading should be 120 kN divided by the skew width of a notional lane when the knife edge loading is in a skew position with respect to the notional lane. Hence, the total load is always 120 kN per notional lane.

*Wheel load*   The wheel load is used mainly for local effect calculations and the nominal load is a single load of 100 kN with a contact pressure of 1.1 N/mm²: the contact area could thus be a circle of diameter 340 mm or a square of side 300 mm. The wheel load is considered to disperse through asphalt at a spread to depth ratio of 1 to 2 and through concrete at 1 to 1 down to the neutral axis.

This loading is thus different to the BE 1/77 load which consists of two 112 kN wheel loads, with a contact pressure of 1.4 N/mm², and a 45° angle of dispersion through both asphalt and concrete.

The change from two wheel loads to a single wheel load may seem drastic; however, there is a requirement in the Code that all bridges be checked under 25 units of HB loading. It is envisaged that the worst effects of the single 100 kN wheel load, or at least 25 units of HB loading, will be at least as onerous as those of the two 112 kN wheel loads.

The reduction in contact pressure results in a greater contact area, but the reduced dispersal through asphalt offsets this somewhat when the effective area at the neutral axis is considered.

## HB loading

HB loading is intended to represent an abnormally heavy vehicle. The nominal loading consists of a single vehicle with 16 wheels arranged on four axles, as shown in Fig. 3.6, which also shows the BE 1/77 HB vehicle. It can be seen that the transverse spacing of the wheels on an axle has been rounded-off to 1.0 m, and that the overall width of the vehicle is now given as 3.5 m. The latter point means that it is not necessary to specify a minimum distance of the vehicle from a kerb, as is necessary in BE 1/77. However, the most significant difference is that the longitudinal spacing of the centre pair of axles is no longer constant at 6.1 m, but can be any of five values between 6 and 26 m inclusive. The reason for this is that the BE 1/77 vehicle originated in BS 153, which was intended for simply supported bridges (although, in practice, it was also applied to continuous bridges) and the worst effects in a simply supported bridge occur with the axles as close together as possible. However, since the Code is intended to be applied to any span configuration, a variable spacing

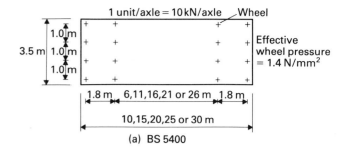

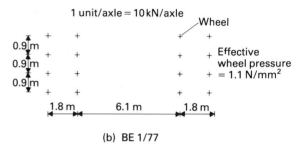

**Fig. 3.6(a),(b)** HB loading

has been specified in order to calculate the worst effects at all design points. As an example, the worst effects at an interior support of a continuous bridge could occur with a wide axle spacing.

The comments on the contact pressure of the HA wheel load and its dispersal are also pertinent to the wheels of the HB vehicle.

The nominal HB loading is specified, as in BE 1/77, in terms of units of loading, with one unit being equivalent to a total vehicle weight of 40 kN. The number of units for *all* roads can vary from 25 to 45, and this is at variance to BE 1/77 in which no minimum is specified; but, unlike BE 1/77, the number of units to be adopted for different types of road are not specified. Presumably, the Department of Transport will issue a memorandum containing guidance on this point.

### Application

*HA loading*   The full uniformly distributed and knife edge loads are applied to two notional lanes and one-third of these loads to all other notional lanes. The wheel load is applied anywhere on the carriageway. The applications are thus identical to those of BE 1/77.

*HB loading*   Only one HB vehicle is, in general, required to be considered on any one superstructure, or any substructure supporting two or more superstructures. The vehicle can be either wholly in a notional lane, or can straddle two notional lanes. If it is wholly in a notional lane, then the knife edge component of HA loading for that lane is completely removed, and the uniformly distributed component is removed for 25 m in front to 25 m behind the vehicle; the remainder of the lane is loaded with the uniformly distributed loading component of HA having an intensity appropriate to a loaded length which includes the displaced length. The vehicle is thus considered to displace part of the HA loading in one lane, but the adjacent lane is still assumed to carry full HA loading. This is more severe than the BE 1/77 loading, where the adjacent lane carries only one-third HA loading.

If the vehicle straddles two lanes, then it is considered to straddle *either* the two lanes loaded with full HA, *or* one lane with full and one with one-third HA; in each case the rules for omitting parts of the HA loading, which were described in the last paragraph, are applied. These arrangements of load are different to those of BE 1/77.

The reason for the more severe arrangement of the accompanying HA loading is that, in practice, queues of heavy vehicles accumulate behind abnormal loads and, when they overtake, they do so in a platoon [18].

It should be noted from Table 3.1 that, when HA loading is applied with HB loading, the $\gamma_{fL}$ values for HA are the same as those for HB.

*Verges, central reserves, etc.*  The accidental wheel loading of BE 1/77, for the loading of verges and central reserves, has been replaced in the Code by 25 units of HB loading: outer verges need only to be able to support any four wheels of 25 units of HB loading.

Transverse cantilever slabs should be loaded with the appropriate number of units of HB loading for the type of road in one notional lane *plus* 25 units of HB in one other notional lane. The latter is intended to be a substitute for HA loading and has been introduced because the HA loading no longer increases for spans less than 6.5 m, as it does in BS 153. This is the only occasion when more than one HB vehicle can act on a structure.

# Secondary highway loading

## General

The secondary effects of highway loading are loads parallel to the carriageway due to changes in speed or direction of the traffic. One should note that each of the following secondary loads is considered separately, and not in combination with the others. An associated primary load is applied with each of the secondary loads.

## Centrifugal load

This is a radial force applied at the surface of the road of a curved bridge. The nominal load is the same as that in BE 1/77 and is given by

$$F_c = \frac{30\ 000}{r+150}\ \text{kN} \tag{3.1}$$

where $r$ is the radius of the lane in metres. Any number of these loads at 50 m centres should be applied to any two notional lanes. Each load $F_c$ can be divided into two parts of $F_c/3$ and $2F_c/3$ at 5 m centres if these give a lesser effect.

The loading was based upon tests carried out at the Transport and Road Research Laboratory [104].

The nominal primary load associated with each load $F_c$ is a vertical load of 300 kN distributed uniformly over the notional lane for a length of 5 m. If the centrifugal load is divided, then the vertical load is divided in the same proportion.

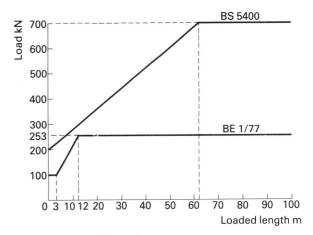

**Fig. 3.7** HA braking load

## Longitudinal braking

The longitudinal forces due to braking are applied at the level of the road surface. The nominal HA braking load is 8 kN per metre of loaded length plus 200 kN, with a maximum value of 700 kN. This load is much greater than the BE 1/77 loading with which it is compared in Fig. 3.7. The new loading is based upon the work of Burt [105] and is much greater than the BE 1/77 loading because of the greater efficiency of modern brakes, which can achieve decelerations which approach 1 g on dry roads [18]. The large increase in braking load could be significant in terms of substructure design. The HA braking load is applied, in one notional lane, over the entire loaded length, and in combination with full primary HA loading in that lane.

It is assumed that abnormally heavy vehicles can only develop a deceleration of 0.25 g and, thus, the HB braking force is taken to be 25% of the primary HB loading and to be equally distributed between the eight wheels of either the front two or the back two axles. This load is thus very similar to that in BE 1/77 for 45 units of loading, but is less severe for a smaller number of units.

## Skidding

This is a new loading which has been introduced because a coefficient of friction for lateral skidding of nearly 1.0 can be developed under dry road conditions [18]. A single nominal point load of 250 kN is considered to act in one notional lane, in any direction, and in combination with primary HA loading.

## Collision with parapets

The Code is not concerned with the design of the parapets, which will presumably still be covered by BE 5 [106], but only with the load transmitted to the member supporting the parapets. The nominal load is thus similar to the BE 1/77 loading, and is defined as the load to cause collapse of the parapet or its connection to the supporting member, whichever is the greater. The additional primary loading assumed to be acting adjacent to the point of collision consists of any four wheels of 25 units of the HB vehicle.

41

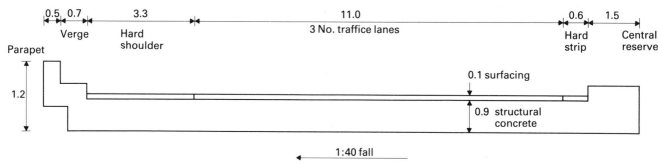

**Fig. 3.8** Loading example

### Collision with supports

In general, the Code recommends the provision of protection of bridge supports from possible vehicle collision. The nominal loads for highway bridge supports which should oe considered are the loads transmitted by the guard rail of 150 and 50 kN, normal and parallel to the carriageway respectively, at 0.75 m above the carriageway level or at the bracket attachment point. In addition, at the most severe point between 1 and 3 m above the carriageway, residual loads of 100 kN should be considered both normal and parallel to the carriageway. The normal and parallel loads should not be considered to act together. The above loads are only two-thirds of those in BE 1/77 and the normal and parallel components of the latter have to be applied together. Thus, even allowing for the fact that, at the ultimate limit state, the product $\gamma_{fL}\,\gamma_{f3}$ is about 1.44, as opposed to the BE 1/77 safety factor of 1.15, the Code loading is less severe than the BE 1/77 loading.

For a foot or cycle track bridge, the nominal collision load is a single load of 50 kN applied in any direction up to a height of 3 m above the carriageway. In view of the safety factors of 1.44 and 1.15 mentioned above, this loading is more severe than its BE 1/77 equivalent.

### Fatigue and dynamic loading

Fatigue loading is considered in Chapter 12.

It has been found [107] that the stress increments due to the dynamic effects of highway loading are within the allowance made for impact in the nominal loading, and thus it is not necessary to consider the effects of vibration.

## Footway or cycle track loading

If the bridge supports only a footway or a cycle track, the nominal load is 5 kN/m² for loaded lengths up to 30 m, above which the load of 5 kN/m² is reduced in the ratio of the HA uniformly distributed load, for the loaded length under consideration, to that for 30 m. The loading for loaded lengths greater than 30 m is thus less severe than the BE 1/77 loading.

The loading on elements supporting footways or cycle tracks, in addition to a highway or railway, is 80% of that mentioned above. However, if the footpath is wider than 2 m, the loading may be reduced further.

Greater reductions in loading are permitted if a main structural member supports two or more highway traffic lanes or railway tracks, in addition to foot or cycle track

loading. It is not obvious whether a slab bridge was intended to come under this category, but it would seem more reasonable to apply the previous '80% rule' to slab bridges.

It is necessary to consider the vibrations of foot and cycle track bridges as explained in Chapter 12.

## Railway loading

The railway loading was derived by a committee of the International Union of Railways and its derivation is fully explained in an appendix to Part 2 of the Code.

# Example

In the following example, the notation is in accordance with Part 2 of the Code, to which the various figure and table numbers refer. Numbers in brackets are the Part 2 clause numbers.

It is required to calculate the nominal transient loads which should be considered for a highway underbridge of composite slab construction and having the cross-section of Fig. 3.8, zero skew and a span of 15 m. The bridge is situated in the Birmingham area at a site which is 150 m above sea level and there are no special funnelling, gust or frost conditions. The anticipated effective bridge temperature at the time of setting the bearings is 16°C. Assume open parapets.

## Lanes

Carriageway width (3.2.9.1) = 3.3 + 11.0 + 0.6
$$= 14.9 \text{ m}$$
Number of notional lanes (3.2.9.3.1) = 4
Width of each notional lane = 14.9/4 = 3.725 m

## Wind

Since the bridge is less than 20 m span and greater than 10 m wide, it would be possible to ignore the effects of wind on the superstructure; however, the wind loads will be calculated in order to illustrate the calculation steps.

Mean hourly wind speed (Fig. 2) = $v$ = 28 m/s

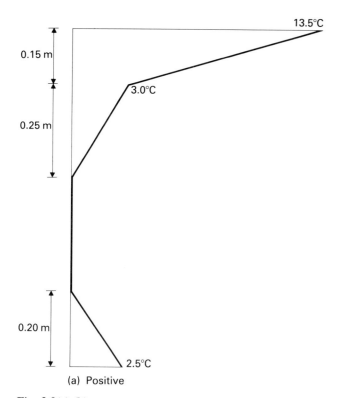

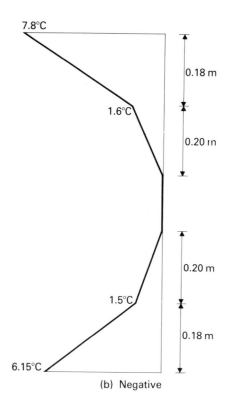

(a) Positive        (b) Negative

**Fig. 3.9(a),(b)**

$K_1$ (5.3.2.1.2) = 1.0
$S_1$ (5.3.2.1.3) = 1.0
$S_2$ (Table 2) = 1.49 for 6 m above ground and loaded length of 15 m.
Maximum gust speed (5.3.2.1) = $v_c = v K_1 S_1 S_2$
∴ $v_c$ = 41.7 m/s
This value of $v_c$ applies when there is no live load on the bridge; 5.3.2.3 states that $v_c \not> 35$ m/s with live load.
Area $A_1$ is calculated for the unloaded and loaded conditions, with $L$ = 15 m for both cases.
Unloaded (5.3.3.1.2(a)(1)(i)), $d$ = 1.2 m (Table 4)
∴ $A_1$ = 15 × 1.2 = 18 m²
Loaded (5.3.3.1.2(b)), d = 0.9 + 0.1 + 2.5 = 3.5 m
∴ $A_1$ = 15 × 3.5 = 52.5 m²
Drag coefficient ($C_D$) is calculated for the unloaded and loaded conditions, with $b$ = 17.6 m.
Unloaded, $d$ = 1.2 m (Table 5 (b))
$b/d$ = 17.6/1.2 = 14.7
$C_D$ = 1.0 (Fig. 5)
Superelevation = 1:40 = 1.43°
Increase $C_D$ by (Note 4 to Fig. 5) 3 × 1.43 = 4%
$C_D$ = 1.04
Loaded, $d$ = 2.5 m (Table 5 (b))
$b/d$ = 17.6/2.5 = 7.04
$C_D$ = 1.24 (Fig. 5)
Increase for superelevation, $C_D$ = 1.29
Dynamic pressure head (5.3.3) = $q = 0.613\ v_c^2$ N/m²
Unloaded, $q$ = 0.613 × 41.7²/1000 = 1.07 kN/m²
Loaded, $q$ = 0.613 × 35²/1000 = 0.751 kN/m²
Nominal transverse wind load (5.3.3) = $P_t = q A_1 C_D$
Unloaded, $P_t$ = 1.07 × 18 × 1.04 = 20.0 kN
Loaded, $P_t$ = 0.751 × 52.5 × 1.29 = 50.9 kN
Nominal longitudinal wind load (5.3.4) is the more severe of:

$P_{Ls} = 0.25\ q\ A_1\ C_D$
with $q$ = 1.07 kN/m², $A_1$ = 18 m², $C_D$ = 1.3 (5.3.4.1)
∴ $P_{Ls}$ = 0.25 × 1.07 × 18 × 1.3 = 6.26 kN
or $P_{Ls} + P_{LL}$
where $P_{Ls} = 0.25\ q\ A_1\ C_D$
with $q$ = 0.751 kN/m², $A_1$ = 18 m², $C_D$ = 1.3 (5.3.4.1)
and $P_{LL} = 0.5\ q\ A_1\ C_D$
with $q$ = 0.751 kN/m², $A_1$ = 2.5 × 15 = 37.5 m², $C_D$ = 1.45(5.3.4.3)
Thus $P_{Ls} + P_{LL}$ = (0.25 × 0.751 × 18 × 1.3) + (0.5 × 0.751 × 37.5 × 1.45)
= 24.8 kN
Thus $P_L$ = 24.8 kN
Nominal vertical wind load (5.3.5) = $P_v = q A_3 C_L$
$A_3$ = 15 × 17.6 = 264 m²
$C_L$ = ± 0.75
Unloaded, $P_v$ = 1.07 × 264 × (± 0.75) = ± 212 kN
Loaded, $P_v$ = 0.751 × 264 × (± 0.75) = ± 149 kN

## Temperature range

Minimum shade air temperature (Fig. 7) = −20°C
Maximum shade air temperature (Fig. 8) = 35°C
Height corrections (5.4.2.2) are (−0.5) (150/100) = −0.8°C
and (1.0) (150/100) = 1.5°C respectively.
Corrected minimum shade air temperature = −20.8°C
Corrected maximum shade air temperature = 36.5°C
Minimum effective bridge temperature (Table 10) = −12°C
Maximum effective bridge temperature (Table 11) = 36°C
Take the coefficient of expansion (5.4.6) to be 12 × 10⁻⁶/°C.
Nominal expansion = (36 − 16) 12 × 10⁻⁶ × 15 = 3.6 mm
Nominal contraction = (16 + 12) 12 × 10⁻⁶ × 15 = 5.04 mm

## Temperature differences

The temperature differences obtained from Fig. 9 are shown in Fig. 3.9. The positive differences can coexist (5.4.5.2) with effective bridge temperatures in the range 15° to 36°C and the negative differences with effective bridge temperatures in the range −12° to 34°C.

## HA

Uniformly distributed load for a loaded length of 15 m is 30.0 kN/m of notional lane (6.2.1). Thus the intensity is 30.0/3.725 = 8.05 kN/m². Knife edge load (6.2.2) = 120 kN/notional lane. Thus the intensity is 120/3.725 = 32.2 kN/m.

The wheel load (6.2.5) would not be considered for this bridge, but it is a 100 kN load with a circular contact area of 340 mm diameter. It can be dispersed (6.2.6) through the surfacing at a spread-to-depth ratio of 1 : 2, and through the structural concrete at 1 : 1 down to the neutral axis. Thus, diameter at neutral axis is 0.34 + 0.1 + 0.9 = 1.34 m.

## HB

Assume 45 units, then load per axle is 450 kN (6.3.1). For a single span bridge, the shortest axle spacing of 6 m is required. The contact area of a wheel is circular with an effective pressure of 1.1 N/mm² (6.3.2) Contact circle diameter = $[(450/4)(1000)(4)/(1.1\pi)]^{1/2}$ = 361 mm
Disperse (6.3.3) as for HA wheel load.
Diameter at neutral axis = 0.361 + 0.1 + 0.9 = 1.36 m
It should be noted that both HA and HB wheel loads can be considered to have square contact areas (6.2.5 and 6.3.3).

## Load on central reserve and verge

Load is (6.4.3) 25 units of HB, i.e. 62.5 kN/wheel.

Diameter of contact circle at neutral axis
= $[(62.5)(1000)(4)/(1.1\pi)]^{1/2}/1000 + 0.1 + 0.9$
= 1.27 m

## Longitudinal

For a loaded length of 15 m, the HA braking load (6.6.1) is 8 × 15 + 200 = 320 kN applied to one notional lane in combination with primary HA loading.

The total HB braking load (6.6.2) is 0.25 (4 × 450) = 450 kN equally distributed between the eight wheels of two axles 1.8 m apart. Applied with primary HB loading.

## Skidding

Point load of 250 kN (6.7.1) in one notional lane, acting in any direction, in combination with primary HA loading (6.7.2).

## Collision with parapet

Parapet collapse load (6.8.1) in combination with any four wheels of 25 units of HB loading (6.8.2).

## Footway

In order to demonstrate the calculation of footway loading, it will be assumed that the 1.5 m wide central reserve of Fig. 3.8 is replaced by a 3 m wide footway. The bridge supports both a footway and a highway and the nominal load for a 15 m loaded length is (7.2.1) 0.8 × 5.0 = 4.0 kN/m². However, because the footway width is in excess of 2 m, this loading may be reduced as follows.
Load intensity on first 2 m   = 4.0 kN/m²
Load intensity on other 1 m   = 0.85 × 4.0
    = 3.4 kN/m²
Average intensity     = (2 × 4.0 + 1 × 3.4)/3
    = 3.8 kN/m²

*Chapter 4*

# Material properties and design criteria

## Material properties

### Concrete

#### Characteristic strengths

As indicated in Chapter 1 material strengths are defined in terms of characteristic strengths. In Part 4 of the Code the characteristic cube compressive strength ($f_{cu}$) of a concrete is referred to as its grade, e.g. grade 40 concrete has a characteristic strength of 40 N/mm². Grades 20 to 50 may be used for normal weight reinforced concrete and 30 to 60 for prestressed concrete.

#### Stress–strain curve

The general form of the short-term uniaxial stress–strain curve for concrete in compression is shown by the solid line of Fig. 4.1(a). For design purposes, it is assumed that the descending branch of the curve terminates at a strain of 0.0035, and that the peak of the curve and the descending branch can be replaced by the chain dotted horizontal line at a stress of $0.67 f_{cu}$. The resulting Code short-term characteristic stress-strain curve is shown in Fig. 4.1(b).

The elastic modulus shown on Fig. 4.1(b) is an *initial tangent* value, and the Code also tabulates *secant* values which are used for elastic analysis (see Chapter 2) and for serviceability limit state calculations as explained in Chapter 7.

#### Other properties

Poisson's ratio is given as 0.2, and the coefficient of thermal expansion as $12 \times 10^{-6}/°C$ for normal weight concrete, with a warning that it can be as low as $7 \times 10^{-6}/°C$ for lightweight and limestone aggregate concrete. These values are reasonable when compared with published data [96, 108].

Certain other properties, such as tensile strengths, are required for the design of prestressed and composite members, but these properties are included as allowable stresses rather than as explicit characteristic values.

The shrinkage and creep properties of concrete can be evaluated from data contained in an appendix to Part 4 of

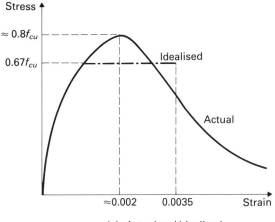

(a) Actual and idealised

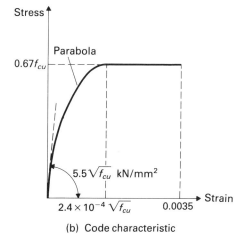

(b) Code characteristic

**Fig. 4.1(a),(b)** Concrete stress–strain curves

the Code. Approximate properties for use in the design of prestressed and composite members are discussed in Chapters 7 and 8.

### Reinforcement

#### Characteristic strengths

The quoted characteristic strengths of reinforcement ($f_y$) are 250 N/mm² for mild steel; 410 N/mm² for hot rolled

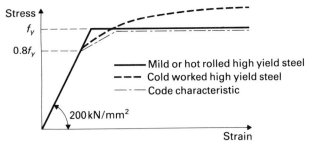

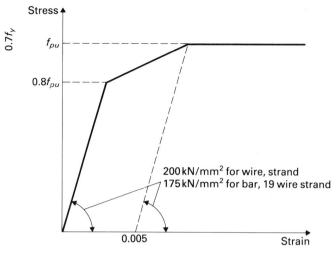

**Fig. 4.2** Reinforcement stress–strain curves

high yield steel; 460 N/mm² for cold worked high yield steel, except for diameters in excess of 16 mm when it is 425 N/mm²; and 485 N/mm² for hard drawn steel wire.

### Stress–strain curve

The general forms of the stress–strain curves for mild or hot rolled high yield steel and for cold worked high yield steel are shown by the solid lines of Fig. 4.2. The Code characteristic stress–strain curve is the tri-linear lower bound approximation to these curves, which is shown chain dotted in Fig. 4.2.

## Prestressing steel

### Characteristic strengths

Tables are given for the characteristic strengths of wire, strand, compacted strand and bars of various nominal size. Each tabulated value is given as a force which is the product of the characteristic strength $(f_{pu})$ and the area $(A_{ps})$ of the tendon.

### Stress–strain curve

The tri-linear characteristic stress–strain curves for normal and low relaxation tendons and for 'as drawn' wire and 'as spun' strand are shown in Fig. 4.3. They are based upon typical curves for commercially available products.

## Material partial safety factors

## Values

As was explained in Chapter 1, design strengths are obtained by dividing characteristic strengths by appropriate partial safety factors $(\gamma_m)$. The $\gamma_m$ values appropriate to the various limit states are summarised in Table 4.1. The sub-divisions of the serviceability limit state are explained later in this chapter when the design criteria are discussed.

The concrete values are greater than those for steel because of the greater uncertainty associated with concrete properties.

The explanation of the choice of 1.0 for both steel and concrete for analysis purposes, at both the serviceability and ultimate limit state, is as follows.

When analysing a structure, its overall response is of interest and, strictly, this is governed by material proper-

(a) Normal and low relaxation products

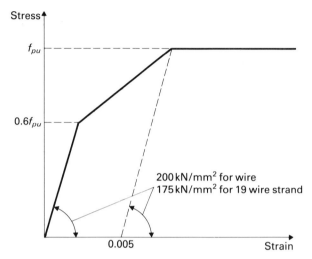

(b) 'As drawn' wire and 'as spun' strand

**Fig. 4.3(a),(b)** Prestressing steel stress–strain curves

**Table 4.1** $\gamma_m$ **values**

| Limit state | Concrete | Steel |
|---|---|---|
| Serviceability | | |
| Analysis of structure | 1.0 | 1.0 |
| Reinforced concrete cracking | 1.0 | 1.0 |
| Prestressed concrete cracking | 1.3 | 1.0 |
| Stress limitations | 1.3 | 1.0 |
| Vibration | 1.0 | 1.0 |
| Ultimate | | |
| Analysis of structure | 1.0 | 1.0 |
| Section design | 1.5 | 1.15 |
| Deflection | 1.0 | 1.0 |
| Fatigue | 1.3 | 1.0 |

ties appropriate to the mean strengths of the materials. If there is a linear relationship between loads and their effects, the values of the latter are determined by the relative and not the absolute values of the stiffnesses. Consequently the same effects are calculated whether the material properties are appropriate to the mean or characteristic strengths of materials. Since the latter, and not the mean strengths, are used throughout the Code it is simpler to use them for analysis; hence $\gamma_m$ values of 1.0 are specified.

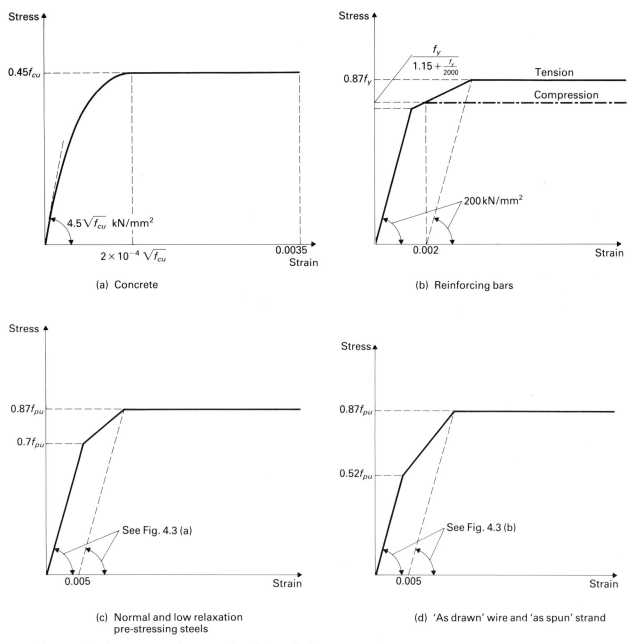

**Fig. 4.4(a)–(d)** Design stress–strain curves for ultimate limit state

For the same reason as above, $\gamma_m$ is taken to be 1.0 when analysing a section if the effect under consideration is associated with deformations; examples of this are cracking in reinforced concrete, deflection and vibration. However, when the effect under consideration is associated with a limiting stress, then a value of $\gamma_m$ should be adopted which should reflect the uncertainty associated with the particular material and the importance of the particular limit state.

The values of 1.5 for concrete and 1.15 for steel originated in the Comité Europeen du Beton, now the Comité Euro-International du Beton (CEB), which chose these values because, when used with the CEB partial safety factors for loads, they led to structures which were sensibly the same as those designed using the European national codes [109]. It should be noted that, although $\gamma_m$ for concrete is partially intended to reflect the degree of control over the production of concrete, a single value of 1.5 is adopted in the Code irrespective of the control, whereas the CEB

Recommendations [110] state that $\gamma_m$ can vary from 1.4, for accurate batching and control, to 1.6, for concrete made without strict supervision.

When carrying out stress calculations at the serviceability limit state, $\gamma_m$ values of 1.3 for concrete and 1.0 for steel are used. These values also originated in the CEB [111]. Cracking in prestressed concrete is considered to be a limiting stress effect because, as explained later in this chapter, it involves a limiting tensile stress calculation. However, it is emphasised that the $\gamma_m$ value of 1.3 for concrete for limiting stress calculations never has to be used by a designer because the stress limitations, which are given in the Code as design criteria, are design values which include the $\gamma_m$ value of 1.3.

A value of 1.3 for concrete is given for fatigue calculations because it is the strength of a section which is of interest. However, this value need never be used by a designer because there is not a requirement in the Code to check the fatigue strength of concrete.

## Design stress–strain curves

The $\gamma_m$ values referred to above are applied to the characteristic strengths whenever they appear in a calculation. Hence design stress–strain curves are obtained from the characteristic curves (see Figs. 4.1 to 4.3) by dividing the characteristic strengths ($f_{cu}, f_y, f_{pu}$), whenever they occur, by the appropriate values of $\gamma_m$. The design curves at the ultimate limit state are of particular interest and are shown in Fig. 4.4.

It should be noted from Fig. 4.1 that the concrete reaches its peak compressive stress, and then starts to crush, at a strain of about 0.002. Once the concrete starts to crush it is less effective in providing lateral restraint to any compression reinforcement, and there is thus a possibility of the latter buckling. Hence the design stress of compression reinforcement is restricted to the stress equivalent to a strain of 0.002 on the design stress–strain curve. This stress is

$$\frac{f_y}{1.15 + \dfrac{f_y}{2000}} \tag{4.1}$$

It lies in the range $0.718 f_y$ to $0.784 f_y$ for $f_y$ in the range 485 to 250 N/mm².

## Design criteria

In this chapter the design criteria are presented and discussed, but methods of satisfying the criteria are presented in subsequent chapters.

The criteria are given in Part 4 of the Code under the headings: ultimate limit state, serviceability limit state and other considerations. The latter includes those criteria which are not specified in the Code but which are, nonetheless, important in design terms. The criteria are listed in Table 4.2 from which it can be seen that there are a great number of criteria to be satisfied and, if calculations had to be carried out for each, the design procedure would be extremely lengthy. Fortunately, as explained later in this chapter, some of the criteria can be checked by 'deemed to satisfy' clauses. Furthermore, as experience of

**Table 4.2** Design criteria

| |
|---|
| *Ultimate limit state* |
| Rupture |
| Buckling |
| Overturning |
| Vibration |
| *Serviceability limit state* |
| Steel stress limitations |
| Concrete stress limitations |
| Cracking of prestressed concrete |
| Cracking of reinforced concrete |
| Vibration |
| *Other considerations* |
| Deflections |
| Fatigue |
| Durability |

the Code is gained, it will be possible to identify those criteria which would not be critical for a particular situation.

Each criterion is now discussed.

## Ultimate limit state

The criterion for rupture of one or more sections, buckling or overturning is simply that these events should not occur.

A vibration criterion, which would be concerned with vibrations to cause collapse of a bridge, is not given, but, instead, compliance with the serviceability limit state vibration criterion is deemed to satisfy the ultimate limit state requirements.

## Serviceability limit state

### Steel stress limitations

*Reinforcement* It is explained in Chapter 7 that it is generally only necessary to check cracks widths in highway bridges under HA loading for load combination 1. This means that there is an indirect check on reinforcement stresses under primary HA loading but not under other loads. In view of the fact that it is desirable to ensure that the steel remains elastic under *all* serviceability conditions, so that cracks which open under the application of occasional loading will close when the loading is removed, it was decided to introduce a separate steel stress criterion. It can be seen from Fig. 4.2 that the steel stress–strain curve becomes non-linear at a stress of $0.8 f_y$ and this stress is, thus, the Code criterion. It is unfortunate that the introduction of this criterion complicates the design procedure for reinforced concrete for two reasons:

1. As shown later in this Chapter, different $\gamma_{f3}$ values are given for the crack width and steel stress calculations and this could cause confusion.
2. As explained in Chapter 7, deemed to satisfy rules for crack control in slab bridges are given in the Code, but the fact that steel stresses have to be calculated counteracts to a large extent the advantages of the deemed to satisfy rules.

*Prestressing steel* Since a stress limitation existed for reinforcement, it was considered logical to specify similar criteria for prestressing steel. Hence, with reference to Fig. 4.3, stress limitations of $0.8 f_{pu}$, for normal and low relaxation products, and $0.6 f_{pu}$, for 'as drawn' wire and 'as spun' strand, are given in the general section of Part 4 of the Code. These criteria imply that tendon stress increments should be calculated under live load. Since such a calculation is not generally carried out in prestressed concrete design, a clause in the prestressed concrete section of Part 4 of the Code effectively states that the criteria can be ignored. Hence, to summarise, although prestressing steel limiting stress criteria are stated in the Code, they can be ignored in practice.

## Concrete stress limitations

It should be noted that this heading does not cover tensile stresses in prestressed concrete, which are covered by 'cracking of prestressed concrete'. The stress limitations referred to as 'concrete stress limitations' include compressive stresses in reinforced and prestressed concrete, and compressive, tensile and interface shear stresses in composite construction.

*Compressive stresses in reinforced concrete* In order to prevent micro-cracking, spalling and unacceptable amounts of creep occurring under serviceability conditions, compressive stresses are limited to $0.5 f_{cu}$.

*Compressive stresses in prestressed concrete* The limiting stresses for the serviceability limit state and at transfer are given in Table 4.3. It is *not* necessary to apply the $\gamma_m$ value of 1.3 to the stresses. The stresses are identical to those in BE 2/73 except that a stress of $0.4 f_{cu}$ is now permitted in support regions because such a region is subject to a triaxial stress system, due to vertical restraint to the compression zone from the support. In addition, the actual flexural stress at a support of finite width is less than the stress calculated assuming a concentrated support.

*Compressive stresses in composite construction* The compression flange of a prestressed precast beam with an in-situ concrete slab is restrained by the latter and placed under a triaxial stress condition. It is thus permitted, under such conditions, to increase the limiting stresses of Table 4.3(a) by 50%, but the increased stress should not exceed $0.5 f_{cu}$. (The Code clause actually refers to the stresses of Table 4.3(b) but it should be those of Table 4.3(a).) The resulting stresses are of the same order as those of BE 2/73. However, the Code implies that the increased stresses may be used both when a precast flange is completely encased in in-situ concrete (such as a composite slab) and for sections formed by adding an in-situ topping to precast beams. This is contrary to BE 2/73, which only permits increased stresses to be used for the first of these situations. The author would suggest that the approach suggested in the CP 110 handbook [112] could be adopted for flanges which have a topping but are not encased in in-situ concrete: the suggestion is that, in such circumstances, an increase of only 25% of the Table 4.3(a) values should be permitted.

**Table 4.3** Limiting concrete compressive stresses in prestressed concrete
(a) Serviceability limit state

| Loading | Allowable stress |
| --- | --- |
| Bending | $0.33 f_{cu}$ ($0.4 f_{cu}$ at supports) |
| Direct compression | $0.25 f_{cu}$ |

(b) Transfer

| Stress distribution | Allowable stress |
| --- | --- |
| Triangular | $0.5 f_{ci}$ |
| Uniform | $0.4 f_{ci}$ |

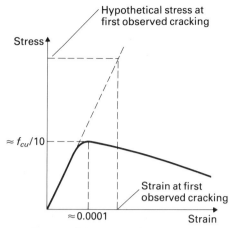

**Fig. 4.5** Tensile stress–strain curve of restrained concrete

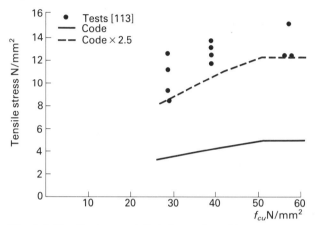

**Fig. 4.6** Tensile stresses at first observed cracking

*Tensile stresses in composite construction* When flexural tensile stresses are induced in the in-situ concrete of a composite member consisting of precast prestressed units and in-situ concrete, the precast concrete adjacent to the in-situ concrete restrains the latter and controls any cracks which may form. Hence, the descending branch of the tensile stress–strain curve can be made use of, and tensile strains in excess of the cracking strain of concrete tolerated. In the Code this is achieved by specifying allowable stresses which are in excess of the ultimate tensile strength of concrete and are, thus, hypothetical stresses as illustrated in Fig. 4.5.

The Code stresses are identical to those in BE 2/73 and are given in Table 4.4

**Table 4.4** Limiting concrete flexural tensile stresses in in-situ concrete

| In-situ concrete grade | 25 | 30 | 40 | 50 |
| --- | --- | --- | --- | --- |
| Tensile stress (N/mm²) | 3.2 | 3.6 | 4.4 | 5.0 |

The Table 4.4 values are extremely conservative, as can be seen from Fig. 4.6 where they are compared with inferred stresses at which cracking was first observed in tests on composite planks reported by Kajfasz, Somerville and Rowe [113]. The Code values could be obtained from the lower bound to the experimental values by applying a partial safety factor of 2.5. It is thus extremely unlikely that cracking of the in-situ concrete would be observed, even at the ultimate limit state, in a composite member designed in accordance with the Code.

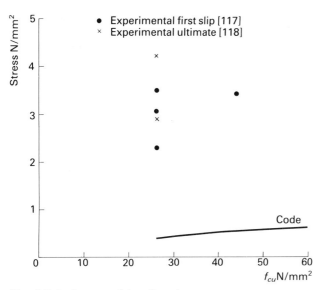

**Fig. 4.7** Surface type 1 interface shear stresses

It is permissible to increase the stresses in Table 4.4 by 50%, provided that the permissible tensile stress in the prestressed unit is reduced by the same numerical amount. This is because more prestress is then required, and it is known [114] that the enhancement of the tensile strain capacity of the adjacent in-situ concrete increases as the level of prestress at the contact surface increases.

*Interface shear in composite construction*   Three types of surface are defined as follows:

1.   Rough and no steel across the interface.
2.   Smooth and at least 0.15% steel across the interface.
3.   Rough and at least 0.15% steel across the interface.

These surface types originated in CP 116 and will lead to considerable problems for bridge engineers because the rough surface of types 1 and 3 requires the laitence to be removed from the surface by either wet brushing or tooling, and this is not the usual practice for precast bridge beams: the top surfaces of the latter are usually left 'rough as cast'.

The minimum link area of 0.15% and a Code detailing rule, which states that the link spacing in composite T-beams should not exceed four times the in-situ concrete thickness, nor 600 mm, were based upon American Concrete Institute recommendations [115].

The allowable interface shear stresses for beam and slab construction are given in Table 4.5; however, it should be emphasised that surface type 1 is not permitted for beam and slab bridge decks because it is always considered

**Table 4.5**   Limiting interface shear stresses

| In-situ concrete grade | Limiting interface shear stress (N/mm²) for | | |
|---|---|---|---|
| | Type 1 | Type 2 | Type 3* |
| 25 | 0.38 | 0.36 | 1.22 |
| 30 | 0.45 | 0.38 | 1.25 |
| 40 | 0.54 | 0.42 | 1.32 |
| 50 | 0.59 | 0.46 | 1.38 |
| 60 | 0.64 | 0.50 | 1.45 |

\* Increase by 0.5 N/mm² per 1% of links in excess of 0.15%

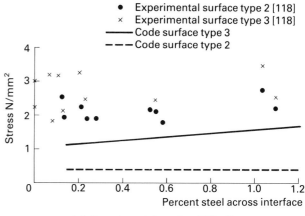

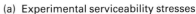

(a) Experimental serviceability stresses

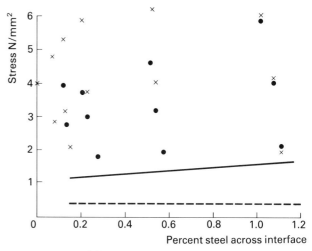

(b) Experimental ultimate stresses

**Fig. 4.8(a),(b)** Surface types 2 and 3 interface shear stresses ($f_{cu} = 25$ N/mm²)

necessary to provide links. This is because the calculated shear capacity of an unreinforced interface cannot be relied upon under conditions of repeated loading as occur on bridges. It can be seen that the Code approach to interface shear design is very different to that of BE 2/73 which is based upon an adaptation of the CP 117 approach [116]. In contrast, the Table 4.5 values were essentially chosen to be a little less conservative than the CP 116 values. However, they are still extremely conservative for surface types 1 and 2.

In Fig. 4.7 the Code surface type 1 stresses are compared with some test results of Hanson [117] and Saemann and Washa [118]. It can be seen that the Code stresses are very conservative.

The Code surface types 2 and 3 stresses can be considered by examining the experimental results of Saemann and Washa [118], who tested composite I-beams in which the steel area across the interface and the shear span to effective depth ratio were varied. In addition, three surfaces were tested, and two of these were equivalent to the Code types 2 and 3. In Fig. 4.8 the test data for a cube strength of about 25 N/mm² and the Code stresses are compared at a serviceability criterion of a slip of 0.127 mm (as suggested by Hanson [117]) and at failure. It can be seen that the Code type 3 stresses are reasonable; but the Code type 2 stresses are very conservative and should be more dependent on steel area than the type 3 stresses, whereas the Code allows no increase of the type 2 stresses for a steel area in excess of 0.15%. In view of this the author would suggest that surface type 3 be considered to be applicable to the 'rough as cast' surface used in bridge practice.

Finally, it is emphasised that it was not intended that interface shear should be checked in composite slabs formed from precast inverted T-beams with solid infill. It is understood that the limiting stresses given in the Code for composite slabs were intended for *shallow* slabs. The type of situation where they would be applicable in bridges is where the top slab of a deck consists of precast units spanning between longitudinal beams, and the units act as permanent formwork for in-situ concrete to form a composite slab.

### Cracking of prestressed concrete

The criteria for the control of cracking in prestressed concrete are presented in terms of limiting flexural tensile stresses for three classes of prestressed concrete.

*Class 1* No tensile stresses are permitted except for 1 N/mm² under prestress plus dead load, and at transfer. These criteria are thus identical to those of BE 2/73.

*Class 2* Tensile stresses are permitted but visible cracking should not occur. Beeby [119] has suggested that the flexural tensile strength of concrete is equal to $0.556 \sqrt{f_{cu}}$. The appropriate partial safety factor to be applied is 1.3 and thus design values of the flexural tensile strength should be given by $0.428 \sqrt{f_{cu}}$ which compares favourably with the Code value of $0.45 \sqrt{f_{cu}}$ for pre-tensioned members. However, only 80% of this value (i.e. $0.36 \sqrt{f_{cu}}$) should be taken for post-tensioned members, because tests reported by Bate [120] indicated that cracks in a grouted post-tensioned member widen at a greater rate than those in a pre-tensioned member, and thus the design stress for the former should be less than that for the latter. No reference is made in the Code to unbonded tendons and, thus, the author would suggest the adoption of the Concrete Society recommendations of $0.15 \sqrt{f_{cu}}$ and zero in sagging and hogging moment regions respectively [121].

It is necessary to check that, under dead and superimposed dead load, a Class 2 member satisfies the Class 1 criterion in order to ensure that large span bridges, for which dead load dominates, have an adequate factor of safety against cracking occurring under the permanent loading which actually occurs in practice.

At transfer of a Class 2 member, flexural tensile stresses of $0.45 \sqrt{f_{ci}}$ and $0.36 \sqrt{f_{ci}}$ are permitted for pre- and post-tensioned members respectively, where $f_{ci}$ is the concrete cube strength at transfer.

*Class 3* Cracking is permitted provided that the crack widths do not exceed the design values for reinforced concrete, given later in Table 4.7. However, it is not necessary to carry out a true crack width calculation, because the Code gives limiting hypothetical flexural tensile stresses which are deemed to be equivalent to the limiting crack

**Table 4.6** Hypothetical flexural tensile stresses for Class 3 members
(a) Basic stresses

| Tendon type | Limiting crack width (mm) | Stress (N/mm²) for concrete grade | | |
|---|---|---|---|---|
| | | 30 | 40 | ≥ 50 |
| Pre-tensioned | 0.1 | – | 4.1 | 4.8 |
| | 0.2 | – | 5.0 | 5.8 |
| | 0.25 | – | 5.5 | 6.3 |
| Grouted post-tensioned | 0.1 | 3.2 | 4.1 | 4.8 |
| | 0.2 | 3.8 | 5.0 | 5.8 |
| | 0.25 | 4.1 | 5.5 | 6.3 |
| Pre-tensioned, close to tension face | 0.1 | – | 5.3 | 6.3 |
| | 0.2 | – | 6.3 | 7.3 |
| | 0.25 | – | 6.8 | 7.8 |

(b) Depth factors

| Depth (mm) ≤ 200 | 400 | 600 | 800 | ≥1000 |
|---|---|---|---|---|
| Factor 1.1 | 1.0 | 0.9 | 0.8 | 0.7 |

widths. The basic stresses are given in Table 4.6(a), and it can be seen that they are referred to as hypothetical stresses because they exceed the tensile strength of concrete and so cannot actually occur. The basic stresses were derived from tests on beams by Bate [120] and Abeles [122], who calculated the hypothetical tensile stresses present at different maximum crack widths observed in the tests.

Beeby and Taylor [123] have shown that the hypothetical tensile stresses should decrease with an increase in depth and, thus, the basic stresses have to be multiplied by a depth factor from Table 4.6(b).

The presence of additional reinforcement in a prestressed member increases the crack control properties and a higher hypothetical tensile stress may be adopted. The Code increases of 4 N/mm² per 1% of additional steel, for pre-tensioned and grouted post-tensioned tendons, and of 3 N/mm² per 1% of additional steel, for pre-tensioned tendons close to the tension face, are based upon the tests of Abeles [122]. It should be noted that the steel percentages are based upon the area of tensile concrete and not the gross section area.

A Class 3 member has to be checked as a Class 1 member under dead and superimposed dead load for the same reason as that given previously for Class 2.

At transfer, the flexural tensile stresses in a Class 3 member should not exceed the limiting stress appropriate to a Class 2 member. This is in order to avoid cracking at the ends of members.

Finally, the Code gives no guidance on unbonded tendons, and the author would suggest that [123] be consulted for such members.

### Cracking of reinforced concrete

The design surface crack widths were assigned from considerations of appearance and durability, and were based partly upon the 1964 CEB recommendations [111]. They

**Table 4.7** Design crack widths

| Conditions of exposure | Design crack width (mm) |
|---|---|
| Moderate<br>Surface sheltered from severe rain and against freezing while saturated with water, e.g.<br>(1) Surfaces protected by a waterproof membrane<br>(2) Internal surfaces, whether subject to condensation or not<br>(3) Buried concrete and concrete continuously under water | 0.25 |
| Severe<br>(1) Soffits<br>(2) Surfaces exposed to driving rain, alternate wetting and drying, e.g. in contact with backfill and to freezing while wet | 0.20 |
| Very severe<br>(1) Surfaces subject to the effects of de-icing salts or salt spray, e.g. roadside structures and marine structures excluding soffits<br>(2) Surfaces exposed to the action of seawater with abrasion or moorland water having a pH of 4.5 or less | 0.10 |

are summarised in Table 4.7, and it should be noted that different design crack widths are assigned for different conditions of exposure; unlike BE 1/73, which differentiates only between different types of loading. In addition, for bridge decks, different design crack widths are assigned for soffits, and for top slabs if the latter are protected by waterproof membranes. The fact that both of these design crack widths are as, or more, onerous than the BE 1/73 values of 0.25 and 0.31 mm is counteracted by the fact that different crack width formulae are adopted, as explained in Chapter 7. However, as discussed in Chapter 9, the very severe exposure condition for roadside structures could prove to be exceptionally onerous, and to lead to impracticably large areas of reinforcement if applied to piers and abutments.

The philosophy of relating the design crack width to the condition of exposure and, thus, indirectly to the amount of corrosion, must now be viewed with some scepticism in the light of research at Munich, from which Schiessl [124] concluded that there was no significant relationship between corrosion and crack width or cover. This work has been discussed by Beeby [125].

### Vibration

It is only necessary to consider foot and cycle track bridges, and the criterion is discomfort to a user of the bridge. The derivation of the Code criterion is described fully by Blanchard, Davies and Smith [107], and a summary follows.

The effects of vibrations on humans are closely related to acceleration and, thus, maximum tolerable accelerations were plotted against frequency in order to derive a criterion in terms of natural frequency. The maximum tolerable accelerations were assessed from two criteria: discomfort while standing on a vibrating bridge, and the impairance of normal walking due to large amplitude vibrations. The Code criterion lies approximately midway between these two criteria and is given in an appendix to Part 2 as an acceleration of $0.5 \sqrt{f_o}$ m/s$^2$ where $f_o$ is the fundamental natural frequency of the unloaded bridge.

## Other considerations

### Deflection

A specific criterion is not given for deflection in terms of an absolute limiting deflection, or of span to depth ratios. However, it is obviously necessary to calculate deflections in order to ensure that clearance specifications are not violated and adequate drainage will obtain. Deflection calculations are also important where the method of construction requires careful control of levels, and for bearing design.

### Fatigue

The relevant criterion is essentially that there should be a fatigue life of 120 years. When considering unwelded bars an equivalent criterion in terms of a stress range is given in the Code The criterion is that the stress range should not exceed 325 N/mm$^2$ for high yield bars nor 265 N/mm$^2$ for mild steel bars. These ranges are identical to those of BE 1/73 except that the range for mild steel bars is independent of bar diameter.

It is not clear why stress ranges which are dependent upon bar type have been adopted, and it would appear more logical for the stress ranges to be dependent upon the type of loading and the loaded length: indeed, such a dependence was considered during the drafting stages of the Code.

### Durability

A durability criterion is not defined but, provided that the requirements of the Code with regard to limiting crack widths, minimum covers and minimum cement contents are complied with, durability should not be a problem.

## $\gamma_{f_3}$ values

In Chapter 3 the nominal loads and the values of the partial safety factor $\gamma_{fL}$, by which these loads are multiplied to give design loads, are presented. Furthermore, it is explained in Chapter 1 that the effects of the latter have to be multiplied by a partial safety factor $\gamma_{f3}$ in order to obtain design load effects. The values of $\gamma_{f3}$ are dependent upon the material of the bridge and hence, for concrete bridges, are given in Part 4 of the Code. It is thus convenient to introduce the $\gamma_{f3}$ values and the design criteria together in this chapter. It should be noted that some of the values given in the following are not necessarily stated in the Code, but are either implied or intended by the drafters.

## Ultimate limit state

A value of 1.15 for dead and superimposed dead load is stated for all methods of analysis since, for loads which are essentially uniformly distributed over the entire structure, any analysis should predict the effects with reasonable accuracy. It is suggested in Chapter 1 that $\gamma_{f3}$ could be considered to be an adjustment factor which ensures that designs to the Code would be similar to designs to existing documents. It can be shown that, on this basis, a value of 1.15 for dead and superimposed dead load is a reasonable average value.

The Code states that, for imposed loads, $\gamma_{f3}$ should be related to the method of analysis and quotes a value of 1.1 for all methods of analysis (including yield line theory) except for methods involving redistribution, in which case a value of $[1.1 + (\beta - 10)/200]$ is quoted, where $\beta$ is the percentage redistribution. These values do not seem entirely logical since the $\gamma_{f3}$ value for an upper bound method should be greater than that for a lower bound method, because the former is theoretically unsafe and the latter theoretically safe. The drafters' reason for including yield line theory with lower bound methods was that, although it is theoretically an upper bound method, tests [86, 87] show that it predicts safe estimates of the strengths of actual slabs; but this is also true of other methods of analysis and, indeed, lower bound methods predict even safer estimates of the strengths of actual slabs [126]. Furthermore, it seems illogical to have a $\gamma_{f3}$ value of 1.1 for the extreme cases of no redistribution (elastic analysis) and what might be considered as full redistribution (yield line theory), yet values greater than 1.1 are permitted for redistribution in the range 10 to 30%.

Since the maximum permitted value of $\beta$ is 30%, $\gamma_{f3}$ cannot be greater than 1.2 and it thus always lies in the range 1.1 to 1.2 for imposed load and is always 1.15 for dead load. In order to simplify the calculations, the Code allows, as an alternative, the adoption of 1.15 for all loads and all types of analysis, provided that $\beta$ does not exceed 20%. The reason for the proviso is that the value of $\gamma_{f3}$ calculated from the formula is greater than 1.15 for $\beta$ greater than 20%.

It should be noted that the formula for $\gamma_{f3}$ gives values less than 1.1 when $\beta$ is less than 10%, but it was intended that a value of 1.1 should be used for these cases.

When using yield line theory, $\gamma_{f3}$ should be applied to the load effects (the required moments of resistance) in accordance with equation (1.2). However, different values of $\gamma_{f3}$ have to be applied to dead loads and imposed loads. This causes problems because, when using yield line theory, the effects of different types of load cannot be calculated separately and then added together. Thus, although strictly incorrect, it is necessary to apply $\gamma_{f3}$ as indicated in equation (1.3) when using yield line theory.

## Serviceability limit state

The $\gamma_{f3}$ values at the serviceability limit state are all unity except for the following: a value of 0.83 is applied to the effects of HA loading and of 0.91 to the effects of HB loading (or of HA combined with HB) when checking the cracking limit state under load combination 1. These values are not stated in the Code, but are implied because the Code states a value of 1.0 for the product $\gamma_{f1} \cdot \gamma_{f2} \cdot \gamma_{f3}$; and, since $\gamma_{fL} (= \gamma_{f1} \cdot \gamma_{f2})$ is 1.2 for HA loading and 1.1 for HB loading at the serviceability limit state for load combination 1 (see Table 3.1), the implied values of $\gamma_{f3}$ are 0.83 and 0.91 respectively.

In the previous paragraph, it is implied by the author that $\gamma_{f3}$ should be taken to be unity when checking the stress limitation limit state. In fact, an appropriate $\gamma_{f3}$ value is not explicitly stated in the Code but it was the intention of the drafters that it should be unity.

The fact that, for load combination 1, different $\gamma_{f3}$ values are specified for the cracking limit state, to those for the stress limitation limit state, causes problems in the calculations for reinforced and prestressed concrete members for the following reasons:

1. In the case of reinforced concrete, stress limitation calculations are carried out for a load of, essentially, 1.2 HA or 1.1 HB, whilst crack width calculations are carried out for a load of 1.0 HA or 1.0 HB. This could obviously cause confusion and it also complicates the calculations.

2. In the case of prestressed concrete, tensile stress calculations are considered at the cracking limit state and, hence, under a load of 1.0 HA or 1.0 HB; whilst compressive stress calculations are considered at the stress limitation limit state under a load of 1.2 HA or 1.1 HB. Hence stresses of different signs on the same member are checked under different loadings. In view of the anomaly so created, the prestressed concrete section of Part 4 of the Code states that *compressive stresses should be checked under the same load as that specified for tensile stresses*: in other words, *loads of 1.0 HA and 1.0 HB are used for calculating both compressive and tensile stresses in prestressed concrete*.

It should be stated that the calculations are not necessarily as complicated as implied above because, strictly, stresses and crack widths should be calculated under the design loads, which are 1.2 HA and 1.1 HB, and the stresses and crack widths so calculated should then be multiplied by the appropriate $\gamma_{f3}$ value (1.0 or 0.83 or 0.91) to give the design load effects as explained in Chapter 1. However, the adoption of different $\gamma_{f3}$ values seems to be an unnecessary complication, when the same final result, in design terms, could have been obtained by modifying the design criteria.

# Summary

An attempt is now made to summarise the implications of the Code $\gamma_{f3}$ values and design criteria by comparing the number of calculations required for designs in accordance with current documents and with the Code.

## Reinforced concrete

In accordance with BE 1/73, a modular ratio design is carried out at the working load, and crack widths are checked at the same load. However, in accordance with the Code, stresses, crack widths and strength have to be checked at different load levels; and thus three calculations, each at a different load level, have to be carried out, as opposed to two calculations, at the same load level, when designing in accordance with BE 1/73.

## Prestressed concrete

The number of calculations for designs in accordance with the Code and in accordance with BE 2/73 are identical because, in both cases, stresses have to be checked at one load level and strength at another.

## Composite construction

In addition to the comments made above regarding reinforced and prestressed concrete construction, the design of composite members is complicated by the interface shear calculation. The latter calculation is considered at the stress limitation limit state ($\gamma_{f3} = 1.0$) and the load level is thus different to that adopted for checking the stresses in, and the strength of, the, generally, prestressed precast members. Thus, three load levels have to be considered for a composite member designed in accordance with the Code, as opposed to two load levels for a design in accordance with BE 2/73.

## Chapter 5

# Ultimate limit state – flexure and in-plane forces

## Reinforced concrete beams

### Assumptions

The following assumptions are made when analysing a cross-section to determine its ultimate moment of resistance:

1. Plane sections remain plane.
2. The design stress-strain curves are as shown in Fig. 4.4.
3. If a beam is reinforced only in tension, the neutral axis depth is limited to half the effective depth in order to ensure that an over-reinforced failure involving crushing of the concrete, before yield of the tension steel, does not occur. This is because such a failure can be brittle, and there is little warning that it is about to take place. A balanced design, in which the concrete crushes and the tension steel yields simultaneously, is given by considering the strain diagram of Fig. 5.1 in which $\varepsilon_y$ is the strain at which the steel commences to yield in tension. This strain is given, by reference to Fig. 4.4(b), by

$$\varepsilon_y = 0.002 + \frac{0.87 f_y}{200\ 000} \qquad (5.1)$$

$\varepsilon_y$ is thus in the approximate range of 0.003 to 0.004 and, for a balanced design, the neutral axis depth $(x)$ is approximately half of the effective depth since, from Fig. 5.1, the neutral axis depth is given by

$$x = \frac{0.0035d}{0.0035 + \varepsilon_y} \qquad (5.2)$$

4. The tensile strength of concrete is ignored.
5. Small axial thrusts, of up to $0.1 f_{cu}$ times the cross-sectional area, are ignored, because they increase the calculated moment of resistance [112].

### Simplified concrete stress block

The parabolic-rectangular distribution of concrete compressive stress, implied by the stress-strain curve of

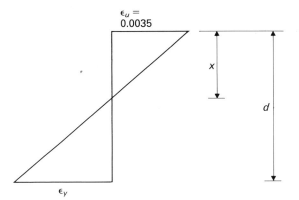

**Fig. 5.1** Strains for a balanced design

Fig. 4.4(a), is tedious to use in practice for hand calculations. The Code thus permits the approximate rectangular stress block, with a constant stress of $0.4 f_{cu}$, as shown in Fig. 5.2, to be adopted. Beeby [127] has demonstrated that, for beams, the adoption of the rectangular stress block results in steel areas which are essentially identical to those using the parabolic-rectangular curve.

### Strain compatibility

The ultimate moment of a section can be determined by using the strain compatibility approach which involves the following steps.

1. Guess a neutral axis depth and, hence, determine the strains in the tension and compression reinforcement by assuming a linear strain distribution and an extreme fibre strain of 0.0035 in the compressive concrete.
2. Determine from the design stress–strain curves the steel stresses appropriate to the calculated steel strains.
3. Calculate the net tensile and compressive forces at the section. If these are not equal, to a reasonable accuracy, adjust the neutral axis depth and return to step 1.
4. If the net tensile and compressive forces are equal, take moments of the forces about a common point in the section to obtain the ultimate moment of resistance.

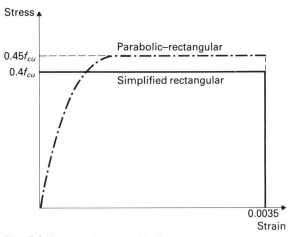

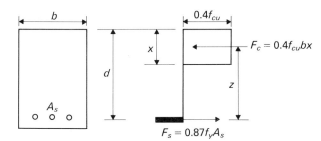

**Fig. 5.2** Rectangular stress block

# Design charts

The strain compatibility method described above is tedious for analysis and is not amenable to direct design. Thus design charts are frequently used and the CP 110 design charts [128] are appropriate.

# Design formulae

As an alternative to using strain compatibility or design charts, the Code gives simplified formulae for hand calculations. The formulae are based upon the simplified rectangular stress block discussed previously and their derivations are now presented.

## Singly reinforced rectangular beam

The stresses and stress resultants at failure are as shown in Fig. 5.3.

For equilibrium
$F_c = F_s$
$0.4f_{cu}bx = 0.87f_yA_s$
$\therefore \; x \simeq \dfrac{2.2f_yA_s}{f_{cu}b}$

Since a rectangular stress block is assumed, the lever arm ($z$) is given by

$$z = d - x/2 = d - \frac{1.1f_yA_s}{f_{cu}b}$$

or $z = \left(1 - \dfrac{1.1f_yA_s}{f_{cu}bd}\right)d$       (5.3)

However, the Code restricts $z$ to a maximum value of $0.95d$. It is not clear why the Code has this restriction, but Beeby [127] has suggested that it could be either that there is evidence that the concrete at the top of a member tends to be less well compacted than that in the rest of the member, or that it is felt desirable to limit the steel strain at failure (a maximum lever arm of $0.95d$ implies a maximum steel strain of 0.0315).

The ultimate moment of resistance ($M_u$) can be obtained by taking moments about the line of action of the resultant concrete force; hence

**Fig. 5.3** Singly reinforced rectangular beam at failure

$M_u = 0.87\,f_yA_sz$       (5.4)

However, the Code restricts the neutral axis depth to a maximum value of $0.5d$, in which case the ultimate moment of resistance can be obtained by taking moments about the reinforcement; hence

$M_u = 0.4f_{cu}b(0.5d)\,(0.75d) = 0.15f_{cu}bd^2$   (5.5)

The ultimate moment of resistance should be taken as the lesser of the values given by equations (5.4) and (5.5).

Equations (5.3) to (5.5) are given in the Code as design equations, but they are obviously more suited to analysing a given section, rather than to designing a section to resist a given moment. In view of this it is best, for design purposes, to rearrange the equations as follows. From (5.4)

$$A_s = \frac{M_u}{0.87f_yz}$$       (5.6)

Substitute in (5.3) and solve the resulting equation for $z$

$$z = 0.5d\left(1 + \sqrt{1 - \frac{5\,M_u}{f_{cu}bd^2}}\right)$$   (5.7)

Thus the lever arm can be calculated from equation (5.7) and, then, the steel area from (5.6).

The Code limits the application of the design equations to situations in which less than 10% redistribution has been assumed. This is because the neutral axis depth is limited to $0.5d$ and, for this value, the relationship between neutral axis depth and amount of redistribution, which is discussed in Chapter 2, implies a maximum redistribution of 10%. However, it is possible to derive the following more general version of equation (5.5), which is appropriate for any amount of redistribution ($\beta$)

$M_u = 0.4f_{cu}bd(0.6-\beta)\,(0.7 + 0.5\beta)$   (5.8)

## Doubly reinforced rectangular beam

The procedure for deriving the Code equations for doubly reinforced beams is to assume that the neutral axis is at the same depth ($0.5d$) as that for balanced design with no compression reinforcement. Compression reinforcement is then provided to resist the applied moment which is in excess of the balanced moment given by equation (5.5), and tension reinforcement provided such that equilibrium is maintained. The strains, stresses and stress resultants are as shown in Fig. 5.4.

It is mentioned in Chapter 4 that the design stress of compression reinforcement is in the range $0.718f_y$ to $0.784f_y$: it is thus conservative always to take a value of $0.72f_y$, as in the Code and Fig. 5.4. It should be noted, from the strain diagram of Fig. 5.4, that, in order that the

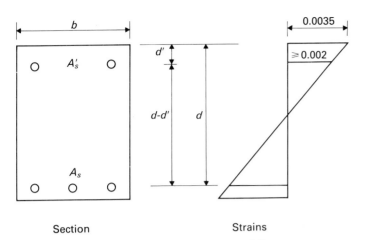

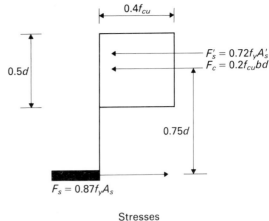

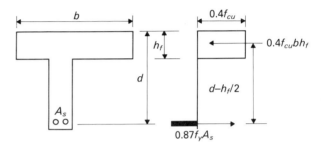

Fig. 5.4 Doubly reinforced rectangular beam at failure

compression reinforcement may develop its yield strain of 0.002, and hence its design strength of $0.72f_y$, it is necessary that

$0.0035\,(0.5d - d')/0.5d \geqslant 0.002$

or $d'/d \leqslant 0.214$

Hence the Code states that

$d'/d \leqslant 0.2$  (5.9)

The ultimate moment of resistance can be obtained by taking moments about the tension reinforcement; hence

$M_u = 0.2f_{cu}bd(0.75d) + 0.72f_yA_s'\,(d-d')$

or $M_u = 0.15f_{cu}bd^2 + 0.72f_yA_s'\,(d-d')$  (5.10)

From which $A_s'$ can be calculated directly. For equilibrium

$0.87f_yA_s = 0.2f_{cu}bd + 0.72f_yA_s'$  (5.11)

From which $A_s$ can be calculated directly.

Equations (5.10) and (5.11) are given in the Code, but it should be noted that the final term of equation (5.10) is incorrectly printed in the Code. The equations are again restricted to less than 10% redistribution, but can be written more generally as

$\dfrac{d'}{d} = \dfrac{3}{7}\,(0.6-\beta)$  (5.12)

$M_u = 0.4f_{cu}bd^2\,(0.6-\beta)\,(0.7 + 0.5\beta) +$
$\qquad\qquad\qquad 0.72f_yA_s'\,(d-d')$  (5.13)

$0.87f_y\,A_s = 0.4f_{cu}bd(0.6-\beta) + 0.72f_yA_s'$  (5.14)

*Flanged beams*

It is assumed in the Code that any compressive stresses in the web concrete can be conservatively ignored: this is valid provided that the flange thickness does not exceed half the effective depth. The stresses and stress resultants at failure are then as shown in Fig. 5.5.

The ultimate moment of resistance is taken to be the lesser of the value calculated assuming the reinforcement to be critical:

$M_u = 0.87f_yA_s\,(d-h_f/2)$  (5.15)

and that assuming the concrete to be critical:

Fig. 5.5 Flanged beam at failure

$M_u = 0.4f_{cu}bh_f\,(d-h_f/2)$  (5.16)

These equations are given in the Code and can be used for design purposes by calculating the steel area from (5.15), and checking the adequacy of the flange using (5.16).

The breadth $b$ shown in Fig. 5.5 is the effective flange width, which is given by the lesser of (a) the actual width and (b) the web width plus one-fifth of the distance between points of zero moment, for T-beams, or the web width plus one-tenth of the distance between points of zero moment, for L-beams. The distance between points of zero moment may be taken as 0.7 times the span for continuous spans, and it would seem reasonable to take a value of 0.85 times the span for an end span of a continuous member.

The resulting effective widths are of the same order as those calculated in accordance with CP 114. In addition, a comparison with values obtained from Table 2 of Part 5 of the Code indicates that, at mid-way between points of zero moment, the Code is generally conservative.

# Prestressed concrete beams

## Assumptions

The assumptions made for reinforced concrete are also made for prestressed concrete; in addition it is assumed that:

1. The stresses at failure in bonded tendons can be obtained from either the appropriate design stress-strain curve of Fig. 4.4 or from a table in the Code

which gives the tendon stress and neutral axis depth at failure as functions of the amount of prestress. The table is based upon the test data of Bate [120], and is very similar to the equivalent table of CP 115. However, the Code neutral axis depths are 87% of the CP 110 values because the Code adopts a design tendon strength of $0.87f_{pu}$, whilst CP 115 uses an ultimate strength of $f_{pu}$. Hence, to maintain equilibrium, a smaller neutral axis depth has to be adopted because the same concrete compressive stress is used in both CP 115 and the Code.

2. The stresses at failure in unbonded tendons are obtained from a table in the Code which gives the tendon stress and neutral axis depth as functions of the amount of prestress and the span to depth ratio. The table is based upon the results of tests carried out by Pannell [129], who concluded that unbonded beams remain elastic up to failure except for a plastic zone, the extent of which depends upon the length of the tendon. Hence, the failure stress and neutral axis depth depend upon the span to depth ratio.

3. In order to give warning of failure it is desirable that cracking of the concrete should occur prior to either the steel yielding and fracturing, or the concrete crushing in the compression zone. This can be checked by ensuring that the strain at the tension face exceeds the tensile strain capacity of the concrete. The latter can be obtained by multiplying the design limiting tensile stress of $0.45 \sqrt{f_{cu}}$ for a Class 2 member (see Chapter 4) by 1.3 (the partial safety factor incorporated in the formula), and dividing by the appropriate elastic modulus given in the Code.

## Strain compatibility

The strain compatibility method described for reinforced concrete can be applied to prestressed concrete, but the prestrain in the tendons should be added to the strain, calculated from the strain diagram at failure, to give the total strain. The latter strain is used to obtain the tendon stress from the stress-strain curve.

## Design charts

Design charts are given in CP 110 [130] for rectangular prestressed beams, but these are of limited use to bridge engineers, who are normally concerned with non-rectangular sections. To the author's knowledge design charts for non-rectangular sections are not generally available, but Taylor and Clarke [131] have produced some typical charts for T-sections. These should be useful to bridge engineers, because they can be applied to a number of standard bridge beams.

## Design formula

The Code gives a formula for calculating the ultimate moment of resistance of a rectangular beam, or of a

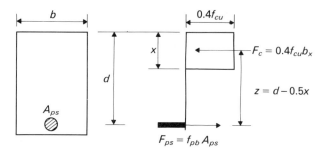

**Fig. 5.6** Rectangular prestressed beam at failure

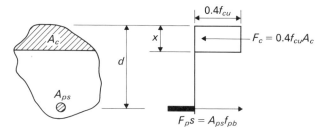

**Fig. 5.7** General prestressed beam at failure

flanged beam with the neutral axis within the flange. The formula is obtained by taking moments of the tendon forces at failure about the line of action of the resultant concrete compressive force. Hence, with reference to Fig. 5.6,

$$M_u = f_{pb}A_{ps} (d - 0.5x) \qquad (5.17)$$

The tendon stress ($f_{pb}$) and neutral axis depth ($x$) at failure are obtained from the tables mentioned previously. Although these tables, and hence equation (5.17), are intended for rectangular sections, it is possible to adapt them to non-rectangular sections [112]. This is achieved by writing down an equilibrium equation in terms of the unknown tendon stress and unknown neutral axis depth. Hence, with reference to Fig. 5.7,

$$A_{ps}f_{pb} = 0.4 f_{cu}A_c$$

where $A_c$ is, generally, a linear function of $x$: thus $f_{pb}$ is also, generally, a linear function of $x$. A graph can be plotted of $f_{pb}$ against $x$ for the section under consideration and, on the same graph, the Code tabulated values of $f_{pb}$ and $x$ can be plotted; the required values of $f_{pb}$ and $x$ can be read off where the two lines cross.

## Reinforced concrete plates

## General

The design of reinforced concrete plates for bridges is more complicated than for buildings because, in bridges, the principal moment directions are very often inclined to the reinforcement directions (e.g. skew slabs), and plates are often subjected to both bending and in-plane effects (e.g. the walls of box girders).

The Code states that allowance should be made for the fact that principal stress resultant directions and reinforcement directions do not generally coincide by calculating required 'resistive stress resultants' such that adequate strength is provided in all directions at a point in a plate. No guidance is given in the Code on the calculation procedure, but the Code statement implies that it is necessary to satisfy the relevant yield criterion. In the following, it is shown how this can be done for plates designed to resist bending, in-plane or combined bending and in-plane effects. It should be noted that all stress resultants are in terms of values per unit length.

## Bending

### Orthogonal reinforcement

In general, it is required to reinforce in the $x$ and $y$ directions a plate element which is subjected to the bending moments $M_x$, $M_y$ and the twisting moment $M_{xy}$ shown in Fig. 2.9.

The yield criterion for a plate element subjected to bending only is simply a relationship between the amounts of reinforcement in the element and the applied moments $(M_x, M_y, M_{xy})$ which would cause yield of the element. It can be shown [132] that the yield criterion is

$$(M^*_x - M_x)(M^*_y - M_y) M^2_{xy} \tag{5.18}$$

$M^*_x$ and $M^*_y$ are the moments of resistance per unit length, of the reinforcement in the $x$ and $y$ directions respectively, calculated in the reinforcement directions. These moments of resistance can be calculated by means of the methods previously described for reinforced concrete beams. A combination of $M_x$, $M_y$, $M_{xy}$ satisfying equation (5.18) would cause the slab element to yield.

A designer is generally interested in determining values of $M^*_x$ and $M^*_y$ to satisfy equation (5.18) for known values of $M_x$, $M_y$, $M_{xy}$. This could be done by choosing either $M^*_x$ or $M^*_y$, and then calculating the other from the yield criterion. However, it is more convenient to make use of equations which give $M^*_x$ and $M^*_y$ directly. Such equations can be derived by noting that the following expressions for $M^*_x$ and $M^*_y$ satisfy the yield criterion

$$M^*_x = M_x - M_{xy}K \tag{5.19}$$

$$M^*_y = M_y - M_{xy}K^{-1} \tag{5.20}$$

Any value of $K$ can be chosen by the designer and thus there is an infinite number of possible combinations of $M^*_x$ and $M^*_y$ capable of resisting a particular set of applied moments. However, a solution which minimises the total amount of steel at a point is generally sought and, to a first order approximation, the total amount of steel is proportional to $(M^*_x + M^*_y)$. Hence the value of $K$ required to give a minimum steel consumption can be obtained by differentiating $(M^*_x + M^*_y)$ with respect to $K$ and equating to zero, thus

$$M^*_x + M^*_y = M_x + M_y - M_{xy}(K + K^{-1})$$

$$\frac{\partial(M^*_x + M^*_y)}{\partial K} = -M_{xy}(1 - K^{-2}) = 0$$

$$K = \pm 1$$

The second derivative is

$$\frac{\partial^2(M^*_x + M^*_y)}{\partial K^2} = -2M_{xy}K^{-3}$$

When considering bottom reinforcement, the sign convention is such that $M^*_x$ and $M^*_y$ must be positive, hence minimum steel consumption coincides with a mathematical minimum and the second derivative should be positive. Hence, $M_{xy}$ and $K$ must be of opposite sign and

$$M_{xy}K = M_{xy}K^{-1} = -|M_{xy}|$$

Thus, for bottom reinforcement

$$\left.\begin{array}{l} M^*_x = M_x + |M_{xy}| \\ M^*_y = M_y + |M_{xy}| \end{array}\right\} \tag{5.21}$$

When considering top reinforcement, the sign convention is such that $M^*_x$ and $M^*_y$ must be negative, hence minimum steel consumption coincides with a mathematical maximum and the second derivative should be negative. Hence, $M_{xy}$ and $K$ must be of the same sign and

$$M_{xy}K = M_{xy}K^{-1} = |M_{xy}|$$

Thus, for top reinforcement

$$\left.\begin{array}{l} M^*_x = M_x - |M_{xy}| \\ M^*_y = M_y - |M_{xy}| \end{array}\right\} \tag{5.22}$$

Equations (5.21) and (5.22) are for the optimum amounts of reinforcement with reinforcement in both the $x$ and $y$ directions, but it is possible for a value of $M^*_x$ and $M^*_y$ so calculated to have the wrong sign. This implies that no reinforcement is required in the appropriate direction, and another set of equations should be used which can be derived as follows.

If $M_x < -|M_{xy}|$, so that a negative value of $M^*_x$ is calculated from the first of equations (5.21), then no reinforcement is required in the bottom in the $x$ direction. Hence, $M^*_x = 0$ can be substituted into equation (5.19) to obtain a value of $K$

$$0 = M_x - M_{xy}K$$

$$K = M_x/M_{xy}$$

This value of $K$ is then substituted into equation (5.20) to give

$$M^*_y = M_y - M^2_{xy}/M_x$$

$M_x$ must be negative and thus this equation is generally written

$$M^*_y = M_y + |M^2_{xy}/M_x| \tag{5.23}$$

Similar equations can be derived for the other possibilities and the complete set of equations, including (5.21) and (5.22), are known to bridge engineers as Wood's equations [133]; although they were originally proposed by Hillerborg [134]. The complete set of equations is given in Appendix A to this book as equations A1 to A8.

### Skew reinforcement

A similar set of equations – (A9) to (A16) of Appendix A – can be derived for skew reinforcement in the $x$ direction and in a direction at an angle α measured clockwise from the *x* direction. These equations are known to bridge engineers as Armer's equations [135].

### Practical considerations

*Experimental verification*  The validity of the theory, upon which equations A1 to A16 is based, has been confirmed by tests on slab elements; in addition, model skew slab bridges have been designed by the equations and successfully tested by Clark [126], Uppenberg [136] and Hallbjorn [137].

*Failure direction*  The direction in which failure (i.e. yield) of a slab element occurs can be determined from theoretical considerations. Strictly, once $M^*_x$, etc., have been calculated from equations (A1) to (A16), they should be resolved into the failure direction and the section design carried out in this direction. However, in practice, it is usual to carry out the section designs independently in each reinforcement direction. Theoretically, this can mean that the concrete is overstressed because the principal concrete stress occurs in the failure direction and not necessarily in a reinforcement direction. The author would suggest that, in practice, this error can be ignored because under-reinforced sections, in which the concrete is not critical, are generally adopted, and the greater ductility of a slab compared with a beam is neglected in design. Slabs are more ductile than beams because the ultimate strain capacity of the concrete in the compression zone increases as the section breadth increases [138]. Thus, although Morley [139] and Clark [126] discuss the correct section design procedure, the author would suggest that the existing practice of designing the sections in the individual steel directions be continued.

*Minimum reinforcement*  When reinforcement is proportioned in accordance with the required moments of resistance, calculated from equations (A1) to (A16), it will sometimes be found that the reinforcement areas are less than the Code minima discussed in Chapter 10. In such situations it is necessary to increase the reinforcement area to the minimum specified in the Code. When this is done it is often theoretically possible to decrease the reinforcement area in another direction. As an example, if the value of $M^*_x$ from equation (A1) implies a reinforcement area in the *x* direction which is less than the minimum, then the minimum area should obviously be adopted in this direction, and the reinforcement area in the *y* direction can then be less than that implied by equation (A2). The equations for carrying out such calculations have been presented by Morley [139]. However, it is obviously conservative, in the above example, to provide the area of reinforcement in the *y* direction implied by equation (A2).

*Multiple load combinations*  Bridges have to be designed for a number of different load positions and combinations and, hence, at each design point, for a number of different moment triads ($M_x$, $M_y$, $M_{xy}$). Many computer programs are available which calculate $M^*_x$, $M^*_y$ or $M^*_\alpha$ for each triad and then output envelopes of maximum values of $M^*_x$, etc. Such an approach ignores the interaction of the multiple triads; but it is simple and conservative. However, it is possible to reduce the total amount of reinforcement required at a particular point by considering the interaction of the multiple triads. The interested reader is referred to the work of Morley [139] and Kemp [140] for further information on such procedures.

## In-plane forces

Equations, very similar to those discussed previously for bending and twisting moments, have been derived for calculating the forces required to resist an in-plane force triad consisting of two in-plane forces per unit length ($N_x$,$N_y$) and an in-plane shear force per unit length ($N_{xy}$). The sign convention adopted for the forces is shown in Fig. A1.

The required resistive forces are designated $N^*_x$ etc., and are equivalent to the appropriate reinforcement area per unit length multiplied by the design stress of the reinforcement. The equations – (A17) to (A30) – for deriving the values of the required resistive forces are given in Appendix A, together with equations for the principal concrete forces per unit length. The principal concrete stresses can be obtained from the latter by dividing by the plate thickness.

The equations for orthogonal reinforcement are generally referred to as Nielson's equations [141] and the equations for skew reinforcement have been presented by Clark [142]. It should be noted that it is required that the values of $N^*_x$, etc., in equations (A17) to (A30) should always be zero or positive, which implies that the reinforcement is always in tension. An extended set of equations which includes the possibility that compression reinforcement may be required has been presented by Clark [142].

The validity of the equations have been confirmed experimentally only for situations in which all of the reinforcement yields in tension [141, 143].

Morley and Gulvanessian [144] have considered the problem of providing a minimum area of reinforcement in a specified direction but have not presented explicit equations, although they do describe a suitable computer program.

Multiple load combinations could be considered by the approaches of [139] and [140].

## Combined bending and in-plane forces

The provision of reinforcement to resist combined bending and in-plane forces is extremely complex. A design solution is usually obtained by adopting a sandwich approach in which the six stress resultants are resolved into two sets of in-plane stress resultants acting in the two outer shells of the sandwich. Such an approach has been suggested by Morley [145]. If the centroids of the outer shells are chosen to coincide with the centroids of the reinforcement layers, as shown in Fig. 5.8, then equations (A17) to (A30) for in-plane forces can be applied to the above two sets of in-plane stress resultants. Such an approach is valid because both equilibrium and the in-plane force yield

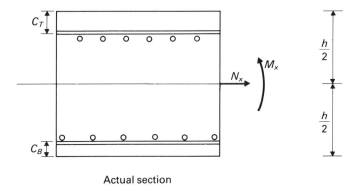

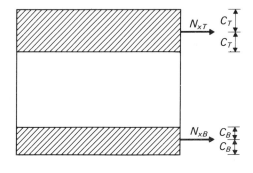

Actual section                    Equivalent sandwich

**Fig. 5.8** Equivalent sandwich plate

criteria are satisfied: thus a safe lower bound design results.

With reference to Fig. 5.8, it can be seen that the in-plane force $N_x$ and the bending moment $M_x$ are statically equivalent to forces $N_{xB}$ and $N_{xT}$ applied at the centroids of the bottom and top outer shells respectively. The values of the latter forces are

$$N_{xB} = \frac{M_x + N_x\,(h/2 - c_T)}{h - c_B - c_T} \tag{5.24}$$

$$N_{xT} = \frac{-M_x + N_x\,(h/2 - c_B)}{h - c_B - c_T} \tag{5.25}$$

Similarly the other stress resultants are

$$N_{yB} = \frac{M_y + N_y\,(h/2 - c_T)}{h - c_B - c_T} \tag{5.26}$$

$$N_{yT} = \frac{-M_y + N_y\,(h/2 - c_B)}{h - c_B - c_T} \tag{5.27}$$

$$N_{xyB} = \frac{M_{xy} + N_{xy}\,(h/2 - c_T)}{h - c_B - c_T} \tag{5.28}$$

$$N_{xyT} = \frac{-M_{xy} + N_{xy}\,(h/2 - c_B)}{h - c_B - c_T} \tag{5.29}$$

Equations (A17) to (A30) can be used to design reinforcement, in the bottom, to resist $N_{xB}$, $N_{yB}$, $N_{xyB}$ and, in the top, to resist $N_{xT}$, $N_{yT}$, $N_{xyT}$.

It should be noted that the core of the sandwich is assumed to make no contribution to the strength of the section. However, Morley and Gulvanessian [144] have extended the method to include the possibility of the core contributing to the strength.

## Prestressed concrete slabs

The Code states that prestressed concrete slabs should be designed in accordance with the clauses for prestressed concrete beams. In addition, 'due allowance should be made in the distribution of prestress in the case of skew slabs'. The latter point is intended to emphasise the fact that, when a skew slab is prestressed longitudinally, some of the longitudinal bending component of the prestress is distributed in the form of transverse bending and twisting moments. The result is that the prestress is less than that calculated on the basis of a simple beam strip in the direc-

**Fig. 5.9** Example 5.1

tion of the prestress. Clark and West [146] have given guidance on the resultant prestress to be expected in skew slab bridges.

Regarding section design for prestressed slabs, it is difficult to imagine how the beam clauses can be applied to a general case. The author would suggest that the prestress should be considered as an applied load at the ultimate limit state, and a set of bending moments and in-plane forces, due to the prestress, calculated and added algebraically to those due to the applied loads. Conventional reinforcement could then be designed to resist the resulting 'out-of-balance' stress resultants by using the equations given in the Appendix. Clark and West have designed, and successfully tested, model skew solid [146] and voided [147] slab bridges by such an approach.

## Examples

### 5.1 Prestressed beam section strength

It is required to calculate the ultimate moment of resistance of the pre-tensioned composite section shown in Fig. 5.9. The initial prestress is 70% of the characteristic strength and the losses amount to 30%. The precast and in-situ concretes are of grades 50 and 40 respectively. From Table 21 of the Code

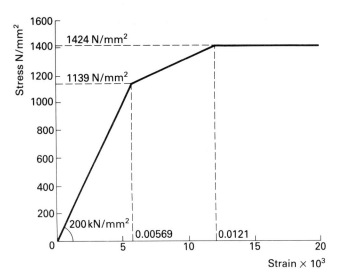

**Fig. 5.10** Design stress–strain curve for 15.2 mm low relaxation strand

$A_{ps} = 138.7$ mm$^2$

$A_{ps}f_{pu} = 227.0$ kN

$\therefore f_{pu} = 227.0 \times 10^3/138.7 = 1637$ N/mm$^2$

Effective prestress $= (0.7)(0.7)(1637) = 802$ N/mm$^2$

$\therefore$ prestrain in tendons $= 802/200 \times 10^3 = 4.01 \times 10^{-3}$

The design stress-strain curve for the tendons at the ultimate limit state is given in Fig. 5.10.

### Strain compatibility approach

By trial and error the neutral axis depth has been found to be 330 mm. The strain distribution is thus as shown in Fig. 5.11, where the total strains at the tendon levels and the tendon stresses are

$\varepsilon_1 = -(0.0035 \times 49/339) + (4.01 \times 10^{-3})$
$\quad = 0.0035$

$f_1 = (0.0035)(200 \times 10^3) = 700$ N/mm$^2$

$\varepsilon_2 = (0.0035 \times 921/339) + (4.01 \times 10^{-3}) = 0.0135$

$f_2 = 1424$ N/mm$^2$

$\varepsilon_3 = (0.0035 \times 971/339) + (4.01 \times 10^{-3}) = 0.0140$

$f_3 = 1424$ N/mm$^2$

The tensile forces in the tendons are

$T_1 = (2)(138.7)(700 \times 10^{-3}) = 194$

$T_2 = (14)(138.7)(1424 \times 10^{-3}) = 2765$

$T_3 = (15)(138.7)(1424 \times 10^{-3}) = 2963$

$\qquad\qquad\qquad\qquad \Sigma T = 5922$ kN

The compressive forces in the concrete are calculated for zones 1 to 4 of Fig. 5.11 (the effective breadth is the actual breadth since the distance between points of zero moment would be at least 20 m).

$C_1 = (1200 \times 200)(0.4 \times 40)10^{-3} = 3840$

$C_2 = (300 \times 170)(0.4 \times 40)10^{-3} = 816$

$C_3 = (300 \times 45)(0.4 \times 50)10^{-3} = 270$

$C_4 = (400 \times 124)(0.4 \times 50)10^{-3} = 992$

$\qquad\qquad\qquad\qquad \Sigma C = 5918$ kN

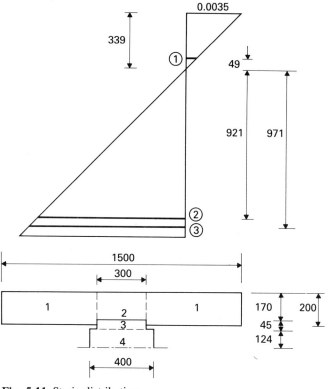

**Fig. 5.11** Strain distribution

$\Sigma T = \Sigma C$, thus take moments about neutral axis

$T_1 \quad (194)(-49)10^{-3} \quad = -10$

$T_2 \quad (2765)(921)10^{-3} \quad = 2547$

$T_3 \quad (2963)(971)10^{-3} \quad = 2877$

$C_1 \quad (3840)(239)10^{-3} \quad = 918$

$C_2 \quad (816)(254)10^{-3} \quad = 207$

$C_3 \quad (270)(146.5)10^{-3} = 40$

$C_4 \quad (992)(62)10^{-3} \quad = 62$

$\qquad\qquad\qquad \Sigma = 6641$ kN m

### Code table approach

Centroid of tendons in tension zone is at $d$ from top of slab, where

$d = (14 \times 1260 + 15 \times 1310)/29 = 1286$ mm

For equilibrium, and ignoring $T_1$

$T_2 + T_3 = C_1 + C_2 + C_3 + C_4$

or $A_{ps}f_{pb} = (3840 + 816 + 270)10^3 +$
$\qquad\qquad\qquad (0.4 \times 50)(400)(x - 215)$

where $x$ is the unknown neutral axis depth and

$A_{ps} = 29 \times 138.7 = 4022.3$ mm$^2$

$\therefore f_{pb} = 797 + 1.99x$

$\therefore (0.87f_{pu})(f_{pb}/0.87f_{pu}) = 797 + 1.99(x/d)d$

$\therefore 1424(f_{pb}/0.87f_{pu}) = 797 + 1.99(x/d)1286$

$\therefore f_{pb}/0.87f_{pu} = 0.560 + 1.80(x/d)$

In Fig. 5.12, the latter expression is plotted together with the Code Table 29 values. It can be seen that the intersection occurs at $f_{pb}/0.87f_{pu} = 1.0$ and $x/d = 0.244$. Hence $f_{pb} = 1424$ N/mm$^2$ and $x = 314$ mm.

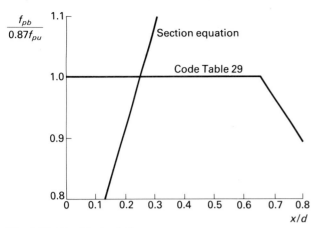

Fig. 5.12 Graphical solution

$T_2$, $T_3$, $C_1$, $C_2$ and $C_3$ are as calculated for the strain compatibility approach and

$C_4 = (400 \times 99)(0.4 \times 50)10^{-3} = 792$ kN

Moments about the neutral axis

| | | | |
|---|---|---|---|
| $T_2$ | $(2765)(946)10^{-3}$ | = | 2616 |
| $T_3$ | $(2963)(996)10^{-3}$ | = | 2951 |
| $C_1$ | $(3840)(214)10^{-3}$ | = | 822 |
| $C_2$ | $(816)(229)10^{-3}$ | = | 187 |
| $C_3$ | $(270)(121.5)10^{-3}$ | = | 33 |
| $C_4$ | $(792)(49.5)10^{-3}$ | = | 39 |
| | | $\Sigma =$ | 6648 kN m |

This value is within 0.1% of the value calculated using strain compatibility.

## 5.2 Slab

The design applied moment triad, at the ultimate limit state, in the obtuse corner of a reinforced concrete skew slab bridge is (with the axes shown in Fig. 5.13)

$M_x = -2.484$ MNm/m

$M_y = 1.139$ MNm/m

$M_{xy} = -0.900$ MNm/m

Obtain the required moments of resistance in the reinforcement directions, if the latter are (a) parallel and perpendicular to the abutments and (b) parallel to the slab edges. In the following the equation numbers are those of Appendix A.

*Orthogonal reinforcement*

*Bottom reinforcement*
From equation A1
$M^*_x = -2.484 + |-0.9| = -1.584$

$M^*_x < 0$, $\therefore M^*_x = 0$ and calculate $M^*_y$ from equation (A3)

$M^*_y = 1.139 + |(-0.9)^2/(-2.484)| = 1.465$ MNm/m

*Top reinforcement*
From equation (A5)

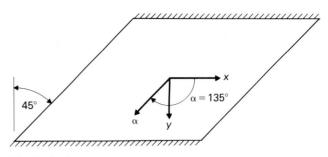

Fig. 5.13 Skew slab axes

$M^*_x = -2.484 - |-0.9| = -3.384$

From equation (A6)

$M^*_y = 1.139 - |-0.9| = 0.239$

$M^*_x > 0$, $\therefore M^*_y = 0$ and calculate $M^*_x$
from equation (A8)

$M^*_x = -2.484 - |(0.9)^2/1.139| = -3.195$ MNm/m

*Skew reinforcement*

From Fig. 5.13 it can be seen that $\alpha = 135°$

*Bottom reinforcement*
From equation (A9)

$M^*_x = -2.484 + 2(-0.9)(-1) + 1.139(-1)^2 + \left| \dfrac{-0.9 + 1.139(-1)}{1/\sqrt{2}} \right|$

$= 3.336$ MNm/m

From equation (A10)

$M^*_\alpha = \dfrac{1.139}{(1/\sqrt{2})^2} + \left| \dfrac{-0.9 + 1.139(-1)}{1/\sqrt{2}} \right|$

$= 5.159$ MNm/m

*Top reinforcement*
From equation (A13)

$M^*_x = -2.484 + 2(-0.9)(-1) + 1.139(-1)^2 - \left| \dfrac{-0.9 + 1.139(-1)}{1/\sqrt{2}} \right|$

$= -2.427$ MNm/m

From equation (A14)

$M^*_\alpha = \dfrac{1.139}{(1/\sqrt{2})^2} - \left| \dfrac{-0.9 + 1.139(-1)}{1/\sqrt{2}} \right|$

$= -0.605$ MNm/m

It should be noted that reinforcement is required in each direction in both the top and bottom of the slab when skew reinforcement is used. However, reinforcement is required only transversely, in the bottom, and longitudinally, in the top of the slab, when orthogonal reinforcement is used.

## 5.3 Box girder wall

A wall of a box girder, 250 mm thick, and with the centroid of the reinforcement in each face at a distance of

60 mm from the face, is subjected to the following design stress resultants at the ultimate limit state

| | | | |
|---|---|---|---|
| $N_x$ | $= -240$ kN/m | $M_x$ | $= -166.0$ kNm/m |
| $N_y$ | $= 600$ kN/m | $M_y$ | $= 24.0$ kNm/m |
| $N_{xy}$ | $= 340$ kN/m | $M_{xy}$ | $= 1.6$ kNm/m |

Design reinforcement in the $x$ and $y$ directions if $f_y = 250$ N/mm$^2$ and $f_{cu} = 50$ N/mm$^2$

The design strengths are

Reinforcement $= 250/1.15 = 217$ N/mm$^2$

Concrete $= 0.4 \times 50 = 20$ N/mm$^2$

In equations (5.24) to (5.29)

$c_T = c_B = 60$ mm

$h/2 - c_T = h/2 - c_B = 65$ mm

$h - c_B - c_T = 130$ mm

The statically equivalent stress resultants in the outer shells of the sandwich plate of Fig. 5.8 are calculated, from equations (5.24) to (5.29), as, in N/mm units

| | | | |
|---|---|---|---|
| $N_{xB}$ | $= -1397$ | $N_{xT}$ | $= 1157$ |
| $N_{yB}$ | $= 485$ | $N_{yT}$ | $= 115$ |
| $N_{xyB}$ | $= 182$ | $N_{xyT}$ | $= 158$ |

### Bottom reinforcement

From equation (A17)

$N^*_{xB} = -1397 + |182| = -1215$

$N^*_{xB} < 0$, $\therefore N^*_{xB} = 0$ and calculate $N^*_{yB}$ from equation (A20)

$N^*_{yB} = 485 + |(182)^2/(-1397)| = 509$ N/mm

$A_{yB} = (509/217)10^3 = 2340$ mm$^2$/m

From equation (A21)

$F_{cB} = -1397 + (182)^2/(-1397) = -1421$ N/mm

Bottom concrete compressive stress $= 1421/120$
$= 11.8$ N/mm$^2$

This is less than the design stress of 20 N/mm$^2$

### Top reinforcement

From equation (A17)

$N^*_{xT} = 1157 + |158| = 1315$ N/mm

$A_{xT} = (1315/217)10^3 = 6060$ mm$^2$/m

From equation (A18)

$N^*_{yT} = 115 + |158| = 273$ N/mm

$A_{yT} = (273/217)10^3 = 1260$ mm$^2$/m

From equation (A19)

$F_{cT} = -2|158| = -316$ N/mm

Top concrete compressive stress $= 316/120 = 2.6$ N/mm$^2$
This is less than the design stress of 20 N/mm$^2$.

## Chapter 6

# Ultimate limit state – shear and torsion

## Introduction

In this chapter, the Code methods for designing against shear and torsion for reinforced and prestressed concrete construction are discussed. The particular problems which arise in composite construction are not dealt with in this chapter but are presented in Chapter 8.

It should be noted that, in accordance with the Code, all shear and torsion calculations, with the exception of interface shear in composite construction, are carried out at the ultimate limit state.

With regard to shear, calculations have to be carried out, as at present, for both flexural shear and, where appropriate, punching shear. However, it should be noted that BE 1/73 requires shear calculations for reinforced concrete to be carried out under working load conditions as opposed to the ultimate limit state as required by the Code. In addition, as explained later, the procedures for designing shear reinforcement in beams differ between BE 1/73 and the Code.

The design of prestressed concrete to resist shear is carried out at the ultimate limit state in accordance with both BE 2/73 and the Code, and the calculation procedures for each are very similar.

Design against torsion is not covered in the Department of Transport's current design documents and, in practice, either CP 110 or the Australian Code of Practice is often referred to for design guidance. The latter document is written in terms of permissible stresses and working loads, and thus differs from the Code approach of designing at the ultimate limit state.

## Shear in reinforced concrete

### Flexural shear

#### Background

The design rules for flexural shear in beams are based upon the work of the Shear Study Group of the Institution

of Structural Engineers [148]. The background to the rules, which are identical to those of CP 110, has been described by Baker, Yu and Regan [149] and by Regan [150].

The general approach adopted by the Shear Study Group was, first, to study test data from beams without shear reinforcement and, then, to study test data from beams with shear reinforcement.

#### Beams without shear reinforcement

The data from beams without shear reinforcement indicate that, for a constant concrete strength and longitudinal steel percentage, the relationship between the ratio of the observed bending moment at collapse ($M_c$) to the calculated ultimate flexural moment ($M_u$) and the ratio of shear span ($a_v$) to effective depth ($d$) is of the form shown in Fig. 6.1(a).

This diagram has four distinct regions within each of which a different mode of failure occurs: region 1, corbel action or crushing of a compression strut which runs from the load to the support; region 2, diagonal tension causing splitting along a line joining the load to the support; region 3, a flexural crack develops into a shear failure crack; region 4, flexure. These modes are illustrated in Fig. 6.1 (b–e).

From the design point of view, it is obviously safe to propose a design method which results in the observed bending moment at collapse ($M_c$) always exceeding the lesser of the calculated ultimate flexural moment and the moment when the section attains its calculated ultimate shear capacity. Such an approach can be developed as follows: if the shear force at failure of a point loaded beam is $V_c$, then

$$M_c = V_c a_v$$

Now, for a particular concrete strength and steel percentage, the ultimate flexural moment is given by

$$M_u = K b d^2$$

where $K$ is a constant. Thus

$$\frac{M_c}{M_u} = \frac{M_c}{K b d^2} = \frac{V_c a_v}{K b d^2} = \frac{1}{K}\left(\frac{V_c}{bd}\right)\left(\frac{a_v}{d}\right)$$

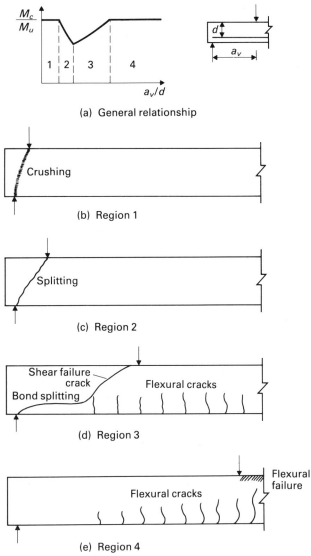

(a) General relationship

(b) Region 1

(c) Region 2

(d) Region 3

(e) Region 4

**Fig. 6.1(a)–(e)** Shear failure modes in reinforced concrete beams

Hence,

$$\frac{M_c}{bd^2} = \left(\frac{V_c}{bd}\right)\left(\frac{a_v}{d}\right) \tag{6.1}$$

It is thus convenient to replot the test data in the form of a graph of $(M_c/bd^2)$ against $(a_v/d)$ as shown by the solid line of Fig. 6.2. The dashed line is that calculated assuming that flexural failures always occur (i.e. $M_u/bd^2$), and the chain dotted line is a line that cuts off the unsafe side of the graph (those beams which fail in shear) and can thus be considered to be an 'allowable shear' line. The significance of equation (6.1) can now be seen because the term $V_c/bd$ is the slope of any line which passes through the origin. Hence if $V_c/bd$ is *chosen* to be the slope of the chain dotted line, then $V_c/bd$ can be considered to define an 'allowable shear' line which separates an unsafe region to its left from a safe region to its right. Furthermore $V_c/bd$ can be considered to be a nominal allowable shear stress ($v_c$) which acts over a nominal shear area ($bd$). It is emphasised that, in reality, a constant shear stress does not act over such an area, but it is merely convenient to choose a shear area of ($bd$) and to then select values of $v_c$ such that the allowable values of the moment to cause collapse fall below the test values.

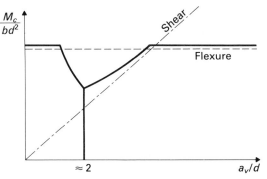

**Fig. 6.2** Shear design diagram

**Table 6.1** Design shear stresses ($v_c$ N/mm²)

| $\dfrac{100\,A_s}{bd}$ | Concrete grade | | | |
|---|---|---|---|---|
| | 20 | 25 | 30 | ≥ 40 |
| ≤ 0.25 | 0.35 | 0.35 | 0.35 | 0.35 |
| 0.50 | 0.45 | 0.50 | 0.55 | 0.55 |
| 1.00 | 0.60 | 0.65 | 0.70 | 0.75 |
| 2.00 | 0.80 | 0.85 | 0.90 | 0.95 |
| ≥ 3.00 | 0.85 | 0.90 | 0.95 | 1.00 |

The code contains a table which gives values of $v_c$ for various concrete grades and longitudinal steel percentages. Reference to the table (see Table 6.1) shows that $v_c$ is only slightly affected by the concrete strength but is greatly dependent upon the area of the longitudinal steel. This is because the latter contributes to the shear capacity of a section in the following two ways:

1. Directly, by dowel action [151] which can contribute 15 to 25% of the total shear capacity [152].
2. Indirectly, by controlling crack widths which, in turn, influence the amount of shear force which can be transferred by the interlock of aggregate particles across cracks. Aggregate interlock can contribute 33 to 50% of the total shear capacity [152].

It should be noted that when using Table 6.1, the longitudinal steel area to be used is that which extends at least an effective depth beyond the section under consideration. The reason for this is given later in this chapter.

It is not necessary to apply a material partial safety factor to the tabulated $v_c$ values because they incorporate a partial safety factor of 1.25 and are thus design values. In fact the $v_c$ values can be considered to be, by adopting the terminology of Chapter 1, design resistances obtained from equation (1.8). An appropriate $\gamma_m$ value would lie between the steel value of 1.15 and the concrete value of 1.5 because the shear resistance of a section is dependent upon both materials. It was decided that a value of 1.25 was reasonable for shear resistance when compared with the usual value of 1.15 for flexural resistance.

## Beams with shear reinforcement

When the nominal shear stress exceeds the appropriate tabulated value of $v_c$ it is necessary to provide shear reinforcement to resist the shear force *in excess of* ($v_c bd$). This approach differs to that of BE 1/73 in which shear rein-

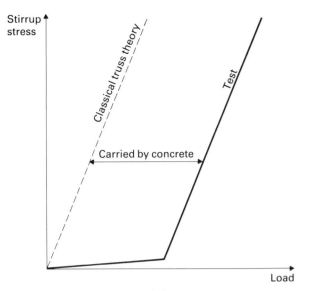

Fig. 6.3 Influence of shear reinforcement

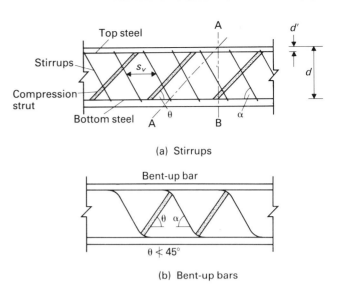

(a) Stirrups

(b) Bent-up bars

**Fig. 6.4(a),(b)** Truss analogy for shear

forcement has to be designed to resist the *entire* shear force when the BE 1/73 allowable concrete shear stress is exceeded, and two-thirds of the shear force when the allowable concrete shear stress is not exceeded. The justification for designing reinforcement to resist only the excess shear force is that tests indicate that the stresses in shear reinforcement are extremely small until shear cracks occur, after which, the stresses gradually increase as shown in Fig. 6.3. Hence, the shear resistance of the shear reinforcement is additive to that of the section without shear reinforcement (i.e. $v_c bd$). This was confirmed by the Shear Study Group who, for vertical stirrups, obtained a good lower bound fit to test data in the form:

$$v = v_c + 1.1 f_{yv} (A_{sv}/bs_v) \qquad (6.2)$$

where $v = V/bd$ and $V$ is the shear force at failures; $f_{yv}$ is the characteristic strength of the shear reinforcement; and $A_{sv}$ and $s_v$ are the area and spacing of the shear reinforcement.

A theoretical expression for $A_{sv}$ can be derived by considering the truss analogy shown in Fig. 6.4(a). Theoretically, the inclination ($\theta$) of the compression struts can be assumed to take any value, provided that shear reinforcement is designed in accordance with the chosen value of $\theta$. However, the greater the difference between $\theta$ and the inclination of the elastic principal stress (45°) when a shear crack first forms, the greater is the implied amount of stress redistribution between initial cracking and collapse. In order to minimise the stress redistribution, $\theta$ is chosen in the Code to coincide with the inclination of the elastic principal stress (45°).

For vertical equilibrium along section A–A and assuming that only the excess shear force is resisted by the shear reinforcement:

$$(v - v_c)bd = A_{sv} f_{yv} (\sin \alpha + \cos \alpha) (d - d')/s_v \qquad (6.3)$$

where $\alpha$ is the inclination of the shear reinforcement (stirrups or bent-up bars). The Code assumes that $d - d' \simeq d$, hence equation (6.3) can be rearranged to give

$$v = v_c + f_{yv} (\sin \alpha + \cos \alpha) (A_{sv}/bs_v) \qquad (6.4)$$

For the case of vertical stirrups, $\alpha = 90°$ and equation (6.4) becomes

$$v = v_c + f_{yv} (A_{sv}/bs_v) \qquad (6.5)$$

It can be seen that equations (6.5) and (6.2) which have been obtained theoretically and from test data respectively are in good agreement. For design purposes it is necessary to apply a material partial safety factor of 1.15 (the value for reinforcement at the ultimate limit state) to $f_{yv}$; equations (6.4) and (6.5) can then be rearranged as:

$$\frac{A_{sv}}{s_v} = \frac{b(v - v_c)}{0.87 f_{yv} (\sin \alpha + \cos \alpha)} \qquad (6.6)$$

$$\frac{A_{sv}}{s_v} = \frac{b(v - v_c)}{0.87 f_{yv}} \qquad (6.7)$$

The latter equation appears in the Code, with $f_{yv}$ restricted to a maximum value of 425 N/mm². This is because the data considered by the Shear Study Group indicated that the yield stress of shear reinforcement should not exceed about 480 N/mm² in order that it could be guaranteed that the shear reinforcement would yield at collapse prior to crushing of the concrete. The Code value of 425 N/mm² is thus conservative. It is implied, in the above derivation, that equation (6.6) can be applied to either inclined stirrups or bent-up bars. Although this is theoretically correct, the Code states that when using bent-up bars the truss analogy of Fig. 6.4(b) should be used in which the compression struts join the centres of the bends of the lower and upper bars. This approach is identical to that of CP 114 and it is not clear why, originally, the CP 110 committee and, subsequently, the Code committee retained it. It is worth mentioning that Pederson [153] has demonstrated the validity of considering the compression struts to be at an angle other than that of Fig. 6.4(b).

In view of the limited amount of test data obtained from beams with bent-up bars used as shear reinforcement and because of the risk of the concrete being crushed at the bends, the Code permits only 50% of the shear reinforcement to be in the form of bent-up bars.

Finally, an examination of Fig. 6.4(a) shows that if the shear strength is checked at section A–B, then the assumed shear failure plane intersects the longitudinal tension reinforcement at a distance equal to the effective depth from the section A–B. Hence the requirement mentioned previously that the value of $A_s$ in Table 6.1 should be the area of the longitudinal steel which extends at least an effective depth beyond the section under consideration. This argument is not strictly correct, but instead, as shown in Chapter 10 when discussing bar curtailment, the area of steel should be considered at a distance of half of the lever arm beyond the section under consideration. Thus the Code requirement is conservative.

## Maximum shear stress ($v_u$)

It is shown in the last section that the shear capacity of a reinforced concrete beam can be increased by increasing the amount of shear reinforcement. However, eventually a point is reached when the shear capacity is no longer increased by adding more shear reinforcement because the beam is then over-reinforced in shear. Such a beam fails in shear by crushing of the concrete compression struts of the truss before the shear reinforcement yields in tension. The Code thus gives, in a table, a maximum nominal flexural shear stress of $0.75 \sqrt{f_{cu}}$ (but not greater than 4.75 N/mm²) which is a design value and incorporates a partial safety factor of 1.5 applied to $f_{cu}$; hence the effective partial safety factor applied to the nominal stress is $\sqrt{1.5}$. Clarke and Taylor [154] have considered data from beams which failed in shear by crushing of the web concrete. They found that the ratios of the experimental nominal shear stress to that given by $0.75 \sqrt{f_{cu}}$ were in the range 1.02 to 3.32 with a mean value of 1.90.

The upper limit of 4.75 N/mm² imposed by the Code is to allow for the fact that shear cracks in beams of very high strength concrete can occur through, rather than around, the aggregate particles. Hence a smooth crack surface can result across which less shear can be transferred in the form of aggregate interlock [152].

## Short shear spans

An examination of Fig. 6.2 reveals that for short shear spans ($a_v/d$ less than approximately 2) the shear strength increases with a decrease in the shear span. Hence, the allowable nominal shear stresses ($v_c$) are very conservative for short shear spans. In view of this, an enhanced value of $v_c$ which is given by $v_c(2d/a_v)$ is adopted for $a_v/d$ less than 2. However, the enhanced stress should not exceed the maximum allowable nominal shear stress of $0.75 \sqrt{f_{cu}}$. The enhanced stress has been shown to be conservative when compared with data from tests on beams loaded close to supports and on corbels [112].

## Minimum shear reinforcement

If the nominal shear stress is less than $0.5 v_c$, the factor of safety against shear cracking occurring is greater than twice that against flexural failure occurring. This level of safety is considered to be adequate and the provision of shear reinforcement is not necessary in such situations.

If the nominal shear stress exceeds $0.5 v_c$ but is less than $v_c$, it is necessary to provide a minimum amount of shear reinforcement. In order that the presence of shear rein-

forcement may enhance the strength of a member, it is necessary that it should raise the shear capacity above the shear cracking load. The Shear Study Group originally suggested, from considerations of the available test data, a minimum value of $0.87 f_{yv} A_{sv}/bs_v$ equal to 60 lb/in² (0.414 N/mm²) in order to ensure that the shear reinforcement would increase the shear capacity. Hence for $f_{yv} = 250$ N/mm² (mild steel) and 425 N/mm² (the greatest permitted in the Code for high yield steel for shear reinforcement), $A_{sv}/s_v = 0.0019b$ and $0.00112b$ respectively: these values have been rounded up to $0.002b$ and $0.0012b$ in the Code so that each is equivalent to $0.87 f_{yv} A_{sv}/bs_v = 0.44$ N/mm². The value of $0.0012b$ for high yield steel is also the minimum value given in CP 114.

It is also necessary to specify a maximum spacing of stirrups in order to ensure that the shear failure plane cannot form between two adjacent stirrups, in which case the stirrups would not contribute to the shear strength. Figure 6.4(a) shows that the spacing should not exceed $[(d-d')(1 + \cot\alpha)]$. This expression has a minimum value of $(d - d')$ when $\alpha = 90°$ and, to simplify the Code clause, it is further assumed that $(d - d') \simeq 0.75d$. This spacing was also shown, experimentally, to be conservative by plotting shear strength against the ratio of stirrup spacing to effective depth for various test data. It was observed that the test data exhibited a reduction in shear strength for a ratio greater than about 1.0 [149].

Finally, it is necessary for the stirrups to enclose all the tension reinforcement because the latter contributes, in the form of dowel action, about 15 to 25% of the total shear strength [151, 152]. If the tension reinforcement is not supported by being enclosed by stirrups, then the dowel action tears away the concrete cover to the reinforcement and the contribution of dowel action to the shear strength is lost because the reinforcement can then no longer act as a dowel.

## Shear at points of contraflexure

A problem arises near to points of contraflexure of beams because the value of $v_c$ to be adopted is dependent upon the area of the longitudinal reinforcement. It is thus necessary to consider whether the design shear force is accompanied by a sagging or hogging moment in order to determine the appropriate area of longitudinal reinforcement.

A situation can arise in which, for example, the area of top steel is less than the area of bottom steel and the maximum shear force ($V_s$) associated with a sagging moment exceeds the maximum shear force ($V_h$) associated with a hogging moment. However, because the area of top steel is less than the area of bottom steel, the value of $v_c$ to be considered with $V_h$ could be less than that to be considered with $V_s$. Thus although $V_s$ is greater than $V_h$, it could be the latter which results in the greater amount of shear reinforcement. It can thus be seen that it is always necessary to consider the maximum shear force associated with a sagging moment *and* the maximum shear force associated with a hogging moment. A conservative alternative procedure, which would reduce the number of calculations, would be to consider only the absolute maximum shear force and to use a value of $v_c$ appropriate to the lesser of the top or bottom steel areas.

If either of these procedures is adopted then, generally, more shear reinforcement is required than when the calculations are carried out in accordance with BE 1/73, in which the allowable shear stress is not dependent upon the area of the longitudinal reinforcement. The increase in shear reinforcement in regions of contraflexure was the subject of criticism during the drafting stages of the Code and in order to mitigate the situation an empirical design rule, which takes account of the minimum area of shear reinforcement which has to be provided, has been included in the Code.

The design rule implies that, for the situation described above, shear reinforcement should be designed to resist (a) $V_s$ with a $v_c$ value appropriate to the bottom reinforcement and (b) the *lesser* of $V_h$ and $0.8\,V_s$ with a $v_c$ value appropriate to the top reinforcement. The greater area of shear reinforcement calculated from (a) and (b) should then be provided. The rule should be interpreted in a similar manner for other relative values of $V_s$, $V_h$ and of bottom and top reinforcement.

The logic behind the above rule is not clear. Furthermore, it was based upon a limited number of trial calculations which indicated that it was conservative. However, it can be shown to be unconservative in some circumstances. In view of this, the author would suggest that it would be safer not to adopt the rule in practice.

## Slabs

*General* The design procedure for slabs is essentially identical to that for beams and was originally proposed for building slabs designed in accordance with CP 110. The implications of this are now discussed.

*$v_c$ values* The values of $v_c$ in Table 6.1 were derived from the results of tests on, mainly, beams, although some one-way spanning slabs with no shear reinforcement were also considered. The slabs had breadth to depth ratios of about 2.5 to 4 and thus were, essentially, wide beams.

It is probably reasonable to apply the $v_c$ values to building slabs because the design loading is, essentially, uniform and the design procedure generally involves considering one-way bending in orthogonal directions parallel to the flexural reinforcement. Hence it is reasonable to consider slabs as wide beams. However, it is not clear whether the same $v_c$ values can be applied to slabs, in more general circumstances, when the support conditions and/or the loading are non-uniform. An additional problem occurs when the flexural reinforcement is not perpendicular to the planes of the principal shear forces because it is not then obvious what area of reinforcement should be used in Table 6.1. Although, strictly, this situation also arises in building slabs, it is ignored for design purposes. It is not certain whether it can also be ignored in bridge slabs, where large principal shear forces can act at large angles to the flexural reinforcement directions. It should also be noted that shear forces in bridge slabs can vary rapidly across their widths. A decision then has to be made as to whether to design against the peak shear force or a value averaged over a certain width.

None of the above problems is considered in the Code.

A possible 'engineering solution' would be to design against shear forces averaged over a width of slab equal to twice the effective depth, and to carry out the shear design calculation for the shear forces acting on planes normal to each flexural reinforcement direction. The latter suggestion of considering beam strips in each of the flexural reinforcement directions can be shown to be, in general, unsafe. This is because it is the stiffness, rather than the strength, of the flexural reinforcement which is of importance in terms of shear resistance. The flexural reinforcement should be resolved into a direction perpendicular to the plane of the critical shear crack, and it is explained in Chapter 7 that, when considering stiffness, reinforcement areas resolve in accordance with $\cos^4\alpha$, where $\alpha$ is the orientation of the reinforcement to the perpendicular to the critical crack. Thus the resolved area and, hence, the appropriate $v_c$ value could be much less than the values appropriate to the steel directions. However, in those regions of slabs where a flexural shear failure could possibly occur, such as near to free edges, the suggested approach should be reasonable. Unfortunately, there is no experimental evidence to justify the approaches suggested above.

*Enhanced $v_c$ values* The basic $v_c$ values of Table 6.1 may be enhanced by multiplying by a tabulated factor, $(\xi_s > 1)$, which increases as the overall depth decreases provided that the overall depth is less than 300 mm. The reason for this is that tests have shown that the shear strength of a member increases as its depth decreases. Relevant test data have been collated by Taylor [155] and are summarised in Fig. 6.5 in terms of the shear strength ($V_u$) divided by the shear strength of an equivalent specimen of 250 mm depth ($V_{250}$); due allowance has been made for dead load shear forces. It should be noted that all of the test specimens were beams, whereas the Code applies the enhancement factor to slabs.

It can be seen that the Code values give a reasonable lower bound to the test data for overall depths less than 500 mm. For greater depths, Taylor observed that there is a reduction in shear strength for large beams but that the minimum stirrups required for beams should take care of this. However, the Code does not require minimum stirrups to be provided for *slabs* unless more than 1% of compression steel is present: this did not cause a problem in the drafting of CP 110 because building slabs are generally thin; however, bridge slabs can be thick. Furthermore, Taylor suggested that code allowable shear stresses should be reduced by 40% if the depth to breadth ratio of a beam exceeds 4. Such a ratio could be exceeded in bridge beams and the webs of box girders. These points are raised here to emphasise that the values of $v_c$ and $\xi_s$ were derived with buildings in mind. It is not possible at present to state whether the values are appropriate to bridges because of the lack of data from tests on slabs subjected to the stress conditions which occur in bridges.

*Shear reinforcement* When the nominal shear stress exceeds $\xi_s v_c$, shear reinforcement should be provided and designed, as for beams, to resist the shear force in excess of that which can be resisted by the concrete ($\xi_s v_c d$ per unit length). The required amount of shear reinforcement

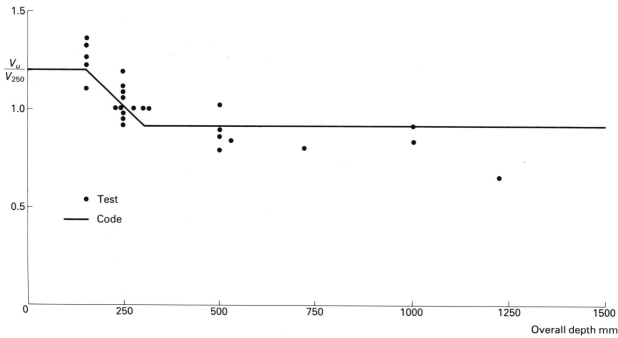

**Fig. 6.5** Slab enhancement factor

should be calculated from equations (6.6) or (6.7), as appropriate, with $v_c$ replaced by $\xi_s v_c$ This design approach is probably reasonable for building slabs for the reasons discussed previously: however, it is not clear whether it is reasonable for slabs subjected to loadings which cause principal shears which are not aligned with the flexural reinforcement. A possible design approach for such situations is suggested earlier in this chapter.

*Maximum shear stress* The maximum nominal shear stress in slabs is limited to $0.375 \sqrt{f_{cu}}$, which is half of that for beams. It is understood that this was originally suggested by the CP 110 committee because it was felt that, in building slabs, the anchorage of stirrups could not be relied upon. The author would suggest that, if this is correct, it would also be the case for top slabs of bridge decks but not necessarily for deeper slab bridges.

Furthermore, it is considered that shear reinforcement cannot be detailed and placed correctly in slabs less than 200 mm thick, and shear reinforcement is consequently considered to be ineffective in such slabs. Hence, the maximum nominal shear stress in a slab less than 200 mm thick is limited to $\xi_s v_c$ and not to $0.375 \sqrt{f_{cu}}$.

*Minimum shear reinforcement* Unlike beams, it is not necessary to provide minimum shear reinforcement if the nominal shear stress is less than $\xi_s v_c$. It is understood that this decision was made by the CP 110 committee because it was considered to be in accordance with normal practice for building slabs. It would also appear to be reasonable for bridge slabs.

The maximum stirrup spacing for slabs is the effective depth. This is greater than the maximum spacing of $0.75d$ for beams because the latter value was considered to be too restrictive for building slabs. The test data referred to when discussing the $0.75d$ value for beams suggest that a spacing of $d$ should be adequate for both beams and slabs.

*Voided or cellular slabs* No specific rules are given in the Code for designing voided or cellular slabs to resist shear. However, when considering longitudinal shear, it is reasonable to apply the solid slab clauses and to consider the shear force to be resisted by the minimum web thickness. With regard to transverse shear, designers, at present, either arrange the voids so that they are at points of low transverse shear force or use their own design rules. Possible approaches to the design of cellular and voided slabs to resist transverse shear forces are given in Appendix B of this book.

## Punching shear

### Introduction

Prior to discussing the Code clauses for punching shear, it should be stated that most codes of practice approach the problem of designing against punching shear failure by considering a specified allowable shear stress acting over a specified surface at a specified distance from the load. It is emphasised that the specified surfaces do not coincide with the failure surfaces which occur in tests. This fact can cause problems when code clauses are applied in circumstances different to those envisaged by those originally responsible for writing the clauses.

The Code clauses are based very much on those in CP 110, which were written with building slabs in mind, and these clauses are now summarised.

### CP 110 clauses

Punching shear in CP 110 is considered under two separate headings: namely, 'Shear stresses in solid slabs under concentrated loadings' and 'Shear in flat slabs'. The former clauses are concerned with the punching of applied loads through a slab, whereas the latter are concerned with punching at columns acting monolithically with a slab.

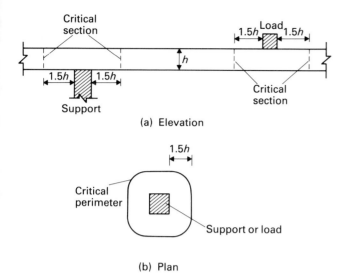

(a) Elevation

(b) Plan

**Fig. 6.6(a),(b)** Punching shear perimeter

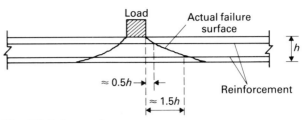

**Fig. 6.7** Failure surface

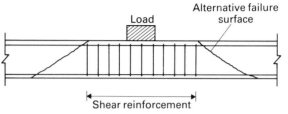

**Fig. 6.8** Influence of shear reinforcement

The critical shear perimeter for both situations is given as 1.5 times the slab *overall* depth from the face of the load or column as shown in Fig. 6.6, and the area of concrete, deemed to be providing shear resistance, is the length of the perimeter multiplied by the slab *effective* depth. A constant allowable design shear stress is assumed to act over this area. The perimeter was *chosen* so that the allowable design shear stress could be taken to be the values of $v_c$ given in Table 6.1. Hence, the perimeter was chosen so that the same $v_c$ values could be used for both flexural and punching shear. However, in the original version of CP 110 it was stated that, for *flat slabs* having lateral stability and with adjacent spans differing by less than 25%, the tabulated values of $v_c$ should be reduced by 20%. This was because the design approach was to take the design shear force to be that acting when all panels adjacent to the column were loaded and, thus, it was necessary to make an allowance for the non-symmetrical shear distribution which would occur if patterned loading were considered. The reduction of $v_c$ by 20% was thus intended to allow for patterned loading [112]. Subsequently, in 1976, the flat slab clause was amended so that, at present, $v_c$ is *not* reduced but the design shear force is increased by 25% to allow for the possible non-symmetrical shear distribution. If the slab does not have lateral stability or if the adjacent spans are appreciably different, it has always been necessary to calculate the moment $(M)$ transmitted by the slab and to increase the design shear force $(V)$ by the factor $(1 + 12.5\,M/Vl)$, where $l$ is the longer of the two spans in the direction in which bending is being considered.

Regan [156] has shown, by comparing with test data, that the original CP 110 clauses were reasonable. It should be noted that most of the tests were carried out on simply supported square slabs under a concentrated load and there are very few data for slabs loaded with a concentrated load near to a concentrated reaction, as occurs near to a bridge pier. Regan quotes only three tests of such a nature and reports satisfactory prediction, by the CP 110 clauses, of the ultimate strength.

It is generally the case that the flexural reinforcement in the vicinity of a concentrated load or a column head is different in the two directions of the reinforcement. The question then arises as to what area of reinforcement to adopt for determining $v_c$ from Table 6.1. CP 110 allows one to adopt the average of the reinforcement areas in the two directions and tests carried out by Nylander and Sundquist [157] in which the ratio of the steel areas varied from 1.0 to 4.1 justify this approach. These tests essentially modelled a pair of columns with line loads on each side as could occur for a bridge.

When averaging the reinforcement areas in the two directions, the area in each direction should include all of the reinforcement within the loaded area and within an area extending to within three times the overall slab depth on each side of the loaded area. The reason for considering the reinforcement within such a wide band is that the actual failure surface extends a large distance from the load as shown in Fig. 6.7. The validity of considering the reinforcement in a large band has been confirmed by the results of tests carried out by Moe [158] in which the same area of reinforcement was distributed differently. It was found that the punching strength was essentially independent of the reinforcement arrangement.

If the actual nominal shear stress $(v)$ on the perimeter exceeds the allowable value of $\xi_s v_c$, it is necessary to design shear reinforcement in accordance with

$$0.4 \text{ N/mm}^2 \leqslant \frac{(\Sigma A_{sv})(0.87f_{yv})}{u_{crit}d} \geqslant v - \xi_s v_c \qquad (6.8)$$

where $(\Sigma A_{sv})$ is the total area of shear reinforcement and $u_{crit}$ is the length of the perimeter. This equation can be derived in a similar manner to equation (6.7). It can be seen from Fig. 6.7 that, in order to ensure that the shear reinforcement crosses the failure surface, it is necessary for the reinforcement to be placed at a distance of about $0.5h$ to $1.5h$ from the face of the load. In fact, CP 110 requires the shear reinforcement calculated from equation (6.8) to be placed at a distance of $0.75h$. However, CP 110 also requires the same amount of reinforcement to be provided at the critical perimeter distance of $1.5h$; hence twice as much shear reinforcement as is theoretically required has to be provided. It appears that such a conservative approach was proposed because of the limited range of shear reinforcement details covered by the available test data [158].

The presence of shear reinforcement obviously strengthens

a slab in the vicinity of the shear reinforcement and can thus cause failure to occur by the formation of shear cracks outside the zone of the shear reinforcement as shown in Fig. 6.8. CP 110 thus requires the shear strength to be checked also at distances, in steps of $0.75h$, beyond $1.5h$ and, if necessary, shear reinforcement should be provided at these distances.

The minimum amount of shear reinforcement implied by equation (6.8) has to be provided only if $v > \xi_s v_c$ and is very similar to the amount originally proposed for beams by the Shear Study Group.

As is also the case for flexural shear in slabs, the maximum nominal punching shear stress should not exceed $0.375 \sqrt{f_{cu}}$.

### BS 5400 clause

The clause in the Code which covers punching shear is identical to that in CP 110 which covers 'Shear stresses in solid slabs under concentrated loadings'. Hence, the modifications in CP 110 which allow for non-uniform shear distributions in flat slabs are not included in the Code. Instead, whether punching of a wheel through a deck or of a pier (integral or otherwise) is being considered the design procedure is to adopt the CP 110 perimeters and the Table 6.1 values of $v_c$ (modified by $\xi_s$ if the depth permits), and to design shear reinforcement using equation (6.8).

Such an approach is probably reasonable when considering wheel loads or piers which are not integral with the deck. However, when dealing with piers which are integral with the deck, and thus non-symmetry of the shear distribution and moment transfer should be considered, it could be that a modification, similar to that for flat slabs in CP 110, should be made to either the $v_c$ values or the design shear force. However, this has not been included in the Code and the implication is that the effects of non-symmetry and moment transfer can be ignored.

Further problems, which are probably of more importance to the bridge engineer than to the building engineer, are those caused by voids running parallel to the plane of a slab and by changes of section due to the accommodation of services. These problems are not considered by the Code.

Some tests have been carried out by Hanson [159] on the influence on shear strength of service ducts having widths equal to the slab thickness and depths equal to 0.35 of the slab thickness. He concluded that provided the ducts were not within two slab thicknesses of the load there was no reduction in shear strength. However, it is not clear whether such a rule would apply to slabs with voids as deep as those which occur in bridge decks.

# Shear in prestressed concrete

## Flexural shear

### Beam failure modes

Two different types of shear failure can occur in prestressed concrete beams as shown in Fig. 6.9.

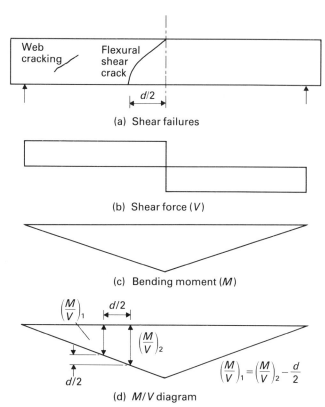

**Fig. 6.9(a)–(d)** Shear failure modes in prestressed concrete beams

In regions uncracked in flexure, a shear failure is caused by web cracks forming when the principal tensile stress exceeds the tensile strength of the concrete. In regions cracked in flexure, a shear failure is caused either by web cracks or by a flexural crack developing into a shear failure. Hence, it is necessary to check both types of failure and to take the lesser of the ultimate loads associated with the two types of failure as the critical load.

### Sections uncracked in flexure

The ultimate shear strength in this condition is designated $V_{co}$, and the criterion of failure for a section with no shear reinforcement is that the principal tensile stress anywhere in the section exceeds the tensile strength of the concrete. If the principal tensile stress is taken as positive and equal to the tensile strength of the concrete, then for equilibrium

$$f_s = V_{co} A\bar{y}/Ib$$

and

$$-f_t = (f_{cp} + f_b)/2 - \sqrt{(f_{cp} + f_b)^2/4 + f_s^2}$$

combining the equations gives

$$V_{co} = \frac{Ib}{A\bar{y}} \sqrt{f_t^2 + (f_{cp} + f_b)f_t}$$

In the above equation,

| | | |
|---|---|---|
| $V_{co}$ | = | shear force to cause web cracking |
| $I$ | = | second moment of area |
| $b$ | = | width |
| $A\bar{y}$ | = | first moment of area |
| $f_{cp}$ | = | compressive stress due to prestress |
| $f_b$ | = | flexural compressive stress |
| $f_s$ | = | shear stress |
| $f_t$ | = | tensile strength of concrete. |

In order to simplify calculations, it is assumed that the principal tensile stress is a maximum at the beam centroid, in which case $f_b = 0$, and, for a rectangular section, $Ib/A\bar{y} = 0.67\,bh$. Hence

$$V_{co} = 0.67\,bh\,\sqrt{f_t^2 + f_{cp}f_t} \qquad (6.8)$$

The above simplification was originally introduced into the Australian Code [160]. It should be noted that it is unsafe for I-beams to consider only the centroid but this is mitigated by the fact that, for such beams, $Ib/A\bar{y} \simeq 0.8\,bh$ as opposed to $0.67\,bh$. However, for flanged beams in which the neutral axis is within the flange, it is considered to be adequate to check the principal tensile stress at the junction of the web and flange. This simplification again originated in the Australian Code [160].

Tests on beams of concretes made with rounded river gravels as aggregate have indicated [160] that $f_t = 5\sqrt{f_{cy1}}$ (in Imperial units). However, the Australian Code adopted $4\sqrt{f_{cy1}}$ in order to allow for strength reductions caused by shrinkage cracking, mild fatigue loading and variations in concrete quality. If the latter value is converted to S.I. units and it is assumed that $f_{cy1} = 0.8f_{cu}$, then $f_t = 0.297\sqrt{f_{cu}}$. A partial safety factor of 1.5 was then applied to $f_{cu}$ to give the design value of $0.24\sqrt{f_{cu}}$ which appears in the Code.

Since a partial safety factor of $\sqrt{1.5}$ is applied to $f_t$, partial safety factors of $(\sqrt{1.5})^2$ and $\sqrt{1.5}$ are implied in the first and second terms respectively under the square root sign of equation (6.8). In order that a partial safety factor of $(\sqrt{1.5})^2$ is implied for both terms, it is necessary to apply a multiplying factor of $1/\sqrt{1.5} \simeq 0.8$ to $f_{cp}$. This results in the following equation, which appears in the Code

$$V_{co} = 0.67\,bh\,\sqrt{f_t^2 + 0.8\,f_{cp}f_t} \qquad (6.9)$$

Reynolds, Clarke and Taylor [161] have compared equation (6.9), without the partial safety factors, with test results and found that the ratios of the observed shear forces causing web cracking to $V_{co}$ were, with the exception of one beam which had a ratio of 0.68, in the range 0.92 to 1.59 with a mean of 1.13.

When inclined tendons are used, the vertical component of the prestress should be added to $V_{co}$ to obtain the total shear resistance. However, the Code only permits 80% of the vertical component to be added in order to be consistent with equation (6.9).

### Sections cracked in flexure

A shear failure can occur in a prestressed beam by a flexural crack developing into an inclined crack which eventually causes a shear failure. The position of the critical flexural crack, relative to the load, varies, but it has been shown [159] that it can be assumed to be at half the effective depth from the load.

Sozen and Hawkins [162] considered the loads at which a flexure-shear crack formed in 190 tests and showed that a good lower bound to the shear force ($V_{cr}$) could be given by the following empirical equation in Imperial units

$$V_{cr} = 0.6\,bd\,\sqrt{f_{cy1}} + M_t/(M/V - d/2) \qquad (6.10)$$

where $M_t$ is the cracking moment and $M$ and $V$ are the moment and shear force at the section under consideration. If this equation is transformed to S.I. units, it is assumed that $f_{cy1} = 0.8\,f_{cu}$ and a partial safety factor of 1.5 applied to $f_{cu}$, then

$$V_{cr} = 0.037\,bd\,\sqrt{f_{cu}} + M_t/(M/V - d/2)$$

It is obviously conservative to ignore the $d/2$ term, in which case the following equation, which appears in the Code, is obtained.

$$V_{cr} = 0.037\,bd\,\sqrt{f_{cu}} + V(M_t/M) \qquad (6.11)$$

It should be noted that, in equation (6.11), $d$ is the distance from the extreme compression fibre to the centroid of the tendons.

If the modulus of rupture of the concrete is $f_r$, then the cracking moment is given by

$$M_t = (f_r + f_{pt})I/y$$

where $f_{pt}$ is the tensile stress due to prestress at an extreme fibre, distance $y$ from the centroid. ACI-ASCE Committee 323 [163] originally suggested that, in Imperial units, $f_r = 7.5\sqrt{f_{cy1}}$ and the Australian Code [160] subsequently reduced this to $6\sqrt{f_{cy1}}$ to allow for shrinkage cracking, repeated loading and variations in concrete quality. If the latter expression is converted to S.I. units, it is assumed that $f_{cy1} = 0.8\,f_{cu}$ and a partial safety factor of 1.5 is applied to $f_{cu}$, then the Code design value of $0.37\sqrt{f_{cu}}$ is obtained. Again, only 80% of the prestress should be taken to give a consistent partial safety factor. Hence, the following design equation, which appears in the Code, is obtained.

$$M_t = (0.37\sqrt{f_{cu}} + 0.8f_{pt})\,I/y \qquad (6.12)$$

A minimum value of $V_{cr}$ of $0.1\,bd\sqrt{f_{cu}}$ is stipulated in the Code. This value originated in the American Code as, in Imperial units, $1.7\,bd\sqrt{f_{cy1}}$. The reason for this value is not apparent but if it is converted to S.I. units, it is assumed that $f_{cy1} = 0.8f_{cu}$ and a partial safety factor of 1.5 is applied to $f_{cu}$, then the Code value is obtained.

The majority of the beams for which equation (6.10) was found to give a good lower bound fit had relatively high levels of prestress, with the ratio of effective prestress to tendon characteristic strength ($f_{pe}/f_{pu}$) in excess of 0.5. These beams were thus representative of Class 1 or Class 2 beams, but not necessarily of Class 3 beams which can have much lower levels of prestress. A modified expression for $V_{cr}$ was thus derived for Class 3 beams which gives a linear transition from the reinforced concrete shear clauses ($f_{pe}/f_{pu} = 0$) to the Class 1 and Class 2 formula (equation (6.11)) when $f_{pe}/f_{pu} = 0.6$. In view of the two terms of equation (6.11) it was proposed [161] that for Class 3 members.

$$V_{cr} = A + B \qquad (6.13)$$

where $A$ depends on material strength and is analogous to the shear force calculated from the $v_c$ values of Table 6.1, and $B$ is the shear force to flexurally crack the beam. Both $A$ and $B$ are to be determined. The term, $A$, was written as a function of $v_c$ and the effective prestress ($f_{pe}$)

$$A = (1 - nf_{pe}/f_{pu})\,v_cbd$$

where $n$ is to be determined. This function was chosen for $A$ because it reduces to the reinforced concrete equation when $f_{pe} = 0$.

Equation (6.11) can be expanded to the following by using equation (6.12)

$$V_{cr} = 0.037\, bd\, \sqrt{f_{cu}} + 0.37\,\sqrt{f_{cu}}\,\frac{I}{y}\,\frac{V}{M} + \frac{M_0 V}{M} \qquad (6.14)$$

where $M_0 = 0.8 f_{pt}\, I/y$ is the moment to produce zero stress at *the level of the steel centroid*. It is thus convenient to write the term, $B$, of equation (6.13) as $M_0\, V/M$ and $V_{cr}$ becomes for reinforced and all classes of prestressed concrete

$$V_{cr} = (1 - n f_{pe}/f_{pu})\, v_c bd + M_0 V/M \qquad (6.15)$$

For reinforced concrete, $f_{pe} = M_o = 0$ and hence $V_{cr} = v_c bd$, which agrees with the reinforced concrete clauses. In order that equations (6.14) and (6.15) for Classes 1 and 2 and Class 3 respectively agree for $f_{pe}/f_{pu} = 0.6$, it is necessary that

$$0.037\, bd\, \sqrt{f_{cu}} + 0.37\,\sqrt{f_{cu}}\,\frac{I}{y}\,\frac{V}{M} = (1 - 0.6n)\, v_c bd$$

A shear failure is unlikely to occur if $M/V > 4h$ and thus it is conservative [161] to put $M/V = 4h$. It is further assumed that $d \simeq h$, $I/y = bh^2/6$ (the value for a rectangular section), $f_{cu} = 50$ N/mm$^2$, $v_c = 0.55$ N/mm$^2$ (i.e. 0.5% steel) and thus $n = 0.55$. Hence, the following equation, which appears in the Code, is obtained

$$V_{cr} = (1 - 0.55\, f_{pe}/f_{pu})\, v_c bd + M_0 V/M \qquad (6.16)$$

In view of the large number of simplifications made in deriving this equation, Reynolds, Clarke and Taylor [161] compared it, with the partial safety factors removed, with observed $V_{cr}$ values from 38 partially prestressed beams. The ratios of the experimental ultimate shear forces to $V_{cr}$ were, with the exception of one beam which had a ratio of 0.77, in the range 0.97 to 1.40 with a mean of 1.18. The exceptional beam had a cube strength of only 20 N/mm$^2$ and a high amount of web reinforcement.

The total area of both tensioned and untensioned steel in the tension zone should be used when assessing $v_c$ from Table 6.1; and, in equation (6.16), $d$ should be the distance from the extreme compression fibre to the centroid of the steel in the tension zone. The total area of tension steel is used because the longitudinal steel contributes to the shear strength by acting as dowel reinforcement and by controlling crack widths, and thus indirectly influencing the amount of aggregate interlock. Thus any bonded steel can be considered. The Code also implies that unbonded tendons should be considered, but it could be argued that they should be excluded because they cannot develop dowel strength and are less effective in controlling crack widths.

When both tensioned steel of area $A_{s(t)}$ and characteristic strength $f_{pu(t)}$ and untensioned steel of area $A_{s(u)}$ and characteristic strength $f_{yL(u)}$ are present, $f_{pe}/f_{pu}$ should be taken as, by analogy, the ratio of the effective prestressing *force* $(P_f)$ divided by the total ultimate *force* developed by both the tensioned and untensioned steels, i.e.

$P_f / (A_{s(t)} f_{pu(t)} + A_{s(u)} f_{yL(u)})$.

It will be recalled that the $d/2$ term which appears in equation (6.10) was ignored in deriving equations (6.11) and (6.16). An examination of the bending moment and shear force diagrams of Fig. 6.9 reveals that the value of $M/V$ at a particular section is equal to the value of $(M/V - d/2)$ at a section distance $d/2$ from the particular section. It is thus reasonable to consider a value of $V_{cr}$ calculated from equations (6.11) and (6.16) to be applicable for a distance of $d/2$ in the direction of increasing moment from the particular section under consideration.

Finally, contrary to the principles of statics, the Code does *not* permit the vertical component of the forces in inclined tendons to be added to $V_{cr}$ to give the total shear resistance. This requirement was based upon the results of tests on prestressed beams, with tendon drape angles of zero to 9.95°, reported by MacGregor, Sozen and Siess [164]. They concluded that the drape *decreased* the shear strength. However, since, except at the lowest point of a tendon, the effective depth of a draped tendon is less than that of a straight tendon, equations (6.11) and (6.16) *do predict a reduction* in shear strength for a draped, as compared with a straight, tendon. It is not clear whether the reduction in strength observed in the tests was due to the tendon inclination or the reduction in effective depth. If it is because of the latter, then the Code effectively allows for the reduction twice by adopting equations (6.11) and (6.16) and excluding the vertical component of the prestress. Hence, the Code, although conservative, does seem illogical in its treatment of inclined tendons.

### Shear reinforcement

The shear force $(V_c)$ which can be carried by the concrete alone is the lesser of $V_{co}$ and $V_{cr}$. If $V_c$ exceeds the applied shear force $(V)$ then, theoretically, no shear reinforcement is required. However, the Code requires nominal shear reinforcement to be provided, such that $0.87 f_{yv} A_{sv}/b s_v \geq 0.4$ N/mm$^2$, if $V \geq 0.5\, V_c$. These requirements were taken directly from the American Code. Thus shear reinforcement need not be provided if $V < 0.5\, V_c$. In addition the Code does not require shear reinforcement in members of minor importance nor where tests have shown that shear reinforcement is unnecessary. The CP 110 handbook [112] defines members of minor importance as slabs, footings, pile caps and walls. However, it is not clear whether such members should be considered to be minor in bridge situations and the interpretation of the Code obviously involves 'engineering judgement'.

If $V$ exceeds $V_c$ then shear reinforcement should be provided in accordance with

$$\frac{A_{sv}}{s_v} = \frac{V - V_c}{0.87\, f_{yv}\, d_t} \qquad (6.17)$$

where $d_t$ is the distance from the extreme compression fibre to the centroid of the tendons or to any longitudinal bars placed in the corners of the links, whichever is the greater. The amount of shear reinforcement provided should exceed the minimum referred to in the last paragraph. Equation (6.17) can be derived in the same way as equation (6.7).

The basic maximum link spacing is 0.75 $d_t$, which is the same as that for reinforced concrete beams. However, if $V > 1.8 V_c$, the maximum spacing should be reduced to 0.5 $d_t$: the reason for this is not clear. In addition, for any value of $V$, the link spacing in flanged members should not exceed four times the web width: this requirement presumably follows from the CP 115 implication that special considerations should be given to beams in which the web depth to breadth ratio exceeds four.

### Maximum shear force

In order to avoid premature crushing of the web concrete, it is necessary to impose an upper limit to the maximum shear force. The Code tabulates maximum design shear stresses which are derived from the same formula (0.75 $\sqrt{f_{cu}}$) as those for reinforced concrete. The shear stress is considered to act over a nominal area of the web breadth, minus an allowance for ducts, times the distance from the extreme compression fibre to the centroid of all (tensioned or untensioned) steel in the tension zone.

Clarke and Taylor [154] have considered prestressed concrete beams which failed by web crushing and found that the ratios of observed web crushing stress to the Code value of 0.75 $\sqrt{f_{cu}}$ were in the range 1.04 to 4.50 with a mean of 2.13.

It has been suggested by Bennett and Balasooriya [165] that beams with a web depth to breadth ratio in excess of ten could exhibit a tendency to buckle prior to crushing: such a ratio could be exceeded in a bridge. However, the test data considered by Clarke and Taylor [154] included specimens with ratios of up to 17; and Edwards [166] has tested a prestressed box girder having webs with slenderness ratios of 33 and did not observe any instability problems. It thus appears that web instability should not be a problem in the vast majority of bridges.

It is mentioned previously that a reduced web breadth, to allow for ducts, should be used when calculating shear stresses. The Code stipulates that the reduced breadth should be the actual breadth less either the duct diameter for ungrouted ducts or two-thirds the duct diameter for grouted ducts. These values were originally suggested by Leonhardt [167] and have subsequently been shown to be reasonable by tests carried out by Clarke and Taylor [154] on prisms with ducts passing through them.

It should be noted that when checking the maximum shear force any vertical component of prestress should be considered only for sections uncracked in flexure. This again defies statics but is consistent with the approach to calculating $V_{co}$ and $V_{cr}$.

### Slabs

The Code states that the flexural shear resistance of prestressed slabs should be calculated in exactly the same manner as that of prestressed beams, except that shear reinforcement is not required in slabs when the applied shear force is less than $V_c$. This recommendation does not appear to be based upon test data and the author has the same reservations about the recommendation as those discussed previously in connection with reinforced slabs.

For design purposes, it would seem reasonable to average shear forces over a width of slab equal to twice the effective depth, and to carry out the shear design calculation for the shear forces acting on planes normal to the tendons and untensioned reinforcement. A similar approach, for reinforced slabs, is discussed elsewhere in this chapter.

## Punching shear

The Code specifies a different critical shear perimeter for prestressed concrete than for reinforced concrete. The perimeter for prestressed concrete is taken to be at a distance of half of the overall slab depth from the load.

The section should then be considered to be uncracked and $V_{co}$ calculated as for flexural shear. In other words, the principal tensile stress at the centroidal axis around the critical perimeter should be limited to 0.24 $\sqrt{f_{cu}}$. It should be noted that Clause 7.4 of the Code refers to values of $V_{co}$ in Table 32 of the Code whereas it should read Table 31.

If shear reinforcement is required, it should be designed in the same way as that for flexural shear.

The above design approach is, essentially, identical to that of the American Code [168] and was originally proposed by Hawkins, Crisswell and Roll [169]. They considered data from tests on slab-column specimens and slab systems, and found that the ratios of observed to calculated shear strength (with material partial safety factors removed) were in the range 0.82 to 1.28 with a mean of 1.06. The data were mainly from reinforced concrete slabs but 32 of the slabs were prestressed. In addition, the specimens had concrete strengths and depths less than those which would occur in bridges. However, the author feels that the Code approach should be applicable to bridge structures.

Finally, the reservations, expressed earlier when discussing reinforced concrete slabs, regarding non-uniform shear distributions and the presence of voids are also applicable to prestressed slabs.

## Torsion – general

### Equilibrium and compatibility torsion

In the introduction to this chapter it is stated that, according to the Code, torsion calculations have to be carried out only at the ultimate limit state.

An implication of this fact is that it is necessary to think in terms of two types of torsion.

### Equilibrium torsion

In a statically determinate structure, subjected to torsional loading, torsional stress resultants must be present in order to maintain equilibrium. Hence, such torsion is referred to as equilibrium torsion and torsional strength *must* be provided to prevent collapse occurring. An example of equilibrium torsion is that which arises in a cantilever beam due to torsional loading.

It is assumed in the Code that the torsion reinforcement provided to resist equilibrium torsion at the ultimate limit

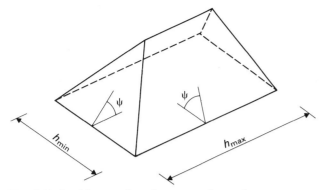

**Fig. 6.10** Sand-heap analogy for rectangular section

state is adequate to control torsional cracking at the serviceability limit state.

### Compatibility torsion

In a structure which is statically indeterminate, it is theoretically possible to provide no torsional strength, and to prevent collapse occurring by designing more flexural and shear strength than would be necessary if torsional strength were provided. The explanation of this is that a stress resultant distribution, within the structure, with zero torsional stress resultants and which satisfies equilibrium can always be found. Since such a distribution satisfies equilibrium it leads to a safe lower bound design [27].

Although a safe design results from a stress resultant distribution with no twisting moments or torques, it is obviously necessary, from considerations of compatibility, for various parts of the structure to displace by twisting. Hence, such torsion is referred to as compatibility torsion.

The torsion which occurs in bridge decks is, generally, compatibility torsion and it would be acceptable, in an elastic analysis, to assign zero torsional stiffness to a deck. This would result in zero twisting moments or torques throughout the deck and bending moments greater than those which would occur if the full torsional stiffness were used.

In the above discussion it is implied that either zero or the full torsional stiffness should be adopted. However, it is emphasised that any value of torsional stiffness could be adopted. As an example, Clark and West [170] have shown that it is reasonable, when considering the end diaphragms of beam and slab bridges, to adopt only 50% of the torsional stiffness obtained by multiplying the elastic shear modulus of concrete by $J_i$ (see equation (2.43)).

Finally, although it is permissible to assume zero torsional stiffness at the ultimate limit state, which implies the provision of no torsion reinforcement, it is necessary to provide some torsion reinforcement to control any torsional cracks which could occur at the serviceability limit state. The Code assumes that the nominal flexural shear reinforcement discussed earlier in this chapter is sufficient for controlling any torsional cracks.

## Combined stress resultants

The Code acknowledges the fact that, at a particular point, the maximum bending moment, shear and torque do not generally occur under the same loading. Thus, when

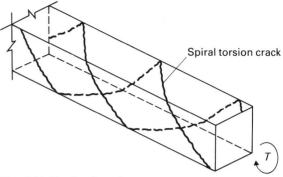

**Fig. 6.11** Torsional cracks

designing reinforcement to resist the maximum torque, reinforcement, which is present and is in excess of that required to resist the other stress resultants associated with the maximum torque, may be used for torsion reinforcement.

## Torsion of reinforced concrete

### Rectangular section

#### Torsional shear stress

Methods of calculating elastic and plastic distributions of torsional shear stress are available for homogeneous sections having a variety of cross-sectional shapes, including rectangular. The calculation of an elastic distribution is generally complex, and that of a plastic distribution is generally much simpler. However, neither distribution is correct for non-homogeneous sections such as cracked structural concrete.

In order to simplify calculation procedures, the Code adopts a plastic distribution of torsional shear stress over the entire cross-section. It is emphasised that such a distribution is assumed not because it is correct but merely for convenience. The Code also gives allowable nominal torsional shear stresses with which to compare the calculated plastic shear stresses. The allowable values were obtained from test data.

It can be seen that the above approach is similar to that adopted for flexural shear, in which allowable nominal flexural shear stresses, acting over a nominal area of breadth times effective depth, were chosen to give agreement with test data.

The plastic torsional shear stress distribution is best calculated by making use of the sand-heap analogy [171] in which the constant plastic torsional shear stress ($v_t$) is proportional to the constant slope ($\psi$) of a heap of sand on the cross-section under consideration. In fact

$$v_t = T\,\psi/K \tag{6.18}$$

where $T$ is the torque and $K$ is twice the volume of the heap of sand.

The plastic shear stress for a rectangular section can thus be evaluated from Fig. 6.10 as follows

$$
\begin{aligned}
\text{Volume of sand-heap} =\ & (h_{min}/2)(h_{max})(\psi h_{min}/2) \\
& - (2)(1/6)(h_{min})(h_{min}/2)(\psi h_{min}/2) \\
=\ & (1/4)\psi\, h^2_{min}\,(h_{max} - h_{min}/3)
\end{aligned}
$$

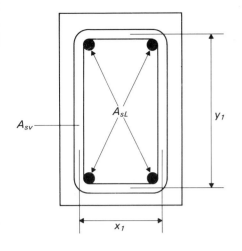

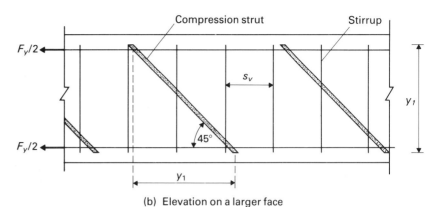

(a) Cross–section           (b) Elevation on a larger face

**Fig. 6.12(a),(b)** Space truss analogy

Thus, from equation (6.18),

$$v_t = \frac{2T}{h^2_{min}\,(h_{max} - h_{min}/3)} \qquad (6.19)$$

This equation is given in the Code.

### Torsion reinforcement

*Design* If the applied torsional shear stress calculated from equation (6.19) exceeds a specified value ($v_{tmin}$) it is necessary to provide torsion reinforcement. One might expect $v_{tmin}$ to be taken as the stress to cause torsional cracking or that corresponding to the pure torsional strength of a member without web reinforcement. In fact, it is taken in the Code to be 25% of the latter value. Such a value was originally chosen by American Concrete Committee 438 [172] because tests have shown [173, 174] that the presence of such a torque does not cause a significant reduction in the shear or flexural strength of a member.

The tabulated Code values of $v_{tmin}$ are given by $0.067\ \sqrt{f_{cu}}$ (but not greater than $0.42$ N/mm²) and are design values which include a partial safety factor of 1.5 applied to $f_{cu}$. The formula is based upon that originally proposed by the American Concrete Institute but has been modified to (a) convert from cylinder to cube strength, (b) allow for partial safety factor and (c) allow for the fact that the American Concrete Institute calculates $v_t$ from an equation based upon the skew bending theory of Hsu [175] instead of the plastic theory.

If $v_t$ exceeds $v_{tmin}$, torsion reinforcement has to be provided in the form of longitudinal reinforcement plus closed links. The reason for requiring both types of reinforcement is that, under pure torsional loading, principal tensile stresses are produced at 45° to the longitudinal axis of a beam. Hence, torsional cracks also occur at 45° and these tend to form continuous spiral cracks as shown in Fig. 6.11. It is necessary to have reinforcement, on each face, parallel and normal to the longitudinal axis in order that the torsional cracks can be controlled and adequate torsional strength developed.

The amounts of reinforcement required are calculated by considering, at failure, a space truss. This is analogous to the plane truss considered for flexural shear earlier in this Chapter. In the space truss analogy, a spiral failure surface

is considered which, theoretically, could form at any angle. However, for design purposes, it is considered to form at the same angle (45°) as the initial cracks in order that the amount of stress redistribution required prior to collapse may be minimised. The failure surface assumed by the Code is shown in Fig. 6.12 together with relevant dimensions.

If the two legs of a link have a total area of $A_{sv}$ and the links are spaced at $s_v$, then $y_1/s_v$ links cross a line parallel to a compression strut on a larger face of the member. If the characteristic strength of the link reinforcement is $f_{yv}$, then the steel force at failure in *each* larger face is

$$F_y = f_{yv}(A_{sv}/2)(y_1/s_v)$$

Similarly, the steel force at failure in *each* smaller face is

$$F_x = f_{yv}(A_{sv}/2)\,(x_1/s_v)$$

The total resisting torque is

$$\begin{aligned} T &= F_y x_1 + F_x y_1 \\ &= f_{yv}A_{sv}\,x_1 y_1/s_v \end{aligned} \qquad (6.20)$$

At the ultimate limit state, $f_{yv}$ has to be divided by a partial safety factor of 1.15 to give a design stress of $0.87\,f_{yv}$. Hence, in the code, $f_{yv}$ in equation (6.20) is replaced by $0.87\,f_{yv}$. Furthermore, the Code introduces an efficiency factor, which is discussed later, of 0.8 in order to obtain good agreement between the space truss analogy and test results. Hence, in the Code, equation (6.20) is presented as

$$\frac{A_{sv}}{s_v} \geq \frac{T}{0.8x_1y_1\,(0.87f_{yv})} \qquad (6.21)$$

The force $F_y$ is considered to be the vertical component of a principal force ($F$) which acts perpendicular to the failure surface and thus $F = F_y\,\sqrt{2}$. The principal force also has a horizontal (longitudinal) component $F/\sqrt{2}$ which, from above, is equal to $F_y$. A force of this magnitude acts in each larger face and, similarly, a horizontal (longitudinal) force of magnitude $F_x$ acts in each smaller face. Thus the *total* horizontal (longitudinal) force, which tends to elongate the section, is $2(F_x + F_y)$. This elongating force has to be resisted by the longitudinal reinforcement: if the *total* area of longitudinal reinforcement is $A_{sL}$

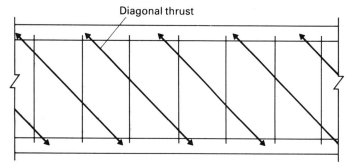

(a) Elevation

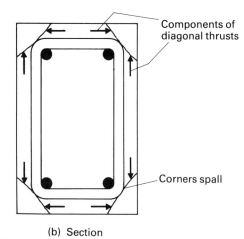

Components of diagonal thrusts

Corners spall

(b) Section

**Fig. 6.13(a),(b)** Diagonal torsional compressive stresses

and it has a characteristic strength of $f_{yL}$

$$A_{sL}f_{yL} = 2(F_x + F_y) = A_{sv}f_{yv}(x_1 + y_1)/s_v$$

Thus, in the Code,

$$A_{sL} \geq \frac{A_{sv}}{s_v}\left(\frac{f_{yv}}{f_{yL}}\right)(x_1 + y_1) \qquad (6.22)$$

*Detailing* The detailing of torsional reinforcement has been considered by Mitchell and Collins [176] and the following Code rules are based very much on their work.

Diagonal compressive stresses occur in the concrete between torsional cracks, and such stresses near the edges of a section cause the corners to spall off as shown in Fig. 6.13. In order to prevent premature spalling it is necessary to restrict the link spacing and the flexibility of the portion of longitudinal bar between the links. Tests reported by Mitchell and Collins [176] indicate that the link spacing should not exceed $(x_1 + y_1)/4$ nor 16 times the longitudinal corner bar diameter. The Code specifies these spacings and also states that the link spacing should not exceed 300 mm. The latter limitation is intended to control cracking at the serviceability limit state in large members where the two other limitations can result in large spacings. In addition, the longitudinal corner bar diameter should not be less than the link diameter.

The characteristic strength of all torsional reinforcement is limited to 425 N/mm² primarily because such a restriction exists for shear reinforcement. However, it is justified by the fact that some beams, which were tested by Swann [177] and reinforced with steel having yield stresses in excess of 430 N/mm², failed at ultimate torques slightly less than those predicted by the Code method of calculation [178]. The reason for this was that the large concrete

strains necessary to mobilise the yield stress of such high strength steel resulted in a reduction in the efficiency factor mentioned earlier and discussed in the next section of this chapter.

### Maximum torsional shear stress

The space truss of Fig. 6.12 consists of the torsion reinforcement acting as tensile ties plus concrete compressive struts. As is also the case for flexural shear, it is necessary to limit the compressive stresses in the struts to prevent the struts crushing prior to the torsion reinforcement yielding in tension. This is achieved by limiting the nominal torsional shear stress. However, the derivation of the limiting values in the Code is connected with the choice of the efficiency factor applied to the reinforcement in equation (6.20).

It is mentioned elsewhere in this chapter that an efficiency factor has to be introduced into equation (6.20) in order to obtain agreement between test results and the predictions of the space truss analogy. Swann [178] found that the efficiency factor decreases with an increase in the nominal plastic torsional shear stress at collapse and also decreases with a decrease in specimen size.

The dependence of the efficiency factor on stress is explained by the fact that, if the nominal shear stress is high, the stresses in the inclined compression struts between the torsional cracks of Fig. 6.11 are also high. Thus, at high nominal stresses, a greater reliance is placed upon the ability of the concrete in compression to develop high stresses at high strains than is the case for low nominal stress. In view of the strain-softening exhibited by concrete in compression (see Fig. 4.1(a)) it is to be expected that the efficiency factor should decrease with an increase in nominal stress.

The size effect is explained by the fact that spalling occurs at the corners of a member in torsion as shown in Fig. 6.13. This alters the path of the torsional shear flow and can be considered to reduce the area of concrete resisting the torque. As shown in Fig. 6.14, this effect is more significant for a small than for a large section.

In order to formulate a simple design method which takes the above points into account, Swann [178] proposed that the efficiency factor should be *chosen* to be a value which could be considered to result in an acceptably high nominal stress level in large sections. The reason for considering large sections was that these are least affected by corner spalling.

Consideration of tests carried out on beams with a maximum cross-sectional dimension of about 500 mm indicated that 0.8 was a reasonable value to take for the efficiency factor. The nominal stress which could be attained with this factor was $0.92 \sqrt{f_{cu}}$. If a partial safety factor of 1.5 is applied to $f_{cu}$, a design maximum nominal torsional shear stress $(v_{tu})$ of $0.75 \sqrt{f_{cu}}$ is obtained. This value is tabulated in the Code with an upper limit of 4.75 N/mm². It should be noted that the stress is the same as the design maximum nominal flexural shear stress. Thus, Swann [178] essentially chose an efficiency factor to give the same design maximum shear stress for torsional shear as for flexural shear. Having established a nominal maximum

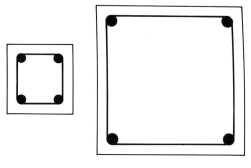

(a) Sections before spalling

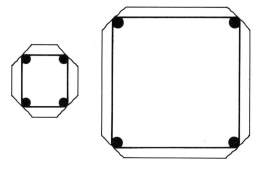

(b) Sections after spalling

**Fig. 6.14(a),(b)** Torsional size effect [178]

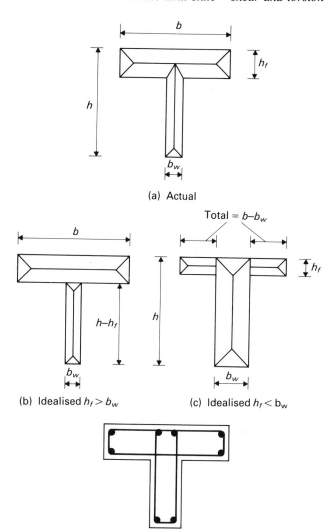

(a) Actual

(b) Idealised $h_f > b_w$    (c) Idealised $h_f < b_w$

(d) Reinforcement

**Fig. 6.15(a)–(d)** Torsion of T-section

stress and an efficiency factor for large sections, it was decided to adopt the same efficiency factor (0.8) for small sections and to determine reduced nominal maximum stresses for the latter by considering test data. It was found that the stress of $0.92\sqrt{f_{cu}}$ had to be modified by multiplying by $(y_1/550)$ for sections where $y_1 < 550$ mm. Hence the design stress of $0.75\sqrt{f_{cu}}$ also has to be multiplied by this ratio.

It is emphasised that the Code requires that both of the following be satisfied:

1. The *total* (flexural plus torsional) shear stress $(v + v_t)$ should not exceed $0.75\sqrt{f_{cu}}$ or 4.75 N/mm²
2. In the case of small sections $(y_1 < 550$ mm), the *torsional* shear stress should not exceed $0.75\sqrt{f_{cu}}(y_1/550)$ or $4.75(y_1/550)$ N/mm².

## T-, L- and I-sections

### Torsional shear stress

The plastic shear stress for a flanged section could be obtained from considerations of the appropriate sand-heap, such as that shown in Fig. 6.15(a) for a T-section. However, the junction effects make the calculations rather tedious and the Code thus permits a section to be divided into its component rectangles which are then considered individually.

The manner in which the section is divided into rectangles should be such that the function $\Sigma(h_{max}h^3_{min})$ is maximised. This implies that, in general, the section should be divided so that the widest of the possible component rectangles is made as long as possible as shown in Fig. 6.15(b) and (c). The total torque $(T)$ applied to the section is then apportioned among the component rectangles such that each rectangle is subjected to a torque:

$$\frac{T(h_{max}h^3_{min})}{\Sigma(h_{max}h^3_{min})}$$

The above considerations imply that an anomalous situation arises in the treatment of flanged sections for the following reason.

An examination of equation (2.43) shows that the division into rectangles and the apportioning of the torque is carried out on the basis of the approximate *elastic* stiffnesses of the rectangles. However, the nominal torsional shear stress in each rectangle is calculated using equation (6.19) based upon plastic theory.

### Torsion reinforcement

If the nominal torsional shear stress in any rectangle is less than the appropriate $v_{tmin}$ value discussed earlier in connection with rectangular sections, then no torsion reinforcement is required in that rectangle. Otherwise, reinforcement for each rectangle should be designed in accordance with equations (6.21) and (6.22), and should be detailed so that the individual rectangles are tied together as, for example, in Fig. 6.15(d).

### Maximum torsional shear stress

The maximum nominal torsional shear stresses discussed

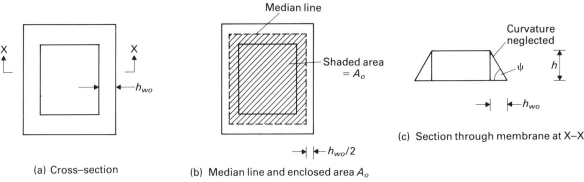

(a) Cross–section      (b) Median line and enclosed area $A_o$

(c) Section through membrane at X–X

**Fig. 6.16(a)–(c)** Torsion of box section

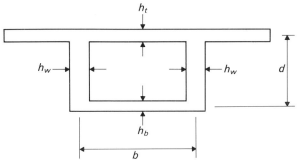

**Fig. 6.17** Box girder notation

in connection with rectangular sections should not be exceeded in any individual rectangle.

## Box sections

### Torsional shear stress

The *elastic* torsional shear stress distribution is easily calculated for a box section. Hence, the Code requires the nominal stress to be calculated from the standard formula for a *thin-walled* closed section [76]:

$$v_t = T/2h_{w0}A_0$$

where $h_{w0}$ is the wall thickness at the point where the shear stress $v_t$ is determined and $A_0$ is the area within the median line of the section as shown in Fig. 6.16.

The above equation can be derived by applying the membrane analogy [76] in which the *variable* elastic torsional shear stress is proportional to the variable slope ($\psi$) of a membrane inflated over the cross-section. The mathematical expression used is identical to that for the sand-heap analogy for plastic torsion (equation (6.18)) but $K$ is now twice the volume under the membrane. With reference to Fig. 6.16 and by ignoring the slight curvature of the membrane (so that a section through the membrane has straight edges), the slope at a particular point is

$$\psi = h/h_{w0}$$

where $h$ = height of membrane, and the volume under the membrane is $\simeq hA_0$. Thus, from equation (6.18),

$$v_t = T\psi/2hA_0 = T/2h_{w0}A_0 \tag{6.23}$$

Equation (6.23) is, strictly, for thin-walled boxes whereas it could be argued that many concrete box sections are not thin. Maisel and Roll [41] have shown that the error ($\triangle v_t$) in calculating $v_t$ for a thick-walled box from

equation (6.23) is, with the notation of Fig. 6.17,

$$\frac{\triangle v_t}{v_t} = \frac{h^2_{w0}}{2A_0}\left(\frac{b}{h_t} + \frac{b}{h_b} + \frac{2d}{h_w}\right) \tag{6.24}$$

### Torsion reinforcement

It can be seen from equation (6.23) that the nominal torsional shear stress at a point is dependent upon the wall thickness at that point and thus varies around a box with non-uniform wall thickness. If the nominal torsional shear stress at any point exceeds the $v_{tmin}$ values discussed earlier in connection with rectangular sections, then torsion reinforcement must be provided.

The design of torsion reinforcement is complicated in the Code by the fact that two sets of equations may be used and the lesser of the amounts so calculated may be provided. The two sets of equations are (a) equations (6.21) and (6.22) which were derived, for solid rectangular sections, from the space truss analogy and (b) equations (6.25) and (6.26) below which were also derived from the space truss analogy, but with the reinforcement assumed to be concentrated along the median line of the box walls and with the efficiency factor taken to be 1.0.

$$\frac{A_{sv}}{s_v} \geqslant \frac{T}{A_0(0.87f_{yv})} \tag{6.25}$$

$$A_{sL} \geqslant \frac{A_{sv}}{s_v}\frac{f_{yv}}{f_{yL}}\frac{(\text{perimeter of }A_0)}{2} \tag{6.26}$$

The reason for having two sets of equations is that (6.21) and (6.22) were originally intended for beams of relatively small cross-section and they can be over-conservative for large thin-walled box sections in which $x_1$ is greater that about 300 mm [179].

Swann and Williams [179] have tested model reinforced concrete box beams under pure torsional loading and found that the observed ultimate torques exceeded the ultimate torques calculated from either set of equations. It was also found that equations (6.21) and (6.22) were more conservative than (6.25) and (6.26): this was to be expected since the models were large in the sense that they were models of large prototype sections.

The consideration of a box girder subjected only to torsion is rather academic since, in practice, flexural loading is also generally present. There is thus an essentially constant compressive stress ($f_{cav}$) due to flexure over the cross-sectional area ($A_c$) of one flange. Thus a compressive in-plane force of $A_cf_{cav}$ acts in conjunction with an in-plane shear force of $A_cv_t$. Hence, from equation (A17), the

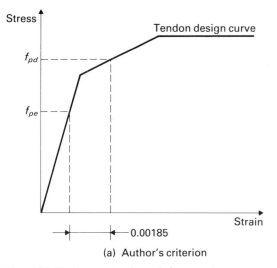

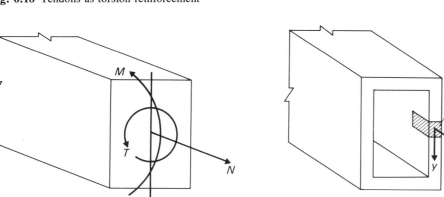

**Fig. 6.18** Tendons as torsion reinforcement

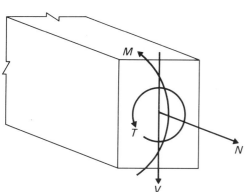

(a) Stress – resultants on entire cross–section      (b) Stress – resultants at a point

**Fig. 6.19(a),(b)** Combined stress-resultants

amount ($A$) of longitudinal reinforcement required is given by

$$(0.87f_{yL})A = -A_c f_{cav} + |A_c v_t|$$

or

$$A = \frac{-A_c f_{cav}}{0.87f_{yL}} + \frac{|A_c v_t|}{0.87f_{yL}}$$

This implies that the amount of reinforcement calculated from equation (6.22) or (6.26) may be reduced by $A_c f_{cav}/0.87f_{yL}$. The validity of this approach to design has been confirmed by the tests of Swann and Williams [179] and is consequently permitted by the Code. It should be noted that the Code adopts for $A_c$ the area of section subjected to flexural compressive stresses instead of simply the flange area, and that the Code also refers to a stress $f_{yc}$ which should read $f_{yL}$.

The reduction in longitudinal steel area due to the effect of flexural compressive stresses could, theoretically, be applied to sections other than boxes but the Code limits the reduction to box sections only.

### *Maximum torsional shear stress*

The maximum nominal torsional shear stresses discussed in connection with rectangular sections have been shown to be reasonable for box sections by the tests of Swann and Williams [179].

## Torsion of prestressed concrete

### General

The code essentially assumes that prestressed concrete members can be designed to resist torsion by ignoring the prestress and designing them as if they were reinforced. Thus values of $v_t$, $v_{tmin}$, $v_{tu}$, $A_{sv}$ and $A_{sL}$ should be calculated in accordance with the equations for reinforced concrete presented earlier in this chapter. The validity of such an approach has been checked, with bridges specifically in mind, by Swann and Williams [179], who carried out tests on model prestressed concrete box beams.

It has been mentioned that the greatest permissible characteristic strength for conventional torsion reinforcement is 425 N/mm². This value was chosen because the large yield strains associated with higher strengths reduce the efficiency factor in equation (6.21) to less than 0.8. Thus, to achieve an efficiency factor of 0.8 when using tendons as torsion reinforcement, it is necessary to ensure that the additional tendon strain, required to mobilise the tendons' ultimate strength, does not exceed the yield strain of other reinforcement in the section having a characteristic strength of 425 N/mm². If the latter is conservatively assumed to be hot-rolled, the *design* yield strain is 0.87 × 425/200 000 = 0.00185. Hence, the *design* ultimate stress

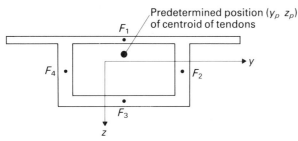

**Fig. 6.20** Segmental box beam

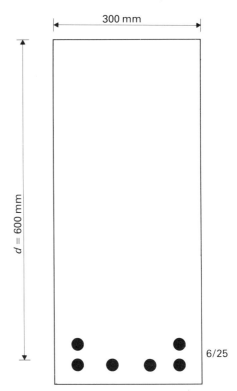

**Fig. 6.21** Example 6.1

$(f_{pd})$ for the tendons should be the lesser of $0.87 f_{pu}$ and (see Fig. 6.18(a)) $f_{pe}$ plus the stress increment equivalent to a strain increment of 0.00185, where $f_{pe}$ is the effective pre-stress in the tendons.

Although the above argument seems logical, the resulting limiting stresses are not included in the Code. Instead, the design stress is limited to the lesser of 460 N/mm² and $(0.87 f_{pu} - f_{pe})$. The specification of such stresses seems to indicate that a criterion suggested by Lampert [180] and by Maisel and Swann [181] was misinterpreted by the drafters. The suggested criterion was that tendons and conventional reinforcement should reach yield at about the same stage of failure of the section, in order that excessively large strains would not develop in one type of reinforcement prior to yield of the other. This implies that (see Fig. 6.18(b))

$$f_{pd} - f_{pe} \simeq f_y$$

where $f_{pd}$ is the tendon design stress at ultimate. The drafters took $f_y$ to be 460 N/mm² (the greatest value likely to be used) although it exceeds the greatest value (425 N/mm²) permitted for torsion reinforcement. Hence, $f_{pd} \simeq f_{pe} + 460$ N/mm². However, $f_{pd}$ cannot exceed $0.87 f_{pu}$, thus, $f_{pd}$ should be taken as the lesser of $0.87 f_{pu}$ and $(f_{pe} + 460$ N/mm²). It thus appears that the limiting stresses given in the Code are not, as stated in the Code, *design* stresses but the stress *increment* necessary to raise the stress from the effective prestress to the design stress. It should also be mentioned that it is inconsistent to apply a partial safety factor to $f_{pu}$ but not to $f_y$, and thus the second limit should be $(f_{pe} + 400$ N/mm²). Hence the stress *increment* in the Code should be 400 N/mm².

In conclusion, the Code stress limits are stress *increments* and not *design* stresses; moreover, the author would suggest that the alternative criterion illustrated in Fig. 6.18(a) is more logical.

## Alternative design methods

The Code method of design of a section subjected to a general loading is to consider the stress resultants acting on the entire cross-section, as shown in Fig. 6.19(a), and then to superpose the effects of any local actions. This method is probably very suitable to small sections but for large sections (and particularly for box beams) it could be considered desirable to vary the reinforcement over, for example, the depth of a web. It is then necessary to consider the stress-resultants acting at various points of the cross-section as shown in Fig. 6.19(b). Each point would generally be subjected to both in-plane and bending stress resultants and thus the sandwich method discussed in Chapter 5

would be appropriate for design. A simplified version of such an approach has been described by Swann and Williams [179] and Maisel and Swann [181].

The Code refers to 'other design methods' without specifically mentioning any; however, the above approach of considering the stress resultants at points of the cross-section would be acceptable.

## Segmental construction

In segmental construction it is not generally convenient to provide continuous longitudinal conventional reinforcement. Thus longitudinal tendons in excess of those needed for flexure have to be provided.

The Code states that the line of action of the longitudinal elongating force should coincide with the centroid of the steel actually provided. If the longitudinal steel capacities required in each flange and web of a segmental box beam are as shown in Fig. 6.20, then they can be replaced by a force $F_t$ situated at $(\bar{y}, \bar{z})$, where:

$$F_t = \Sigma F_i$$
$$\bar{y} = \Sigma F_i y_i / \Sigma F_i$$
$$\bar{z} = \Sigma F_i z_i / \Sigma F_i$$

and each summation is for $i = 1$ to 4.

Thus the total ultimate capacity of the tendons needs to be $F_t$ and their centroid needs to be at $(\bar{y}, \bar{z})$.

In practice, it is often simpler to calculate the necessary total ultimate capacity of tendons situated at predetermined positions such that their centroid is at, say, $(y_p, z_p)$. In this case $F_t$ must be such that the capacity of each flange or web exceeds the appropriate value of $F_i$. Hence, moments should be taken about each web and flange in turn to give four inequalities. For example, by taking moments about web 2

$$(y_2 - y_p)F_t \geqslant F_1(y_2 - y_1) + F_3(y_2 - y_3) + F_4(y_2 - y_4)$$

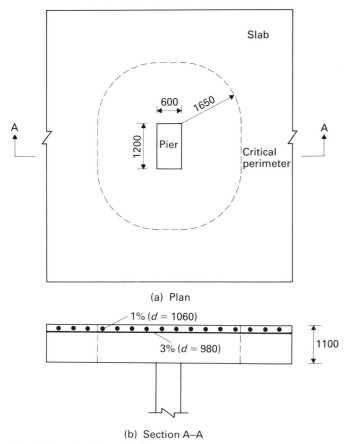

(a) Plan

(b) Section A–A

**Fig. 6.22(a),(b)** Example 6.2

The largest value of $F_t$ obtained from the four inequalities is the total ultimate tendon capacity which must be provided. Swann and Williams [179] give a numerical example of such a calculation, and they also present the results of tests on two model segmental box beams against which the above method of calculation was checked.

# Examples

## 6.1 Flexural shear in reinforced concrete

Design stirrups of 250 N/mm² characteristic strength to resist an ultimate shear force of 540 kN applied to the section shown in Fig. 6.21. The concrete is of grade 40.
Nominal applied shear stress = $v$ =
$$540 \times 10^3/(300 \times 600) = 3 \text{ N/mm}^2$$
From Table 6 of Code or 0.75 $\sqrt{f_{cu}}$, maximum allowable shear stress =

$v_u = 4.75 \text{ N/mm}^2$
$v_u > v$, o.k.

Area of tension reinforcement $A_s = 2950 \text{ mm}^2$
$100A_s/bd = (100 \times 2950)/(300 \times 600) = 1.64$
From Table 5 of Code, allowable shear stress without shear reinforcement

$v_c = 0.88 \text{ N/mm}^2$

From equation (6.7),

$$\frac{A_{sv}}{s_v} = \frac{300(3.00 - 0.88)}{0.87 \times 250} = 2.92 \text{ mm}^2/\text{mm}$$

The stirrup spacing should not exceed $0.75\,d = 450$ mm.
12 mm stirrups (2 legs) at 75 mm centres give 3.02 mm²/mm.

## 6.2 Punching shear in reinforced concrete

Design the slab shown in Fig. 6.22 to resist punching shear if the characteristic strengths of the steel and concrete are 425 and 40 N/mm² respectively and the column reaction at the ultimate limit state is 15 MN.
The critical perimeter is at $1.5h = (1.5)(1100) = 1650$ mm
The critical section is shown on Fig. 6.22.
Perimeter = $(2)(600) + (2)(1200) + (2\pi)(1650)$
$$= 14\,000 \text{ mm}$$
'Average' effective depth, $d = (3 \times 980 + 1 \times 1060)/4$
$$= 1000 \text{ mm}$$
Nominal applied shear stress,
$$v = 15 \times 10^6/(14\,000 \times 1000) = 1.07 \text{ N/mm}^2$$
From Table 6 of Code or 0.75 $\sqrt{f_{cu}}$, maximum allowable shear stress = $v_u = 4.75 \text{ N/mm}^2$
Only take half for slab, $\therefore v_u = 2.38 \text{ N/mm}^2$

$v_u > v$, o.k.
Average $100A_s/bd = (3 + 1)/2 = 2$
From Table 5 of the Code, allowable shear stress without shear reinforcement $v_c = 0.95 \text{ N/mm}^2$
From Table 8 of Code, $\xi_s = 1.00$

$v - \xi_s v_c = 1.07 - 0.95 = 0.12 \text{ N/mm}^2$
Thus shear reinforcement required, but must provide for at least 0.4 N/mm²
From equation (6.8)

$$\Sigma A_{sv} \geqslant (0.4)(14\,000)(1000)/(0.87)(425) = 15\,100 \text{ mm}^2$$

This amount of reinforcement must be provided along a perimeter $1.5h$ from the loaded area and also along a perimeter $0.75h$ from the loaded area.
Length of perimeter at $0.75h = (2)(600) + (2)(1200) + (2\pi)(825) = 8780$ mm
Thus on $1.5h$ perimeter provide $15\,100/14 = 1080$ mm²/m, and, on $0.75h$ perimeter, $15\,100/8.78 = 1720$ mm²/m.
Now check on perimeter at $(1.5 + 0.75)h = 2.25h$.
Length of perimeter = $(2)(600) + (2)(1200) + (2\pi)(2475)$
$$= 19200 \text{ mm}$$
Nominal applied shear stress = $v$
$$= 15 \times 10^6/(19\,200 \times 1000) = 0.78 \text{ N/mm}^2$$
$v < \xi_s v_c$, thus no need to provide more shear reinforcement.

## 6.3 Flexural shear in prestressed concrete

The pretensioned box beam shown in Fig. 6.23 is subjected to a shear force of 0.9 MN with a co-existing moment of 3.15 MNm at the ultimate limit state. Design shear reinforcement with a characteristic strength of 250 N/mm². The initial prestress is 70% of the characteristic strength and the losses amount to 30%. The concrete is of grade 50 and the section has been designed to be class 1.
From Table 32 of Code or 0.75 $\sqrt{f_{cu}}$, maximum allowable shear stress

$v_u = 5.3 \text{ N/mm}^2$

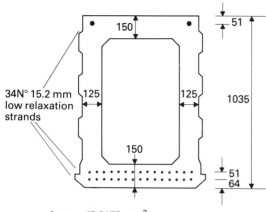

Area = 47 8175 mm²
Bottom fibre modulus = 125.43 × 10⁶ mm³
Neutral axis is 510 mm from bottom fibre

**Fig. 6.23** Example 6.3

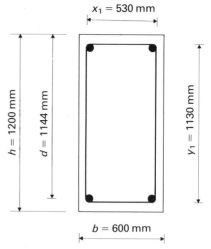

**Fig. 6.24** Example 6.4

maximum allowable shear force $V_u = v_u bd$
where $b$ = total web breadth = 250 mm
$d$ = effective depth of steel in tension zone
= 945.5 mm
$V_u = (5.3)(250)(945.5)10^{-6} = 1.25$ MN
$V_u > V$, ∴ o.k.
From Table 21, $A_{ps} f_{pu} = 227.0$ kN
Effective prestress = (34)(0.7)(0.7)(227) = 3780 kN

### Uncracked in flexure

Compressive stress at centroidal axis
= 3780 × 10³/478 175 = 7.91 N/mm²
Allowable principal tensile stress $f_t = 0.24 \sqrt{f_{cu}}$
= 1.70 N/mm²

From equation 6.9, $V_{co} = (0.67)(250)(1035) \times$
$\sqrt{(1.70)^2 + (0.8)(7.91)(1.70)}$
= 641 × 10³ N
= 0.641 MN

### Cracked in flexure

Distance of centroid of all tendons from bottom fibre
= (16 × 64 + 16 × 115 + 2 × 984)/34 = 142 mm
Eccentricity = 510 − 142 = 368 mm
Prestress at bottom fibre = 7.91 + (3780 × 10³ × 368/
125.43 × 10⁶)
= 19.0 N/mm²
From equation (6.12), cracking moment is

$M_t = (0.37 \sqrt{50} + 0.8 × 19.0) 125.43 × 10^6 × 10^{-9}$
= 2.23 MNm

From equation (6.11),

$V_{cr} = (0.037)(250)(1035 − 142) \sqrt{50} + (0.9 × 10^6)$
(2.23/3.15)
= 696 × 10³ N = 0.696 MN
$V_{cr}$ must be at least $0.1bd \sqrt{f_{cu}} = 0.158$ MN
Thus use $V_{cr} = 0.696$ MN.

### Shear reinforcement

Shear capacity without shear reinforcement $V_c$ is lesser of
$V_{co}$ and $V_{cr}$

∴ $V_c = V_{co} = 0.641$ MN

From equation (6.17),

$A_{sv}/s_v = (0.9−0.641)10^6/(0.87 × 250 × 971) = 1.23$ mm²/mm
$V \not> 1.8V_c$, maximum spacing is lesser of 0.75 × 971 = 728 mm

and 4 × 125 = 500 mm
Provide 12 mm stirrups (2 legs) at 175 mm centres.

$$\frac{A_{sv}}{s_v} \left( \frac{0.87 f_{yv}}{b} \right) = \frac{2 × 113}{175} \left( \frac{0.87 × 250}{250} \right)$$

$$= 1.12 \text{ N/mm}^2$$
$$> 0.4 \text{ N/mm}^2, \text{ O.K.}$$

## 6.4 Torsion in reinforced concrete

An end diaphragm of a beam and slab bridge is, for design purposes, considered as a rectangular reinforced concrete beam, 600 mm wide and 1200 mm deep. The concrete is of grade 40 and the minimum cover is 30 mm. Design torsion reinforcement, having a characteristic strength of 425 N/mm², to resist an ultimate torque of 290 kNm which co-exists with an ultimate shear force of 500 kN. From equation (6.19), the nominal torsional shear stress is

$$v_t = \frac{2 × 290 × 10^6}{600^2 (1200 − 600/3)} = 1.61 \text{ N/mm}^2$$

From Table 7 of Code or $0.067 \sqrt{f_{cu}}$, allowable torsional shear stress without torsion reinforcement $v_{tmin} =$
0.42 N/mm²
$v_t > v_{tmin}$, thus torsion reinforcement required.
Assume $d$ = 1144 mm, $x_1$ = 530 mm, $y_1$ = 1130 mm
(see Fig. 6.24)
Nominal flexural shear stress = $v$
= 500 × 10³/(600 × 1144) = 0.73 N/mm²

$v + v_t = 2.34$ N/mm²

From Table 7 of Code or $0.75 \sqrt{f_{cu}}$, maximum allowable shear stress

$v_{tu} = 4.75$ N/mm²
$v_{tu} > v + v_t$, thus section big enough.

From equation (6.21), links should be provided such that

$A_{sv}/s_v = 290 × 10^6/(0.8 × 530 × 1130 × 0.87 × 425)$
= 1.64 mm²/mm

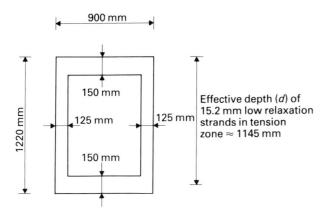

**Fig. 6.25** Example 6.5

From equation (6.22), required area of longitudinal reinforcement is

$$A_{sL} = 1.64\,(425/425)(530 + 1130) = 2720 \text{ mm}^2$$

4 No. 32 mm give 3220 mm² (1 bar in each corner)
Stirrup spacing should not exceed the least of
(a)  $(530 + 1130)/4 = 415$ mm
(b)  $16 \times 32 = 512$ mm
(c)  300 mm
10 mm stirrups (2 legs) at 90 mm centres give 1.74 mm²/mm
If 40 mm cover to main steel then $d$, $x_1$ and $y_1$ are as assumed.

The above reinforcement should be provided in addition to any flexural and shear reinforcement.

Although the Code does not permit one to do so, it would seem reasonable to reduce the area of longitudinal torsion reinforcement in the flexural compression zone in a similar manner to that permitted for box sections.

## 6.5 Torsion in prestressed concrete

The rectangular box section shown in Fig. 6.25 is subjected to an ultimate torque of 610 kNm and an ultimate vertical shear force 230 kN. A co-existing bending moment produces an average flexural compressive stress of 20 N/mm² over the flexural compression zone which extends to a depth of 300 mm below the top of the section. The concrete is of grade 50 and the characteristic strengths of the prestressing steel and the torsional reinforcement are 1637 N/mm² and 425 N/mm² respectively. The effective prestress in the tendons is 860 N/mm². The minimum cover is 25 mm. Design suitable torsion reinforcement.
Area within median line $A_0 = (1220 - 150)(900 - 125)$
$$\qquad\qquad\qquad = 829\,250 \text{ mm}^2$$
From equation (6.23), the nominal torsional shear stress in a flange is

$$v_{tf} = 610 \times 10^6/(2 \times 150 \times 829\,250) = 2.45 \text{ N/mm}^2$$
and in a web is

$$v_{tw} = 610 \times 10^6/(2 \times 125 \times 829\,250) = 2.94 \text{ N/mm}^2$$
From Table 7 of Code allowable torsional shear stress without torsion reinforcement $v_{tmin} = 0.42$ N/mm²
$v_t > v_{tmin}$, thus torsion reinforcement required.
The Code does not, in this context, define $b$ and $d$ to be used to calculate the flexural shear stress which acts only in the webs.
Assume $b = 2 \times 125 = 250$ mm, $d = 1145$ mm.

Flexural shear stress in webs $v_w = 230 \times 10^3/(250 \times 1145) = 0.80$ N/mm² and flexural shear stress in flanges $v_f = 0$.
For each flange, total shear stress $= v_f + v_{tf} = 0 + 2.45$
$$\qquad\qquad\qquad\qquad\qquad\quad = 2.45 \text{ N/mm}^2$$
For one web, total shear stress $= v_w + v_{tw} = 0.80 + 2.94$
$$\qquad\qquad\qquad\qquad\qquad\qquad = 3.74 \text{ N/mm}^2$$
From Table 7, maximum allowable shear stress $v_{tu} = 4.75$ N/mm²

$v_{tu} > v + v_t$, thus section big enough.

Assume $x_1 = 825$ mm, $y_1 = 1140$ mm.
From equation (6.21), links should be provided such that

$$A_{sv}/s_v = 610 \times 10^6/(0.8 \times 825 \times 1140 \times 0.87 \times 425)$$
$$\qquad\quad = 2.19 \text{ mm}^2/\text{mm}$$

From equation (6.25), links should be provided such that
$$A_{sv}/s_v = 610 \times 10^6/(829\,250 \times 0.87 \times 425) = 1.99 \text{ mm}^2/\text{mm}$$
Provide links such that $A_{sv}/s_v = 1.99$ mm²/mm.
The longitudinal torsion reinforcement could be provided by conventional reinforcement or by additional prestressing steel. If the latter were to be used, it would be logical to calculate a design stress as follows (see stress–strain curve in Fig. 5.10).

$f_{pe} = 802$ N/mm²
Strain at $f_{pe} = 0.00401$

Allowable strain increment = 0.00185 (see text), thus total strain = 0.00586 at which the stress is (see Fig. 5.10) 1147 N/mm².

The area of longitudinal conventional reinforcement would be, from equation (6.26),

$$A_{sL} = 1.99(425/425)(2 \times 1070 + 2 \times 775)/2 = 3670 \text{ mm}^2$$

or, the area of longitudinal prestressing steel with an effective prestress of 802 N/mm² would be, from equation (6.26),

$$A_{sL} = 1.99(0.87 \times 425/1147)(2 \times 1070 + 2 \times 775)/2$$
$$\qquad = 1180 \text{ mm}^2$$

Thus provide either 1835 mm² of conventional reinforcement or 590 mm² of extra prestressing steel in each flange. The top flange areas may be reduced by, respectively

$$20(900 \times 150 + 2 \times 125 \times 150)/(0.87 \times 425) = 9330 \text{ mm}^2$$
and

$$20(900 \times 150 + 2 \times 125 \times 150)/1147 = 3010 \text{ mm}^2$$

Thus no torsion reinforcement is required in the top flange. The bottom flange could be reinforced with 4 No. 25 mm bars (giving an area of 1960 mm²) or 5 additional tendons (giving an area of 694 mm²).

12 mm diameter stirrups (2 legs) at 100 mm centres give 2.26 mm²/mm, which is adequate for the transverse torsion reinforcement.

It is emphasised that, as written, the Code would not permit a design stress of 1147 N/mm² to be adopted for the tendons. The Code requires the lesser of 460 N/mm² and $(0.87\,f_{pu} - f_{pe}) = 564$ N/mm² to be adopted, i.e. 460 N/mm². Thus, in accordance with the Code, no advantage could be taken of the larger ultimate strength of the tendons.

## Chapter 7

# Serviceability limit state

## Introduction

As explained in Chapter 4, the criteria which have to be satisfied at the serviceability limit state are those of permissible steel and concrete stress, permissible crack width, interface shear in composite construction and vibration. In this chapter, methods of satisfying the criteria of permissible steel and concrete stress, and of crack width in both reinforced and prestressed concrete construction are presented. The additional criteria, associated with interface shear and tensile stress in the in-situ concrete of composite construction, are presented in Chapter 8.

Compliance with the vibration criterion is discussed, together with other aspects of the dynamic loading of bridges, in Chapter 12.

## Reinforced concrete stress limitations

### General

As discussed in Chapter 4, the concrete compressive stresses in reinforced concrete should not exceed $0.5f_{cu}$ and the reinforcement, tensile or compressive, stresses should not exceed $0.8f_y$.

Although it is not stated in the Code, the above limitations should be applied only to axial or flexural stresses. Thus it is not necessary to consider flexural shear or torsional shear stresses at the serviceability limit state.

In practice, axial and flexural stress calculations will involve the application of conventional elastic modular ratio theory. However, a modular ratio is not explicitly stated in the Code, and it is necessary to calculate a value from the stated modulus of elasticity of the reinforcement (200 kN/mm²) and the modulus of elasticity of the concrete. Short term values of the latter modulus are given in the Code and these vary from 25 kN/mm² for a characteristic strength of 20 N/mm² to 36 kN/mm² for a characteristic strength of 60 N/mm². The Code is not explicit as to whether these short term moduli should be used when calculating stresses or whether they should be adjusted to give long-term moduli. However, the Code does state that a long-term modulus, equal to half of the short term value, should be used when carrying out crack width calculations. In view of this requirement, and the fact that all structural Codes of Practice have hitherto adopted a long-term modulus, the author would suggest that, for stress calculations in accordance with the Code, a long-term modulus should be adopted. Since a relatively weak concrete having a characteristic strength of less than about 40 N/mm² is generally adopted for reinforced concrete, the long-term modular ratio calculated from the Code would be between 13 and 16. These values are of the same order as the value of 15 which is generally adopted for design purposes at present.

## Tension stiffening

In the above discussion, conventional modular ratio theory is mentioned. This theory generally considers a cracked section and ignores the stiffening effect of the concrete in tension between cracks (tension stiffening); hence, it overestimates stresses and strains as shown in Fig. 7.1. The difference between lines AB and OC of Fig. 7.1 is a measure of the tension stiffening effect and can be seen to decrease with an increase in load above the initial cracking load. This decrease results from the development of further cracks and the gradual breakdown of bond between the reinforcement and the concrete. Hence the strain ($\varepsilon_1$) calculated ignoring tension stiffening should be reduced by an amount ($\varepsilon_{ts}$) to give the actual strain ($\varepsilon_m$).

The Code permits tension stiffening to be taken into account in certain crack width calculations (as referred to later in this chapter), but it is not clear whether one is permitted to allow for it in permissible stress calculations. However, test results [183] indicate that, at stresses of the order of the steel stress limitation of $0.8f_y$, the tension stiffening effect is negligible. Therefore, it seems to be reasonable to ignore tension stiffening when carrying out stress calculations.

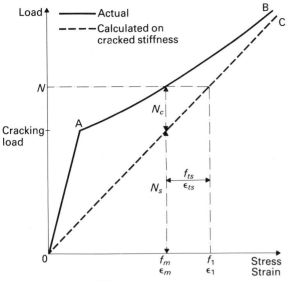

**Fig. 7.1** Tension stiffening

## Slabs

If the principal stresses in a slab do not coincide with the reinforcement directions, it is extremely difficult to calculate accurately the concrete and reinforcement stresses. As the Code gives no guidance, the author would suggest the following procedure:

1. Assume the section to be uncracked and calculate the four principal extreme fibre stresses caused by the stress resultants due to the applied loads.
2. Where a principal tensile stress exceeds the permissible design value, assume cracks to form perpendicular to the direction of that principal stress. The permissible design value could be taken to be the Class 2 prestressed concrete limiting stress of $0.45 \sqrt{f_{cu}}$ (see Chapter 4).
3. Consider each set of cracks in turn and calculate an equivalent area of reinforcement perpendicular to these cracks.
4. Using the equivalent area of reinforcement, calculate the stresses in the direction perpendicular to the cracks by using modular ratio theory.
5. Compare the extreme fibre concrete compressive stress with the allowable value of $0.5 f_{cu}$.
6. If the calculated stress in the equivalent area of reinforcement is $f_n$, then calculate the stress in an $i$-th layer of reinforcement, inclined at an angle $\alpha_i$ to the direction perpendicular to the cracks, from $f_i = \delta f_n$ where $\delta$ is discussed later. Compare $f_i$ with the permissible value of $0.8 f_y$.

The calculation of the equivalent area of reinforcement (step 3 above) is explained by considering a point in a cracked slab where the average direct and shear strains, referred to axes perpendicular and parallel to a crack (see Fig. 7.2), are $\varepsilon_n$, $\varepsilon_t$, $\gamma_{nt}$.

The strain in the direction of an $i$-th layer of reinforcement at an angle $\alpha_i$ to the $n$ direction is, by Mohr's circle

$$\varepsilon_i = \varepsilon_n \cos^2 \alpha_i + \varepsilon_t \sin^2 \alpha_i - \gamma_{nt} \sin \alpha_i \cos \alpha_i$$

The steel stress is thus

$$f_i = E_s \varepsilon_i$$

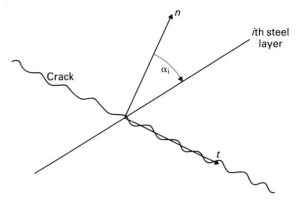

**Fig. 7.2** Cracked slab

where $E_s$ is the elastic modulus of the steel. If the steel area per unit width is $A_i$, the steel force per unit width is

$$F_i = A_i f_i$$

If $N$ such layers are considered, the total resolved steel force in the $n$ direction is

$$F_n = \sum_{i=1}^{N} F_i \cos^2 \alpha_i$$

$$= E_s \sum_{i=1}^{N} A_i (\varepsilon_n \cos^4 \alpha_i + \varepsilon_t \sin^2 \alpha_i \cos^2 \alpha_i - \gamma_{nt} \sin \alpha_i \cos^3 \alpha_i)$$

The force $F_n$ can be considered in terms of an equivalent area $(A_n)$ of reinforcement per unit width in the $n$ direction. Thus $F_n = A_n E_s \varepsilon_n$. Hence, by comparison with the previous equation,

$$A_n = \sum_{i=1}^{N} A_i (\cos^4 \alpha_i + \frac{\varepsilon_t}{\varepsilon_n} \sin^2 \alpha_i \cos^2 \alpha_i - \frac{\gamma_{nt}}{\varepsilon_n} \sin \alpha_i \cos^3 \alpha_i)$$

It is reasonable, at the serviceability limit state, to assume that the $n$ and $t$ directions will very nearly coincide with the principal strain directions. Thus $\gamma_{nt} \approx 0$ and the third term in the brackets of the above equation can be ignored.

There are now three cases to consider for a slab not subjected to significant tensile in-plane stress resultants:

1. If the slab is cracked on one face only and in one direction only, $\varepsilon_n \gg \varepsilon_t$ and the expression for $A_n$ reduces to

$$A_n = \sum_{i=1}^{N} A_i \cos^4 \alpha_i \qquad (7.1)$$

2. If the slab is cracked in two directions on the same face, then $\varepsilon_t$ will be of the same sign as $\varepsilon_n$. If $\varepsilon_t$ is again taken to be zero, the calculated value of $A_n$ will be less than the true value. It is thus conservative to use equation (7.1).
3. If the slab is cracked in two directions on opposite faces, $\varepsilon_t$ will be of opposite sign to $\varepsilon_n$ and could take

any value. The precise value of $\varepsilon_t/\varepsilon_n$ to adopt is very difficult to determine, although some guidance is given by Jofriet and McNeice [184]. As an approximation, it is reasonable to assume that, at the serviceability limit state, it is unlikely that $\varepsilon_n < |\varepsilon_t|$; it is thus conservative to assume $\varepsilon_n/\varepsilon_t = -1$. In which case

$$A_n = \sum_{i=1}^{N} A_i \cos^2\alpha_i \, (\cos^2\alpha_i - \sin^2\alpha_i) \qquad (7.2)$$

It is implied in equation (7.2) that reinforcement inclined at more than 45° to the $n$ direction should be ignored.

By implication the stress transformation factor, $\delta$, referred to earlier in this chapter should be taken as $\cos^2\alpha_i$ if equation (7.1) is used to evaluate $A_n$ and as $(\cos^2\alpha_i - \sin^2\alpha_i)$ if equation (7.2) is used.

Little error is involved in adopting the above approximations for $A_n$ when the reinforcement is inclined at less than about 25° to the perpendicular to the cracks. If, when using the above approximations with reinforcement inclined at more than 25°, it is found that the reinforcement stresses are excessive, it would be advisable to estimate a more accurate value of $\varepsilon_n/\varepsilon_t$ (as opposed to the above approximate values of zero and $-1$) as described in [184].

# Crack control in reinforced concrete

## General

### Statistical approach

The design crack widths given in Table 4.7, and discussed in Chapter 4, are design *surface* crack widths and are derived from considerations of appearance and durability.

In order to calculate crack widths it is necessary to decide on an interpretation of the design crack width: is it a maximum, a mean or some other value? It is not possible to think in terms of a maximum crack width but it is feasible to predict a crack width with a certain probability of exceedence. This can be illustrated by considering two nominally identical beams having zones of constant bending moment. If each beam is subjected to the same loading and the widths of all of the cracks within the respective constant moment zones are measured, then distributions of crack widths can be plotted as shown in Fig. 7.3. It is found that, although the maximum width of crack measured on each of the two nominally identical beams may be very different the crack widths exceeded by a certain percentage of the results are quite similar. For this reason, the design crack widths are defined as those having a certain probability of exceedence.

The probability of exceedence that has been chosen in the Code is 20% (i.e., 1 in 5 crack widths greater than the design value), which is the same percentage as that adopted in the building code, CP 110. At first sight it may appear very liberal to permit 1 in 5 crack widths to exceed

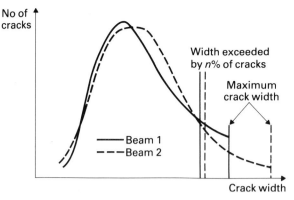

**Fig. 7.3** Crack width distributions for nominally identical beams

the design value; however, the formula used to calculate the '1 in 5' crack width is applicable only to the constant moment zones of specimens tested under laboratory conditions. The exceedence level in practice is much less than 20% and, for buildings, it has been estimated [182] that the chance of a specified width being exceeded by any single crack width will lie in the range $10^{-3}$ to $10^{-4}$. The main reasons for this reduction in probability are:

1. The specified design loading under which cracking should be checked is, essentially, the full nominal loading. This is greater than the 'average' loading which occurs for a significant length of time.

2. In design, lower bound estimates of the material properties are used; thus the probability of the stiffness being as low as is assumed in calculating the strains is low.

3. Structural members are not generally subjected to uniform bending over any great length, and thus the only cracks in a member which have any serious chance of being critical are those close to the critical sections of the member. These will be few in number compared with the total population of cracks.

The above reduction in probability will be less dramatic in bridges because the latter are subjected to repeated loadings which cause crack widths to gradually increase during the life of the bridge. In addition, the design loading is more likely to be achieved on a bridge than on a building. However, considerably less than 20% of the crack widths in a bridge should exceed the design value if the design is carried out in accordance with the Code formulae.

### Crack control in the Code

*General approach*  Although cracking due to such effects as the restraint of shrinkage and early thermal movement is a significant practical problem, it is cracking induced by applied loading that is used as a basis for design in the Code. However, the Code does require at least 0.3% of the gross concrete area of mild steel or 0.25% of high yield steel to be provided to control restrained shrinkage and early thermal movement cracks. These percentages are rather less than those suggested by Hughes [185] and should be used with caution.

The only crack widths that need to be calculated according to the Code are those due to axial and flexural stress resultants. Hence, flexural shear and torsional shear cracks do not have to be considered explicitly since, as discussed in Chapter 6, it is assumed that the presence of nominal links control the widths of such cracks.

Cracks due to applied loadings which cause axial and flexural stress resultants are controlled by limiting the spacing of the bars. This approach is identical to that adopted in BE 1/73. The bar spacing should not generally exceed 300 mm, and should be such that the crack widths in Table 4.7 are not exceeded *midway* between the bars under the specified design loading. It should be noted that the Code gives a different method of ensuring that the crack widths are not exceeded for each type of structural element.

*Loading*   Since the widths of cracks are controlled primarily for durability purposes, it is logical to define the design loading as that which can be considered to be virtually permanent rather than that of occasional but more severe loads. This is because, after the passage of occasional severe loads, the crack widths return to their values under permanent loading provided that the reinforcement remained elastic during the application of the occasional load. Since the limiting reinforcement stress is $0.8f_y$, the reinforcement will remain elastic.

At one time the drafters considered that 50% of HA loading should be taken to be permanent, together with pedestrian loading, dead load and superimposed dead load. When the appropriate partial safety factors were applied, the resulting design load was:

dead load + 1.2 (superimposed dead load) + 0.6 (HA) + 1.0 (pedestrian loading)

Subsequently, it was decided to check crack widths under full HA loading but the partial safety factors were altered so that the design load became:

dead load + 1.2 (superimposed dead load) + 1.0 (HA) + 1.0 (pedestrian loading)

This design loading is given, in the general design section of Part 4 of the Code, with the requirement that the wheel load should be excluded except when considering top flanges and cantilever slabs. In addition, for spans less than 6.5 m, 25 units of HB loading with associated HA loading should be considered. This additional loading was introduced to comply with the requirement of Part 2 of the Code that all bridges should be checked for 25 units of HB loading. It is not clear whether top flanges and cantilever slabs should be loaded with 25 units of HB when they span less than 6.5 m. However, it would seem reasonable to design such slabs for the more severe of the local effects of the HA wheel load or 25 units of HB loading.

The above loadings are very similar to those in BE 1/73, with the exception that the latter document refers to 30 units of HB loading for loaded length less than 6.5 m. The implication in BE 1/73 appears to be that the 30 units of HB loading should be applied to top flanges and cantilever slabs.

Finally, the Code clause concerned with reinforced concrete walls requires that any relevant earth pressure loadings should be considered in addition to dead, superimposed dead and highway loadings.

*Stiffnesses*   Although the design crack widths and loadings, referred to previously, are given in the general design section of Part 4 of the Code, the clauses concerned with crack width calculations appear much later in the Code under the heading: 'Spacing of reinforcement'. Guidance on the stiffnesses to be adopted in the calculations also appears under this heading.

In all crack width calculations the Code requires the elastic modulus of the concrete to be a long-term value equal to half of the tabulated short-term value. Hence a modular ratio of about 13 to 16 would be used to calculate the strains ignoring tension stiffening. In certain situations, which are discussed later in this chapter, the Code permits the strains ignoring tension stiffening to be reduced. It is thus necessary to determine the strain $\varepsilon_{ts}$ in Fig. 7.1. This is best achieved by initially considering an axially reinforced cracked section (having a concrete area of $A_c$ and a steel area of $A_s$) which is axially loaded. At a crack all of the applied force ($N$) is carried by the steel:

$$N = E_s \, \varepsilon_1 \, A_s$$

However, the *average* steel and concrete forces are given by:

$$N_s = E_s \, \varepsilon_m \, A_s$$
$$N_c = A_c \, f_{cm}$$

where $f_{cm}$ is the average tensile stress in the concrete between the cracks.
But

$$N = N_s + N_c$$
$$\therefore E_s \, \varepsilon_1 \, A_s = E_s \, \varepsilon_m \, A_s + A_c \, f_{cm}$$
$$\therefore \varepsilon_m = \varepsilon_1 - A_c \, f_{cm} / E_s \, A_s$$

Or

$$\varepsilon_{ts} = A_c \, f_{cm} / E_s \, A_s$$

At the cracking load, $f_{cm}$ is obviously equal to the tensile strength ($f_t$) of the concrete. At higher loads, tests [183, 186, 187] indicate that $f_{cm}$ reduces in accordance with

$$f_{cm} = f_t \, f_{scr} / f_1$$

where $f_{scr}$ is the steel stress at a crack at the cracking load and $f_1$ is the steel stress at the load corresponding to the strain $\varepsilon_1$.

For an axially reinforced and loaded section it is obvious that $A_c \simeq bh$ where $b$ and $h$ are the breadth and overall depth respectively. However, for a flexural member, it is necessary to define an effective area ($Kbh$) of concrete in tension over which the average stress $f_{cm}$ acts. Hence, for flexure, and considering only surface strains

$$\varepsilon_{ts} = Kbh \, f_{scr} \, f_t / E_s \, A_s \, f_1$$

In order that strains at any depth ($a'$) from the compression face can be considered the above expression is modified to

$$\varepsilon_{ts} = Kbh \, f_{scr} \, f_t \, (a' - x) / E_s \, A_s \, f_1 (h - x)$$

where $x$ is the neutral axis depth.

In the Code, cracking is generally checked under HA loading only and the values of $(\gamma_{fL} \, \gamma_{f3})$ at the ultimate limit state are 1.32 for dead load and 1.73 for HA loading, giving an average of 1.53; hence $f_1 \simeq 0.87 f_y / 1.53 = 0.57 f_y$.

However, implicit in the tension stiffening formula in the Code is the assumption that $f_1 = 0.58f_y$. This is because the formula was originally derived for CP 110, for which $f_1 = 0.58f_y$ is correct. However, the difference between this value of $f_1$ and the 'correct' value for the Code is negligible. Hence

$$\varepsilon_{ts} = Kbh\, f_{scr}\, f_t(a' - x)/E_s\, A_s(0.58\, f_y)\, (h - x)$$

Tests carried out by Stevens [188] indicated that, on average,

$$Kf_{scr}\, f_t/0.58E_s = 1.2 \times 10^{-3}\ \text{N/mm}^2.$$

It is emphasised that this constant has the dimensions of N/mm². Thus

$$\varepsilon_{ts} = 1.2bh(a' - x)\, 10^{-3}/A_s\, (h - x)\, f_y$$

Hence, the tension stiffening formula in the Code is obtained as

$$\varepsilon_m = \varepsilon_1 - 1.2bh(a' - x)\, 10^{-3}/A_s(h - x)\, f_y \qquad (7.3)$$

Beeby [189] has shown that equation (7.3) provides a reasonable lower bound fit to the instantaneous results of Stevens and a reasonable average fit to the latter's results obtained after long term loading of two years' duration. However, test results, under short term loading, reported by Rao and Subrahmanyan [186], Clark and Speirs [183] and Clark and Cranston [187] show tension stiffening values about one-third of those of Stevens and of those predicted by equation (7.3). In addition, the latest CEB tension stiffening equation [110] is in reasonable accord [187] with the data presented in [183], [186] and [187]. There is thus evidence to suggest that equation (7.3) overestimates tension stiffening.

Finally, equation (7.3) was originally derived with buildings in mind and thus the effects of repeated loading were not considered. Bridges are subjected to repeated loading and it is reasonable to assume that such loading reduces the tension stiffening. There is a lack of experimental data in this respect and, as an interim measure, it might be sensible to adopt the CEB recommendation [110] that tension stiffening under repeated loading should be taken as 50% of that under instantaneous loading. Tests on model solid [87, 126] and voided [71] slab bridges indicate that, for HB loading, the tension stiffening, as a proportion of the instantaneous value, is of the order of 60%, 50% and 40% after 1000, 2000 and 4000 load applications respectively. These values are reasonably consistent with the CEB value.

## Crack control calculations

### Beams

Base, Read, Beeby and Taylor [190] have carried out tests on 133 reinforced concrete beams. They found that there was an average difference of only 13% between the crack control performances of plain and deformed bars, and thus it is not necessary to have separate crack width formulae for the two types of bar. This point is particularly valid in

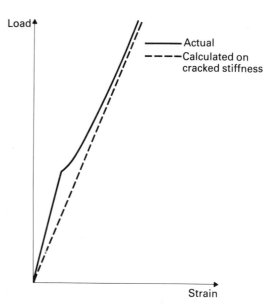

**Fig. 7.4** Tension stiffening: high steel percentage

view of the fact that crack widths are calculated only at positions mid-way between bars where the bar type has the least influence.

Base *et al.* found that the distributions of crack width were Gaussian and that, for deformed bars, the mean crack width ($w_m$) could be predicted by

$$w_m = 1.67\, a_{cr}\, \varepsilon_m$$

where $a_{cr}$ is the perpendicular distance from the point where the crack width is to be predicted to the surface of the nearest reinforcing bar. The standard deviation of the crack width population at any strain was found to be $0.416\, w_m$.

It is mentioned previously in this chapter that the probability of exceedence adopted in the Code is 20%. In a Gaussian distribution, a value of mean plus 0.842 standard deviations is exceeded by 20% of the population; thus the design crack width is given by

$$w = (1 + 0.842 \times 0.416)\, 1.67 a_{cr}\, \varepsilon_m \simeq 2.3 a_{cr}\, \varepsilon_m$$

However, in the Code, tension stiffening is not taken into account for beams and it is thus assumed that $\varepsilon_m = \varepsilon_1$. Hence the following Code equation is obtained

$$w = 2.3 a_{cr}\, \varepsilon_1 \qquad (7.4)$$

The reason for ignoring tension stiffening in beams is that it is envisaged that reinforced concrete bridge beams would be heavily reinforced; in which case the load–strain relationship is as shown in Fig. 7.4. It can be seen that the tension stiffening effect is a small proportion of the actual strain and can be ignored.

Equation (7.4) is very similar to the equation in BE 1/73; however, in the latter document the constant is not 2.3 but is 3.3 for deformed bars and 3.8 for plain bars. The value of 3.3 is appropriate to the 1% exceedence level [190], which is a much more severe criterion than the 20% level adopted in the Code. However, it should be remembered that the design crack widths specified in BE 1/73 and the Code are also different. The BE 1/73 value of 3.8 was obtained by increasing the value of 3.3 for deformed bars by 13% and rounding up [190].

## Solid slab bridges

At one stage in the drafting of the Code, it was hoped to prepare simple bar spacing rules which would obviate the need to carry out crack width calculations. However, it was found that, due to the large number of variables to be considered, it was possible to produce such rules only for single span solid slab bridges. Consequently an exercise was carried out [191] in which a range of solid slab bridges was designed at the ultimate limit state either by yield line theory or in accordance with elastic moment fields. The bar spacings required to control crack widths to the design values were then determined. Bar spacing tables based upon this exercise did appear in some drafts of the Code, but it was eventually decided to simplify the bar spacing rules considerably. Consequently, the Code simply states that the longitudinal bar spacing should not exceed 150 mm and the transverse bar spacing should not exceed 300 mm. These values apply to continuous slabs in addition to single span slabs and, in view of this, the author would suggest that 'longitudinal' be interpreted as primary and 'transverse' as secondary in order to avoid excessive cracking in certain situations. For example, the Code implies that the spacing of the transverse bars in the region of an interior support of a slab bridge continuous over discrete columns could be 300 mm; however, in such situations the transverse bending moments could be large enough to cause excessive cracking if the bar spacing were not considerably less than 300 mm. It is thus suggested that the bar spacing rules should be interpreted with 'engineering judgement'.

From present design experience, it is known that, in some situations, a longitudinal bar spacing of 150 mm does not control the crack widths to the design values unless the reinforcement is used at a stress less than the maximum permitted in BE 1/73 ($\approx 0.56f_y$). The drafters anticipated that when designing in these situations in accordance with the Code, bar spacings less than 150 mm would be forced upon the designer because of the large amount of reinforcement that would be required to satisfy the ultimate limit state criterion.

An examination of [191] shows that if bar spacings were to be calculated for single span slab bridges, they would generally be greater than the Code values of 150 and 300 mm, except for elastically designed slabs having skew angles greater than about 30° and yield line designed slabs having skew angles greater than about 15°. The Code values should thus be used with caution if the skew angle exceeds these values.

Finally, it is worth mentioning that the reason for having bar spacing rules is to avoid carrying out specific crack width calculations. However, since the Code requires stress calculations to be carried out at the serviceability limit state, albeit at a different design load to that for the crack width calculation, a large proportion of the data required for a crack width calculation would already be calculated. Thus, to a certain extent, the advantage of quoting bar spacing rules is lost and little extra effort is involved in carrying out a complete crack width calculation by, for example, applying the general procedure suggested by Clark [192, 193].

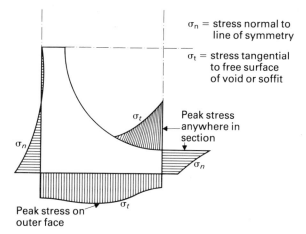

$\sigma_n$ = stress normal to line of symmetry

$\sigma_t$ = stress tangential to free surface of void or soffit

Peak stress anywhere in section

Peak stress on outer face

(a) Elastic stresses in a quadrant

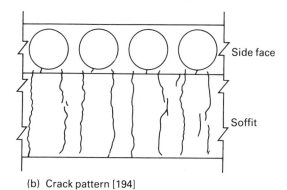

Side face

Soffit

(b) Crack pattern [194]

**Fig. 7.5(a),(b)** Cracks in voided slabs

## Slab bridges with longitudinal circular voids

*Cracks due to longitudinal bending* The spacing and widths of cracks due to longitudinal bending are very similar to those occurring, at the same steel strain, in solid slabs. Thus, the Code limits the spacing of longitudinal reinforcement to 150 mm which is the same as the value for solid slabs. This spacing was not checked for voided slabs by either calculation or experiment; however, the author would suggest that, with the same precautions as discussed for solid slabs, it is a reasonable value to adopt in practice.

*Cracks due to transverse bending* The stress raising effects of the voids result in the response of a voided slab to transverse bending being very different to that of a solid slab. Cracking due to transverse bending in voided slabs has been described in some detail by Clark and Elliott [194] and their findings can be summarised as follows:

1.  Linear elastic analyses indicated that:
    (a) The peak stress occurs at the crown of the void as shown in Fig. 7.5(a). It is here, on the *inside* of the void, that the first crack should initiate.
    (b) The peak stress on the outer face does not occur at the void centreline, but at approximately the quarter points of the void spacing as shown in Fig. 7.5(a). Thus cracks propagating from the outer face should initiate at the quarter points.
2.  The theoretical predictions were confirmed by tests on transverse strips of voided slabs. An actual crack pattern is shown in Fig. 7.5(b).

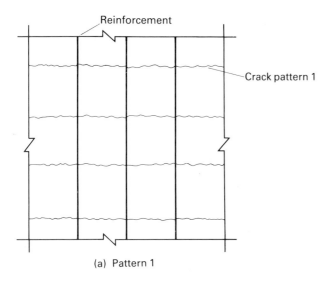

(a) Pattern 1

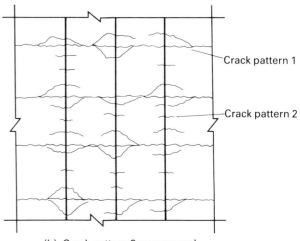

(b) Crack pattern 2 superposed
on crack pattern 1.

**Fig. 7.6(a),(b)** Plan views of crack patterns in solid slab

The tests indicated that in order to obtain a controlled crack pattern, such as that shown in Fig. 7.5(b), with no cracks passing completely through the flange, it was necessary to have at least 1% of transverse reinforcement in the flange. This reinforcement should be calculated as a percentage of the minimum flange area. This minimum reinforcement percentage requirement is given in the Code for predominantly tension flanges, together with an upper limit of 1500 mm²/m. The latter value was introduced to avoid excessive amounts of reinforcement in slabs with very thick flanges. Clark and Elliott [194] have suggested that, in such cases, it may be preferable to consider the minimum flange thickness as having two critical layers: one layer would be adjacent to the outer face and the other adjacent to the crown of the void. The thickness of each layer would be equal to twice the relevant cover plus the bar diameter, and each layer would be provided with a minimum of 1% of reinforcement. This suggestion is similar to that recommended by Holmberg [195].

In the case of predominantly compression flanges, the Code requires that the area of transverse reinforcement should be the lesser of 1000 mm²/m or 0.7% of the minimum flange area. These values were chosen because the tests reported by Clark and Elliott [194] indicated that the moment at which such steel would be stressed to 230 N/mm² would be greater than the cracking moment of the section. Thus, if cracking did occur due to an unexpected severe loading situation, the reinforcement would not yield suddenly and a controlled crack pattern would occur.

In addition to the above limitations on the area of transverse reinforcement, it is necessary to limit the spacing of the reinforcement. The Code states that the spacing should not exceed the solid slab value of 300 mm nor twice the minimum flange thickness. The latter criterion was introduced to discourage the use of large diameter bars if thin flanges were adopted, since such flanges are subjected to particularly large stress concentrations.

### Flanges

In order to discuss crack control in the flanges of beam and slab, cellular slab and box beam construction it is first necessary to consider, in general terms, cracking in slabs.

Beeby [196] has investigated cracking in slabs spanning one way and found that there are two basic crack patterns (see Fig. 7.6):

1. A pattern controlled by the deformation imposed on the section.
2. A pattern controlled by the proximity of the reinforcement.

Beeby [196] proposed a theory which adequately predicts the properties of the two patterns and their interaction. In addition, formulae for predicting the widths of cracks at any point on a slab were derived. These formulae are too complicated for design purposes and thus Beeby [182] reduced them to the following single design formula

$$w = \frac{K_1 a_{cr} \, \varepsilon_m}{1 + K_2 \dfrac{(a_{cr} - c_{min})}{(h - x)}}$$

where $c_{min}$ is the minimum cover to the tension steel and $K_1$ and $K_2$ are constants which depend upon the probability of exceedence of the design crack width. The appropriate values of $K_1$ and $K_2$ are 3 and 2, respectively, for the 20% probability of exceedence adopted in the Code. Hence, the following crack width equation, which is given in the Code, is obtained

$$w = \frac{3 a_{cr} \, \varepsilon_m}{1 + 2 \dfrac{(a_{cr} - c_{min})}{(h - x)}} \tag{7.5}$$

It should be noted that the strain, allowing for tension stiffening, is used because it is considered that slabs are relatively lightly reinforced and have load–strain relationships similar to that shown in Fig. 7.1. In such cases, the tension stiffening effect is significant and should be calculated from equation (7.3); although, as mentioned earlier in this chapter, the tension stiffening could be reduced by repeated loading.

Equation (7.5) was derived from tests in which the reinforcement was perpendicular to the cracks. The general problem of crack control when the reinforcement is not perpendicular to the cracks has been considered by Clark

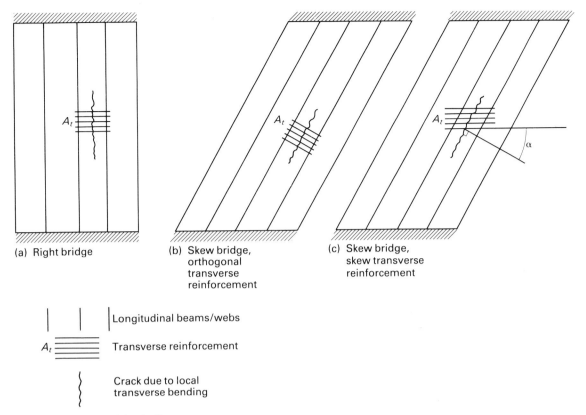

(a) Right bridge

(b) Skew bridge, orthogonal transverse reinforcement

(c) Skew bridge, skew transverse reinforcement

| | Longitudinal beams/webs
$A_t$ ≡≡≡ Transverse reinforcement

Crack due to local transverse bending

**Fig. 7.7(a)–(c)** Cracking in flanges

[192]. This study showed that equation (7.5) is applicable provided that the crack width calculation is carried out in a direction perpendicular to the crack and that all of the reinforcement should be resolved to an equivalent area of steel perpendicular to the crack. However, equation (7.3), which predicts the tension stiffening effect, has not been checked for reinforcement which is not perpendicular to the cracks. In view of the complex stress state in the concrete between the cracks, it would be advisable to ignore tension stiffening when considering such arrangements of reinforcement.

*Longitudinal reinforcement in flanges*  Although it is not clear in the Code, it is unnecessary to calculate the spacing of the longitudinal reinforcement in a flange because a simple bar spacing rule is given.

A bar spacing rule for tension flanges was derived by noting that, for HA loading, the design load at the serviceability limit state is about 70% of that at the ultimate limit state; thus

$$\varepsilon_1 \simeq 0.7 \times 0.87 f_y / 200 \times 10^3$$

If it is assumed that $\varepsilon_m \simeq 0.8\,\varepsilon_1$, then

$$\varepsilon_m \simeq 0.8 \times 0.7 \times 0.87 f_y / 200 \times 10^3$$

Hence, a reasonable upper limit to $\varepsilon_m$ is 0.001. Since the strain is very nearly constant over the depth of a flange, the neutral axis depth ($x$) for the flange tends to infinity. Hence, equation (7.5) reduces to $w = 3a_{cr}\,\varepsilon_m$ and, for $\varepsilon_m = 0.001$ and a design value of $w$ of 0.2 mm (see Table 4.7), $a_{cr} = 67$ mm, which results in a bar spacing of about 150 mm. This maximum bar spacing is given in the Code for predominantly tension flanges together with a value of 300 mm for predominantly compression flanges. The latter

value simply complies with the general maximum bar spacing given in the Code.

*Transverse reinforcement in flanges*  When drafting the clauses for crack control in flanges, it was assumed that the effects of local bending in the flanges would generally dominate the transverse bending effects. Thus the major principal moment in a flange would act very nearly perpendicular to the longitudinal beams or webs. Tests [87, 126, 192] on slab bridges and slab elements indicate that, at the serviceability limit state, it is reasonable to assume that the cracks are perpendicular to the principal moment direction. There is no reason to assume that the cracks in a slab acting as a flange would not form in the same direction and, thus, they would be very nearly parallel to the longitudinal beams or webs.

It can be seen from Fig. 7.7 that, in the case of a right bridge or of a skew bridge in which the transverse reinforcement is perpendicular to the beams or webs, the reinforcement would be perpendicular to the service load cracks. Hence, equations (7.3) and (7.5) can be applied, to a unit width of slab perpendicular to the cracks, with $A_s$ equal to the area of transverse reinforcement per unit width. However, for a skew bridge in which the transverse reinforcement is parallel to the supports and such that it makes an angle ($\alpha$) to the perpendicular to the crack, as shown in Fig. 7.7(c), equations (7.3) and (7.5) cannot be applied directly. It is first necessary to calculate an effective area of reinforcement perpendicular to the crack by using either equation (7.1) or (7.2). In fact, equation (7.1) is adopted in the Code because the drafters had top flanges primarily in mind and it is most likely that these will be cracked on one face only and in one direction. Furthermore, since the longitudinal reinforcement would generally be parallel to the longitudinal beam or webs, equation (7.1) reduces to

$$A_n = A_t \cos^4\alpha$$

where $A_t$ is area of transverse reinforcement per unit width. Hence $A_s$, in equation (7.3), should be replaced by $A_n$ and the latter value used to calculate the neutral axis depth and the strain in a direction perpendicular to the cracks.

Finally, the tension stiffening equation is not dependent upon bar spacing because it was derived from tests on beams; whereas, in the Code, it is used only for slabs. One would expect tension stiffening in slabs to depend upon the bar spacing and, indeed, test data indicate such a dependence for large bar spacings. However, tests reported by Clark and Cranston [187] show that the influence of bar spacing is insignificant provided that the bar spacing does not exceed about 1.5 times the slab depth. Since such large spacings are unlikely to occur in a bridge, the tension stiffening equation can be applied to flanges. However, the reservations expressed earlier, regarding skew reinforcement and repeated loading, should be considered.

### Columns

If tensile stresses occur in a column, then the column should be considered as a beam for crack control purposes and equation (7.4) used.

### Walls

If tensile stresses occur in a reinforced concrete wall, then it is obviously reasonable that the wall should be considered as a slab for crack control purposes. The Code takes such an approach and also distinguishes between the two exposure conditions (see Table 4.7) which are applicable to a wall.

*Severe exposure* This condition includes surfaces in contact with backfill and is thus appropriate to the back faces of retaining walls and wing walls. The design crack width is 0.2 mm which is also the value for soffits. Hence the Code permits the bar spacing rules for slab bridges, of 150 and 300 mm for longitudinal and transverse reinforcement, respectively, to be adopted for walls subjected to a severe exposure condition. However, it should be noted that calculations for walls were not carried out to check specifically that the spacings of 150 and 300 mm would be reasonable.

*Very severe exposure* A leaf pier is an example of a wall subject to the effects of salt spray; hence, its exposure condition is classed by the Code as very severe. The bar spacing rules for slab bridges are appropriate only to the severe exposure condition and it is thus necessary to carry out crack width calculations for walls if the exposure condition is very severe. Thus equation (7.5) should be used, in conjunction with equation (7.3).

### Bases

Except for the statement that 'reinforcement need not be provided in the side faces of bases to control cracking', the Code is rather vague regarding crack control in bases. The relevant clause states that the method of checking crack widths depends on the type of base and the design assumptions. Before discussing this statement, it should be mentioned that, although reinforcement is generally provided in the side faces of deep members to control cracks for aesthetic purposes, it is not necessary to do this in bases because they are generally buried.

In drafting the Code clause on crack control in bases it was intended that the various components of a base should be checked for cracking in accordance with the most appropriate of the procedures given for other structural elements. It was intended that 'beam components' should be checked by applying equation (7.4), and that 'slab components' (e.g., spread footings) should be considered as follows:

1.  If a moderate or severe exposure condition (see Table 4.7) is appropriate, then apply the bar spacing rules, of 150 and 300 mm, for slab bridges.
2.  If a very severe exposure condition is appropriate, then apply equations (7.3) and (7.5).

The reasons for these recommendations are the same as those discussed previously in connection with walls.

It is obvious that 'engineering judgement' is required when checking crack widths in bases.

# Prestressed concrete stress limitations

## Beams

In Chapter 4, limiting values of prestressing steel stresses and concrete compressive and tensile stresses are given.

It is emphasised in Chapter 4 that, although prestressing steel stress criteria are given in the Code, another clause essentially states that these can be ignored because it is not necessary to calculate tendon stress changes due to the effects of applied loadings.

Concrete stresses can be calculated by applying conventional elastic theory; but the calculation of tensile stresses in Class 3 members requires some comment. Although a Class 3 member is, by definition, cracked at the serviceability limit state, it is considered to be uncracked for the purposes of calculating stresses. It is permissible to do this because the hypothetical tensile stresses in Table 4.6(a) were calculated from test data by assuming elastic uncracked behaviour. Hence, for stress calculation purposes, a cracked Class 3 member is considered in exactly the same way as an uncracked Class 1 or 2 member.

It should be noted that a disadvantage of the above approach to design is that, because the section is actually cracked in practice, the *actual compressive* stress exceeds the value calculated for an uncracked section. Thus the *actual compressive* stress could exceed the allowable compressive stress from Table 4.3(a) although the *calculated compressive* stress might not. Thus the author would suggest that, when designing Class 3 members, it would be good practice not to stress the concrete to its allowable limit in compression.

## Slabs

In an uncracked slab (Class 1 or 2), conventional elastic theory can be applied in the usual way to calculate the principal concrete stresses. However, the approach adopted for cracked Class 3 beams, in which hypothetical tensile stresses appropriate to an uncracked beam are calculated, would not, in general, be correct for cracked Class 3 slabs. This is because the hypothetical tensile stresses in Table 4.6 were calculated from test data for beams and, although probably applicable to slabs in which the prestressed and non-prestressed steel are parallel to the principal stress direction, they would not be applicable to slabs in which these directions do not coincide. Test data are not available for such situations and the author would suggest the following interim measures which are based upon consideration of equation (7.1).

1. Beeby and Taylor [123] have studied, theoretically and experimentally, cracking in Class 3 members. Their theoretical expression for the hypothetical tensile stress is a function of the area of prestressing steel perpendicular to the crack. Equation (7.1) suggests that if the area of prestressing steel per unit width is $A_{ps}$ and it is at an angle $\alpha$ to the major principal stress direction, then the equivalent area of prestressing steel perpendicular to the crack is $A_{ps} \cos^4\alpha$. A study of Beeby and Taylor's work indicates that it is reasonable to assume that a reduction of steel area from $A_{ps}$ to $A_{ps} \cos^4\alpha$ results in a reduction of the hypothetical tensile stress from, say, $f_{ht}$ to $f_{ht} \cos^2\alpha$. Thus the author would suggest that, when the prestressing tendons are at an angle $\alpha$ to the major principal stress directions, the limiting hypothetical tensile stresses of Table 4.6(a) should be multiplied by $\cos^2\alpha$.
2. The depth factors of Table 4.6(b) should not be modified.
3. Any additional conventional reinforcement should be considered in terms of an equivalent area of reinforcement perpendicular to the crack by using equation (7.1). This equivalent area should be used to calculate the increase of hypothetical tensile stress which is permitted when additional reinforcement is present.
4. If the final limiting hypothetical tensile stress is less than the appropriate limiting tensile stress for a Class 2 member, the section will be uncracked and should be treated as a Class 2 member.

## Losses in prestressed concrete

### Initial prestress

In order that excessive relaxation of the stress in the tendons will not occur, the normal jacking force should not exceed 70% of the characteristic tendon strength. However, in order to overcome the effects of friction, the jack-ing force may be increased to 80% of the characteristic tendon strength, provided that the stress–strain curve of the tendon does not become significantly non-linear above a stress of 70% of the characteristic tendon strength.

The above requirements are essentially identical to those of BE 2/73.

### Loss due to steel relaxation

If experimental data are available, then the loss of prestress in the tendon should be taken as the relaxation, after 1000 hours duration, for an initial load equal to the jacking force at transfer. This value is taken because it is approximately equal to the relaxation, after four years, for an initial force of 60% of the tendon strength: this force is roughly the average tendon force over four years [197].

In the absence of experimental data, the relaxation loss should be taken as 8% for an initial prestress of 70% of the characteristic tendon strength, decreasing linearly to 0% for an initial prestress of 50% of the characteristic tendon strength. These values were based upon tests on plain cold drawn wire [112].

The Code also refers to losses given in Part 8 of the Code, but this appears to be a mistake because no losses are given in Part 8.

The Code losses are essentially the same as those of BE 2/73 except for the reduced loss which may be adopted if the initial jacking force is less than 70% of the characteristic tendon strength.

### Loss due to elastic deformation of the concrete

The elastic loss may be calculated by the usual modular ratio procedure.

For post-tensioned construction, the elastic loss may be calculated either, exactly, by considering the tensioning sequence, or, approximately, by multiplying the final stress in the concrete adjacent to the tendons by half of the modular ratio. The latter procedure is an approximate method of allowing for the progressive loss of prestress which occurs as the tendon forces are gradually transferred to the concrete.

Elastic losses calculated in accordance with the Code will thus be identical to those calculated in accordance with BE 2/73.

### Loss due to shrinkage of concrete

For a normal exposure condition of 70% relative humidity the Code gives shrinkage strains of 200 and 300 microstrains, for post-tensioning and pre-tensioning, respectively. These values are identical to the CP 115 values. However, for a humid exposure condition of 90% relative humidity, the Code also gives shrinkage strains of 70 and 100 microstrains, for post-tensioning and pre-tensioning, respectively.

### Loss due to creep of concrete

According to Neville [198], the relationship between creep of concrete and the stress to strength ratio is of the form shown in Fig. 7.8. It can be seen that, for a particular concrete, creep is directly proportional to stress for stress

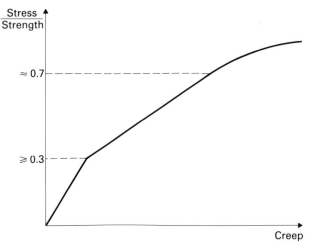

**Fig. 7.8** Creep–stress/strength relationship

to strength ratios less than approximately 0.3. The Code limit of one-third, below which creep can be considered to be proportional to stress, is thus reasonable.

The Code gives values for the specific creep strain (creep strain per unit stress) of:

1. For pre-tensioning; the lesser of $48 \times 10^{-6}$ and $48 \times 10^{-6} \times 40/f_{ci}$ per N/mm².
2. For post-tensioning; the lesser of $36 \times 10^{-6}$ and $36 \times 10^{-6} \times 40/f_{ci}$ per N/mm².

In the above, $f_{ci}$ is the cube strength at the time of transfer. The values of specific creep strain are identical to those in BE 2/73 and CP 115.

If the compressive stress anywhere in the section exceeds one-third of the cube strength at transfer, the specific creep strain should be increased as indicated by Fig. 7.8. The Code gives a factor, by which the above specific creep strains should be multiplied, which varies linearly from unity at a stress-to-strength ratio of one-third to 1.25 at a ratio of one-half (the greatest allowable ratio under any conditions). This factor is less than that indicated by tests [198] on uniaxial compression specimens because, generally, it is parts, rather than the whole, of a cross-section which are subjected to stresses in excess of one-third of the cube strength.

The above requirements are identical to those of BE 2/73 but it should be noted that the Code also states that half of the total creep may be assumed to take place in the first month after transfer and three-quarters in the first six months.

### Loss due to friction in duct

In post-tensioning systems, losses occur due to friction in the duct caused by unintentional variations in the duct profile ('wobble') and by curvature of the duct. The Code adopts the conventional friction equation [199]

$$P_x = P_o \exp(-Kx + \mu x/r_{ps})$$

where

$P_x$ = prestressing force at a distance $x$ from the jack
$P_o$ = prestressing force in the tendon at the jack
$r_{ps}$ = radius of curvature of duct
$K$ = wobble factor
$\mu$ = coefficient of friction of tendon.

In the absence of specific test data, the Code gives a general value of $K$ of $33 \times 10^{-4}$ per metre, but a value of $17 \times 10^{-4}$ may be used when the duct former is rigid or rigidly supported.

The Code gives the following values for $\mu$:

1. 0.55 for steel moving on concrete.
2. 0.30 for steel moving on steel.
3. 0.25 for steel moving on lead.

The above values of $K$ and $\mu$ are identical to those in CP 115 and were originally based upon the test data of Cooley [199]. However, the Construction Industry Research and Information Association has now assessed all of the available experimental data on $K$ and $\mu$ values and has recommended [200]:

1. $\mu$ values of 0.25 and 0.20 for steel moving on steel and lead, respectively.
2. For other than long continuous construction, the $K$ values given in the Code.
3. A $K$ value of $40 \times 10^{-4}$ per metre for long continuous construction because it has been suggested [201] that the Code value of $33 \times 10^{-4}$ underestimates friction losses in such situations. The author presumes that a value of $20 \times 10^{-4}$ would be adopted for rigidly supported ducts.

Finally, where circumferential tendons are used, the following $\mu$ values are recommended in the Code:

1. 0.45 for steel moving in smooth concrete.
2. 0.25 for steel moving on steel bearers fixed to the concrete.
3. 0.10 for steel moving on steel rollers.

These values are the same as those in CP 115 and were originally suggested by Creasy [202].

### Other losses

The Code does not give specific data for calculating the losses due to steam curing nor, in post-tensioning systems, the losses due to friction in the jack and anchorages and to tendon movement at the anchorages during transfer. Instead reference is made, as in CP 115, to specialist advice.

# Deflections

## General

It is explained in Chapter 1 that there is not a limit state of excessive deflection in the Code since a criterion, in the form of an allowable deflection or span-to-depth ratio, is not given. However, in practice, it is necessary to calculate deflections in order, for example, to calculate rotations in the design of bearings. The Code thus gives methods of calculating both short and long term *curvatures* in Appendix A of Part 4.

The procedure is to calculate the curvature assuming,

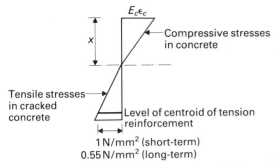

**Fig. 7.9** Tension stiffening for curvature calculations

first, an uncracked section and, second, a cracked section. The larger of the two curvatures is then adopted. Long-term effects of creep are allowed for by using an effective elastic modulus for the concrete which is less than the short-term modulus. Shrinkage is allowed for by separately calculating the curvature due to shrinkage.

## Short-term curvature

The short-term elastic moduli for concrete, which are tabulated in the Code, should be used to calculate the short-term curvature under imposed loading.

The calculation for the uncracked section is straightforward. However, the calculation for the cracked section is more complicated because of the need to allow for tension stiffening.

It is mentioned elsewhere in this chapter, that when calculating crack widths in flanges, tension stiffening is allowed for by subtracting a 'tension stiffening strain' from the reinforcement strain calculated by ignoring tension stiffening. However, when calculating deflections, a different approach is adopted: a triangular distribution of tensile stress is assumed in the concrete below the neutral axis, with a stress of 1 N/mm² at the centroid of the tension reinforcement, as shown in Fig. 7.9.

The stress of 1 N/mm² was derived from the test results of Stevens [188] which are referred to earlier in this chapter.

The CP 110 handbook [112] implies that the neutral axis should be calculated from the stress diagram of Fig. 7.9 by employing a trial-and-error approach; but, as suggested by Allen [203], it is simpler to adopt the following procedure, which involves little error:

1. Calculate the neutral axis depth ($x$) ignoring tension stiffening.
2. Calculate the extreme fibre concrete compressive stress due to the applied bending moment.
3. Calculate the extreme fibre strain ($\varepsilon_c$) by dividing the stress by the elastic modulus of the concrete ($E_c$).
4. Calculate the curvature ($\varepsilon_c/x$).

Allen gives equations which aid the above calculations for csctangular and T-sections.

## Long-term curvature

The curvature calculated above would increase under long-term loading due to creep of the concrete. Creep in compression is allowed for by dividing the short-term elastic modulus of the concrete by $(1 + \phi)$ where $\phi$ is a creep coefficient. It is emphasised that $\phi$ does not, in this chapter, refer to the same coefficient as it does in Chapter 8. In fact, the creep coefficient $\beta$ referred to in Chapter 8 is identical to the creep coefficient $\phi$ referred to above. An appropriate value of $\phi$ can be determined from the data given in Appendix C of the Code, provided that sufficient prior knowledge of the concrete mix and curing conditions are known. Since $\phi$ depends upon many variables, the CP 110 handbook [112] gives a table of $\phi$ values which may be used in the absence of more detailed information. As an alternative, a simplified method of obtaining $\phi$ is given by Parrott [204].

Creep of the concrete in tension is allowed for by reducing the tensile stress in the concrete, at the level of the centroid of the tension reinforcement, from its short-term value of 1 N/mm² to a long-term value of 0.55 N/mm². The latter value was again derived from the test results of Stevens [188].

The long-term curvature can be calculated by following the same procedure as that given earlier for the short-term curvature.

## Shrinkage curvature

The Code gives the following expression for calculating the curvature ($\psi_s$) due to shrinkage.

$$\psi_s = \rho_o \, \varepsilon_{cs}/d \tag{7.6}$$

where $d$ is the effective depth, $\varepsilon_{cs}$ is the free shrinkage strain and $\rho_o$ is a coefficient which depends upon the percentages of tension and compression steel.

The free shrinkage strain can be determined from the data given in Appendix C of the Code. However, as is also the case for $\phi$, the free shrinkage strain depends upon many variables. Thus the CP 110 handbook [112] suggests a value of $300 \times 10^{-6}$ for a section less than 250 mm thick and a value of $250 \times 10^{-6}$ for thicker sections. The handbook also indicates how the shrinkage develops with time. As an alternative, Parrott [204] gives a graph for estimating shrinkage.

Values of the coefficient $\rho_o$ are given in a table in Appendix A of the Code. The tabulated values are based upon empirical equations derived by Branson [205] from two sets of test data. Hobbs [206] has compared the Code values of $\rho_o$ with these data and also with an additional set of data. He found that the Code values are conservative and are particularly conservative, by a factor of about 2, for lightly reinforced and doubly reinforced sections. In view of this, Hobbs suggests an alternative procedure for calculating shrinkage curvatures, which is based upon theory rather than empirical equations, and which gives good agreement with test data [206].

## General calculation procedure

The following general procedure is suggested in the Code for calculating the long-term curvature:

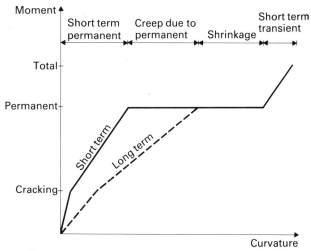

Fig. 7.10 Calculation of long term curvature

1. Calculate the instantaneous (i.e., short-term) curvatures under the total design load and under the permanent design load.
2. Calculate the long-term curvature under the permanent design load.
3. Calculate the difference between the instantaneous curvatures under the total and permanent design loads. This difference is the curvature due to the transient design loads. Add this value to the long-term curvature under the permanent load.
4. Add the shrinkage curvature.

The above procedure is logical and its net effect is illustrated in Fig. 7.10.

# Examples

## 7.1 Reinforced concrete

The top slab of a beam and slab bridge has been designed at the ultimate limit state and is shown in Fig. 7.11. The

**Table 7.1** Example 7.1. Nominal load effects
(a) Local moments (kNm/m)

| Load | Transverse | Longitudinal |
|------|-----------|--------------|
| Dead | 0.45 | 0.0 |
| Superimposed dead | 0.22 | 0.0 |
| HA wheel | 10.8 | 7.20 |
| 45 HB units | 11.7 | 7.65 |
| 25 HB units | 7.56 | 5.26 |

(b) Global effects

| Load | Transverse moment (kNm/m) | Longitudinal compressive stress in slab (N/mm²) | |
|------|------|------|------|
| | | Top | Bottom |
| Superimposed dead | 0.0 | 0.33 | 0.14 |
| HA | 0.0 | 3.90 | 1.60 |
| HA wheel | 6.0 | 0.61 | 0.25 |
| 45 HB units | 43.0 | 6.24 | 2.55 |
| 25 HB units | 23.3 | 3.63 | 1.48 |

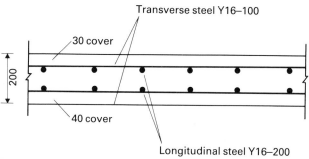

Fig. 7.11 Example 7.1

characteristic strengths of the reinforcement and concrete are 425 and 30 N/mm², respectively.

Mid-way between the beams, the local bending moments given in Table 7.1(a) act together with the global effects given in Table 7.1(b). The effects due to HB include those due to associated HA.

It is required to check that the slab satisfies the serviceability limit state criteria under load combination 1.

### General

From Table 2 of the Code, short-term elastic modulus of concrete = 28 kN/mm². Use a long-term value of 28/2 = 14 kN/mm² for both cracking and stress calculation (Clause 5.3.3.2). Modular ratio = 200/14 = 14.3.

Each design load effect will be calculated as nominal load effect $\times \gamma_{fL} \times \gamma_{f3}$. Details of these partial safety factors are given in Chapters 3 and 4.

### Cracking

*Due to transverse bending*
HA design moment
$$= (0.45)(1.0)(1.0) + (0.22)(1.2)(1.0)$$
$$+ (10.8)(1.2)(0.83) + (6.0)(1.2)(0.83)$$
$$= 17.5 \text{ kNm/m}$$

25 HB design moment
$$= (0.45)(1.0)(1.0) + (0.22)(1.2)(1.0)$$
$$+ (7.56)(1.1)(0.91) + (23.3)(1.1)(0.91)$$
$$= 31.6 \text{ kNm/m}$$
Critical moment = 31.6 kNm/m

Area of bottom steel = area of top steel = 2010 mm²/m

If elastic neutral axis is at depth $x$, then by taking first moments of area about the neutral axis:

$$(1/2)(1000) x^2 + (13.3)(2010) (x - 38)$$
$$= (14.3)(2010)(152 - x)$$
$$\therefore x = 62.2 \text{ mm}$$

Second moment of area, $I$, about the neutral axis is given by:

$$(1/3)(1000)(62.2)^3 + (13.3)(2010)(24.2)^2 +$$
$$(14.3)(2010)(89.8)^2$$
$$= 0.328 \times 10^9 \text{ mm}^4/\text{m}$$

Bottom steel stress =
$$(14.3)(31.6 \times 10^6)(89.8)/0.328 \times 10^9 = 124 \text{ N/mm}^2$$

Soffit strain ignoring tension stiffening is

$$\varepsilon_1 = (124/200 \times 10^3)(137.8/89.8) = 9.51 \times 10^{-4}$$

From equation (7.3), soffit strain allowing for tension stiffening is

$$\varepsilon_m = 9.51 \times 10^{-4} - (1.2)(1000)(200)(10^{-3})/(2010 \times 425)$$
$$= 6.70 \times 10^{-4}$$

Maximum crack width occurs mid-way between bars where the distance ($a_{cr}$) to the nearest bar is given by

$$a_{cr} = \sqrt{50^2 + 48^2} - 8 = 61.3 \text{ mm.}$$

From equation (7.5), crack width is

$$w = \frac{(3)(61.3)(6.70 \times 10^{-4})}{1 + (2)(61.3 - 40)/(200 - 62.2)} = 0.094 \text{ mm}$$

From Table 1 of the Code, allowable crack width is 0.20 mm.

*Due to longitudinal bending*  The longitudinal bars are in a region of predominantly compressive flexural stress and thus the Code maximum spacing is 300 mm. This exceeds the actual spacing of 200 mm.

### Stresses

*Limiting values*  The limiting stresses are $0.5 \times 30 = 15$ N/mm² for concrete in compression, and $0.8 \times 425 = 340$ N/mm² for steel in tension or compression.

*Due to transverse bending*
HA design moment
$$= (0.45)(1.0)(1.0) + (0.22)(1.2)(1.0)$$
$$+ (10.8)(1.2)(1.0) + (6.0)(1.2)(1.0)$$
$$= 20.9 \text{ kNm/m}$$

45 HB design moment
$$= (0.45)(1.0)(1.0) + (0.22)(1.2)(1.0)$$
$$+ (11.7)(1.1)(1.0) + (43.0)(1.1)(1.0)$$
$$= 60.9 \text{ kNm/m}$$
Critical moment $= 60.9$ kNm/m

From crack width calculations, $x = 62.2$ mm,
$$I = 0.328 \times 10^9 \text{ mm}^4/\text{m}$$
Maximum concrete stress
$$= (60.9 \times 10^6)(62.2)/0.328 \times 10^9$$
$$= 11.5 \text{ N/mm}^2 < 15 \text{ N/mm}^2$$
Steel stresses are
$$(14.3)(60.9 \times 10^6)(89.8)/0.328 \times 10^9$$
$$= 238 \text{ N/mm}^2 \text{ tension}$$
and
$$(14.3)(60.9 \times 10^6)(24.2)/0.328 \times 10^9$$
$$= 64 \text{ N/mm}^2 \text{ compression}$$
Both $< 340$ N/mm²

*Due to longitudinal effects*  Maximum compressive stresses occur under 45 units of HB loading. Assume section to be uncracked and ignore reinforcement.
Extreme fibre stresses due to nominal local moment
$$= \pm (7.65 \times 10^3)(6)/(200)^2 = \pm 1.14 \text{ N/mm}^2$$

Net top fibre design stress is
$$(0.33)(1.2)(1.0) + (6.24)(1.1)(1.0) + (1.14)(1.1)(1.0)$$
$$= 8.51 \text{ N/mm}^2 < 15 \text{ N/mm}^2$$

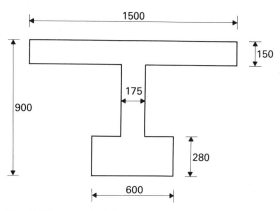

**Fig. 7.12** Example 7.2

Maximum tensile stress in reinforcement occurs under the HA wheel load. The longitudinal global stresses under the HA wheel load and superimposed dead load are:
Top $= 0.61 + 0.33 = 0.94$ N/mm²
Bottom $= 0.25 + 0.14 = 0.39$ N/mm²
These are equivalent to an in-plane force of $200 (0.94 + 0.39)/2 = 133$ kN/m, and a moment of $[(0.94 - 0.39)/2] (200^2) 10^{-3}/6 = 1.83$ kNm/m. Total nominal moment $= 7.2 + 1.83 = 9.03$ kNm/m. If the in-plane force is conservatively ignored, the design load effect is a moment of $(9.03)(1.2)(1.0) = 10.8$ kNm/m. Area of bottom steel = area of top steel = 1010 mm²/m. If elastic neutral axis is at depth $x$ and is above the top steel level

$$(1/2)(1000) x^2 = (14.3)(1010)[(136 - x) + (54 - x)]$$
$$\therefore x = 50.7 \text{ mm and is above the top steel.}$$

Second moment of area $=$

$$(1/3)(1000)(50.7)^3 + (14.3)(1010)(85.3^2 + 3.3^2)$$
$$= 0.149 \times 10^9 \text{ mm}^4/\text{m}$$

Bottom steel stress
$$= (14.3)(10.8 \times 10^6)(85.3)/0.149 \times 10^9$$
$$= 88 \text{ N/mm}^2 < 340 \text{ N/mm}^2$$

## 7.2. Prestressed concrete

A bridge deck is constructed from the pre-tensioned I-beams shown in Fig. 7.12. Each beam is required to resist the moments, due to nominal loads, given in Table 7.2. Determine the prestressing force and eccentricity required to satisfy the serviceability limit state criteria, under load combination 1, for each of the three classes of prestressed concrete. Assume that the losses amount to

**Table 7.2** Example 7.2. Design data

| Load | Moment (kNm) | | Stress (N/mm²) | |
|------|---------|--------|-------|--------|
| | Nominal | Design | Top | Bottom |
| Dead | 477 | 477 | +3.67 | −5.24 |
| Superimposed dead | 135 | 162 | +1.25 | −1.78 |
| HA | 949 | 949 | +7.30 | −10.42 |
| HB + Associated HA | 1060 | 1060 | +8.15 | −11.64 |

11% at transfer and finally amount to 34%. The concrete is of grade 50 and, at transfer, the concrete strength is 40 N/mm².

## General

Area = 475 250 mm²
Centroid is 529.2 mm from bottom fibre.
Second moment of area = $4.819 \times 10^{10}$ mm⁴
Bottom fibre section modulus = $9.11 \times 10^7$ mm³
Top fibre section modulus = $13.0 \times 10^7$ mm³
From Table 24 of the Code, the allowable compressive stress for any class of prestressed concrete is $0.33 \times 50 = 16.5$ N/mm².
From Table 25 of the Code, the allowable compressive stress at transfer for any class of prestressed concrete is $0.5 \times 40 = 20$ N/mm².
The usual sign convention of compressive stresses being positive and tensile stresses being negative is adopted.
The design moments are calculated as nominal moments $\times \gamma_{fL} \times \gamma_{f3}$; and are
Dead load = (477)(1.0)(1.0) = 477 kNm
Superimposed dead load = (135)(1.2)(1.0) = 162 kNm
HA load = (949)(1.2)(0.83) = 949 kNm
HB + associated HA load = (1060)(1.1)(0.91)
= 1060 kNm
Hence HB loading is the critical live load. The design moments, together with the extreme fibre stresses which they induce, are given in Table 7.2.

## Class 1

The allowable tensile stresses in the concrete are zero under the design service load, and 1 N/mm² at transfer or under a service load condition of dead load alone.
The critical stresses, for the given section and loading, are the transfer stresses. If the prestressing force before any losses occur is $P$ and its eccentricity is $e$, then at the top

$$-1 = \frac{0.89P}{475\,250} - \frac{0.89Pe}{13 \times 10^7} + 3.67$$
$$\therefore -2.494 \times 10^6 = P(1 - 3.656 \times 10^{-3}e)$$

and at the bottom

$$\frac{0.89P}{475\,250} + \frac{0.89Pe}{9.11 \times 10^7} - 5.24 = 20$$
$$\therefore P(1 + 5.217 \times 10^{-3}\,e) = 13.478 \times 10^6$$

Solving simultaneously
$P = 4087$ kN and $e = 440$ mm (i.e., 89.2 mm from soffit).
The resulting stresses under the various design loads are given in Table 7.3.

## Class 2

The allowable tensile stresses in the concrete are:

1. Under the design service load, 3.2 N/mm² (from Table 26 of the Code or $0.45 \sqrt{f_{cu}}$).
2. Under dead plus superimposed dead load, zero.
3. At transfer, 2.9 N/mm² (from Table 26 of the Code or $0.45 \sqrt{f_{ci}}$).

The critical stress, for the given section and loading, is the tensile stress at the bottom fibre under the full design service load. Hence

$$-3.2 = \frac{0.66\,P}{475\,250} + \frac{0.66\,Pe}{9.11 \times 10^7} - 5.24 - 1.78 - 11.64$$
$$\therefore P(1 + 5.217 \times 10^{-3}\,e) = 11.132 \times 10^6$$

Assuming the same eccentricity (440 mm) as that adopted for Class 1:
$P = 3378$ kN (i.e., 83% of that for Class 1)
The resulting stresses under the various design loads are given in Table 7.3.

## Class 3

The allowable tensile stresses in the concrete are:

1. Under the design service load, the basic hypothetical tensile stress from Table 27 of the Code or Table 4.6(a) of this book is 5.8 N/mm² for a design crack width of 0.2 mm. The section is 900 mm deep and the stress of 5.8 N/mm² has to be multiplied by 0.75 (from Table 28 of the Code or Table 4.6(b) this book) to give a final stress of 4.35 N/mm². It will be assumed that no conventional reinforcement is present and thus the hypothetical stress cannot be increased.
2. Under dead plus superimposed dead load, zero.
3. At transfer, a Class 3 member must be treated as if it were Class 2 and thus the allowable stress is 2.9 N/mm².

The critical stress, for the given section and loading, is

**Table 7.3** Example 7.2. Design stresses

| Design load | Stresses (N/mm²) | | | | | |
|---|---|---|---|---|---|---|
| | Class 1 | | Class 2 | | Class 3 | |
| | Top | Bottom | Top | Bottom | Top | Bottom |
| Transfer | −1.0 | +20.0 | −0.2 | +15.6 | +0.1 | +14.1 |
| PS+DL | +0.2 | +13.5 | +0.8 | +10.2 | +1.0 | +9.1 |
| PS+DL+SDL | +1.5 | +11.7 | +2.1 | +8.4 | +2.3 | +7.3 |
| PS+DL+SDL+LL | +9.6 | +0.1 | +10.2 | −3.2 | +10.4 | −4.4 |

PS = final prestress
DL = dead load
SDL = superimposed dead load
LL = HB + associated HA load

again the tensile stress at the bottom fibre under the full design service load.
Hence

$$-4.35 = \frac{0.66\,P}{475\,250} + \frac{0.66\,Pe}{9.11 \times 10^7} - 5.24 - 1.78 - 11.64$$
$$\therefore P(1 + 5.217 \times 10^{-3}\,e) = 10.304 \times 10^6$$

Assuming the same eccentricity (440 mm) as that adopted for Classes 1 and 2: $P = 3127$ kN (i.e., 77% and 93% of those for Classes 1 and 2, respectively). The resulting stresses under the various design loads are given in Table 7.3.

### General comments

In an actual design, the section should also be checked at the ultimate limit state. In addition other sections along the length of the beam would need to be checked, particularly with regard to transfer stresses.

The losses have been assumed to be the same for each class but they will obviously be different because of the different prestressing forces.

The different classes of prestressed concrete have been catered for by merely altering the prestress but, in practice, consideration would also be given to altering the cross-section.

## Chapter 8

# Precast concrete and composite construction

## Precast concrete

The design of precast members in general is based upon the design methods for reinforced or prestressed concrete which are discussed in other chapters. Bearings and joints for precast members are considered as part of this chapter.

### Bearings

The Code gives design rules for two types of bearing: corbels and nibs.

#### Corbels

A corbel is defined as a short cantilever bracket with a shear span to depth ratio less than 0.6 (see Fig. 8.1(b)). The design method proposed in the Code is based upon test data reviewed by Somerville [207]. The method assumes, in the spirit of a lower bound design method, the equilibrium strut and tie system shown in Fig. 8.1(a). The calculations are carried out at the ultimate limit state.

In order to assume such a strut and tie system it is first necessary to preclude a shear failure. This is done by proportioning the depth of the corbel, at the face of the supporting member, in accordance with the clauses covering the shear strength of short reinforced concrete beams.

The force ($F_t$) to be resisted by the main tension reinforcement can be determined by considering the equilibrium of the strut and tie system as follows (the notation is in accordance with Fig. 8.1).

$$V = F_c \sin \beta$$

$$F_t = H + F_c \cos \beta = H + V \cot \beta \qquad (8.1)$$

Somerville [207] suggests that $\beta$ can be determined by assuming a depth of concrete $x$ having a constant compressive stress of $0.4 f_{cu}$ and considering equilibrium at the face of the supporting member, as shown in Fig. 8.1(b) and (c). An iterative procedure is suggested in which $x/d$ is first assumed to be 0.4; in which case $\cot \beta = a_v/0.8d$ and the compressive force in the concrete is given by

$$F_c = 0.4 f_{cu} bx \cos \beta$$

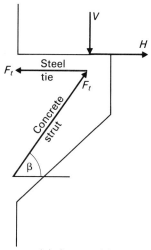

(a) Strut and tie

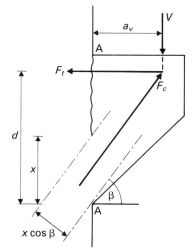

(b) Determination of $\beta$

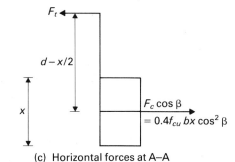

(c) Horizontal forces at A–A

**Fig. 8.1(a)–(c)** Corbel strut and tie system

102

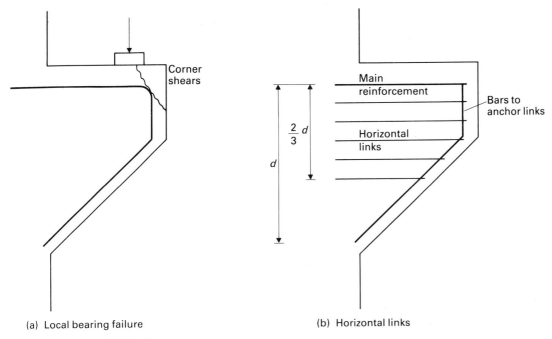

(a) Local bearing failure       (b) Horizontal links

**Fig. 8.2(a),(b)** Corbel detailing

The horizontal component of this force is

$$0.4 f_{cu} bx \cos^2 \beta$$

For equilibrium, the tensile force ($F_t$) in the reinforcement must equal this horizontal component of the concrete compressive force; thus

$$F_t = 0.4 f_{cu} bx \cos^2 \beta. \tag{8.2}$$

A new value of $\beta$ can then be calculated from

$$\cot \beta = a_v/(d - x/2) \tag{8.3}$$

This procedure can be continued iteratively.

It should be noted that both Somerville [207] and the CP 110 handbook [112] give the last term of equation (8.2), incorrectly, as $\cos \beta$ instead of $\cos^2 \beta$.

The area of reinforcement provided should be not less than 0.4% of the section at the face of the supporting member. This requirement was determined empirically from the test data. It is important that the reinforcement is adequately anchored: at the front face of the corbel this can be achieved by welding to a transverse bar or by bending the main bars to form a loop. In the latter case the bearing area of the load should not project beyond the straight portion of the bars, otherwise shearing of the corner of the corbel could occur as shown in Fig. 8.2(a).

Theoretically no reinforcement, other than that referred to above, is required. However, the Code also requires horizontal links, having a total area equal to 50% of that of the main reinforcement, to be provided as shown in Fig. 8.2(b). Horizontal rather than vertical links are required because the tests, upon which the design method is based, showed that horizontal links were more efficient for values of $a_v/d < 0.6$.

The Code states that the above design method is applicable for $a_v/d < 0.6$. The implication is thus that, if $a_v/d \geqslant 0.6$, the corbel should be designed as a flexural cantilever. However, the test data showed that the design

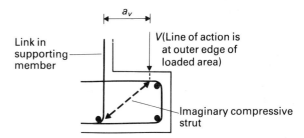

**Fig. 8.3** Nib

method based upon a strut and tie system was applicable to $a_v/d$ values of up to 1.5. In view of this the CP 110 handbook [112] suggests that, as a compromise, the method can be applied to corbels having $a_v/d$ values of up to 1.0.

Finally, in order to prevent a local failure under the load, the test data indicated that the depth of the corbel at the outer edge of the bearing should not be less than 50% of the depth at the face of the supporting member.

### Nibs

The Code requires nibs less than 300 mm deep to be designed as cantilever slabs at the ultimate limit state to resist a bending moment of $V a_v$ (see Fig. 8.3). Clarke [208] has shown by tests that this method is safe, but that an equilibrium strut and tie system of design is more appropriate for nibs which project less than 1.5 times their depth. The distance $a_v$ is taken to be from the outer edge of the loaded area (i.e. the most conservative position of the line of action of $V$) to the position of the nearest vertical leg of the links in the supporting member. The latter position was chosen from considerations of a strut and tie system in which the inclined compressive strut is as shown in Fig. 8.3.

Detailing of the reinforcement is particularly important in small nibs and the Code gives specific rules which are confirmed by the test results of Clarke [208].

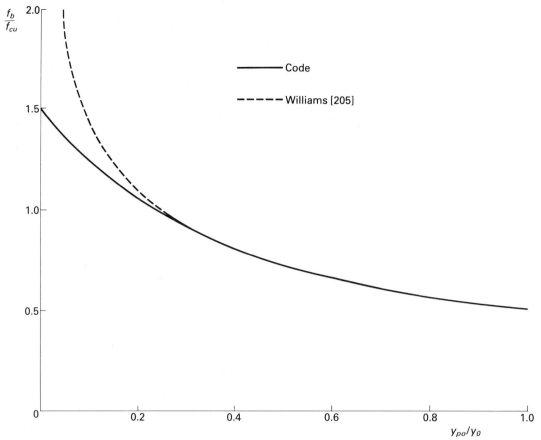

**Fig. 8.4** Bearing stresses

## Bearing stress

Normally the compressive stress at the ultimate limit state between two contact surfaces should not exceed 40% of the characteristic strength of the concrete.

If the bearing area is well-defined and binding reinforcement is provided near to the contact area, then a triaxial stress state is set up in the concrete under the bearing area and stresses much higher than $0.4 f_{cu}$ can be resisted. The Code gives the following equation for the limiting bearing stress $(f_b)$ at the ultimate limit state

$$f_b = \frac{1.5 f_{cu}}{1 + 2 y_{po}/y_o} \tag{8.4}$$

where $y_{po}$ and $y_o$ are half the length of the side of the loaded area and of the resisting concrete block respectively. This equation was obtained from a recommendation of the Comité Europeen du Beton, but the original test data is not readily available. However, Williams [209] has reviewed all of the available data on bearing stresses and found that a good fit to the data is given by

$$f_b = 0.78 f_{cu} (y_{po}/y_o)^{-0.441} \tag{8.5}$$

In Fig. 8.4, the latter expression, with a partial safety factor of 1.5 applied to $f_{cu}$, is compared with the Code equation. It can be seen that the latter is conservative for small loaded areas.

The Code recognises that extremely high bearing stresses can be developed in certain situations: an example is in concrete hinges [210], the design of which will probably be covered in Part 9 of the Code. At the time of writing, Part 9 has not been published and thus the Department of Transport's Technical Memorandum on Freyssinet hinges [211] will, presumably, be used in the interim period. This document permits average compressive stresses in the throat of a hinge of up to $2 f_{cu}$.

## Halving joints

Halving joints are quite common in bridge construction and the Code gives two alternative design methods at the ultimate limit state: one involving inclined links (Fig. 8.5(a)) and the other involving vertical links (Fig. 8.5(b)). Each method is presented in the Code in terms of reinforced concrete, but can also be applied to prestressed concrete.

### Inclined links

When inclined links are used, the Code design method assumes the equilibrium strut and tie system, shown in Fig. 8.5(c), which is based upon tests carried out by Reynolds [212]. The Code emphasises that, in order that the inclined links may contribute to the strength of the joint, they must cross the line of action of the reaction $F_v$.

For equilibrium, the vertical component of the force in the links must equal the reaction $(F_v)$. Hence

$$F_v = A_{sv} f_{yv} \cos \theta$$

where $A_{sv}, f_{yv}$ and $\theta$ are the total area, characteristic strength and inclination of the links respectively. A partial safety factor of 1.15 has to be applied to $f_{yv}$ and hence the Code equation is obtained.

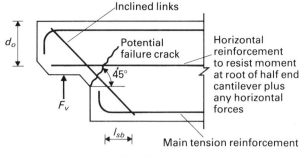

(a) Inclined links

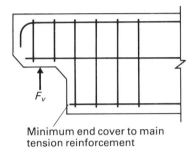

(b) Vertical links

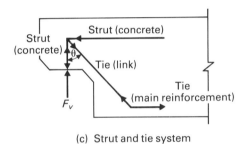

(c) Strut and tie system

**Fig. 8.5(a)–(c)** Halving joint

$$F_v = A_{sv} (0.87 f_{yv}) \cos \theta \qquad (8.6)$$

It should be noted that, theoretically, any value of $\theta$ may be chosen. However, the crack which initiates failure forms at about 45° as shown in Fig. 8.5(a), and thus it is desirable to incline the links at 45°. The adoption of such a value of $\theta$ is implied in the Code.

Reynolds suggested [212] that, although his tests showed that it is possible to reinforce a joint so that the maximum allowable shear force for the *full* beam section could be carried, it would be prudent in practice to limit the reaction at the joint to the maximum allowable shear force for the *reduced* section (with an effective depth of $d_0$). This limit was suggested in order to prevent over-reinforcement of the joint and, hence, to ensure a ductile joint.

One might expect that the above limit would imply $F_v = v_u b d_o$; instead, the Code gives a value of $F_v = 4v_c b d_o$. The reason for this is that, in a draft of CP 110, $4v_c$ was the maximum allowable shear stress in a beam and was thus equivalent to $v_u$, which occurs in the final version of CP 110 and the Code. However, the clause on halving joints was written when $4v_c$ was in CP 110 and was not subsequently altered when $v_u$ ($= 0.75 \sqrt{f_{cu}}$) was introduced (see Chapter 6).

Finally, the horizontal component of the tensile force in the inclined links has to be transferred to the bottom main

tension reinforcement. Normally one would assume that such transfer occurs by bond, but the Code assumes that the transfer occurs in two ways: half by bond with the concrete, and half by friction between the links and bars. For the latter to occur the links must be wired tightly to the main bars. The division of the force transfer into bond and friction was not based upon theory but was an interpretation of the test results. Since only half of the force has to be transferred by bond, the anchorage length of the main tension reinforcement ($l_{sb}$) given in the Code is only half of that which would be calculated by applying the anchorage bond stresses discussed in Chapter 10.

### Vertical links

As an alternative to the inclined link system of Fig. 8.5(a), a halving joint may be reinforced with vertical links as shown in Fig. 8.6(b). The vertical links should be designed by the method described in Chapter 6 and anchored around longitudinal reinforcement which extends to the *end* of the beam as shown in Fig. 8.5(b) [212].

### Horizontal reinforcement in half end

The Code does not require flexural reinforcement to be designed in half ends. However, it would seem prudent to design horizontal reinforcement to resist the moment at the root of the half end cantilever. Such reinforcement should also be designed to resist any horizontal forces.

## Composite construction

### General

In the context of this book, *composite construction* refers to precast concrete acting compositely with in-situ concrete. Very often in bridge construction, the precast concrete is prestressed and the in-situ concrete is reinforced.

The design of composite construction is complicated not only by the fact that, for example, shear and flexure calculations have to be carried out for both the precast unit and the composite member, but because additional calculations, such as those for interface shear stresses, have also to be carried out.

The various calculations are now discussed individually.

### Ultimate limit state

#### Flexure

The flexural design of precast elements can be carried out in accordance with the methods of Chapter 5. In the case of a composite member, the methods of Chapter 5 may be applied to the entire composite section provided that horizontal shear can be transmitted, without excessive slip, across the interface of the precast and in-situ concretes. The criteria for interface shear stresses, discussed in Chapter 4, ensure that excessive slip does not occur.

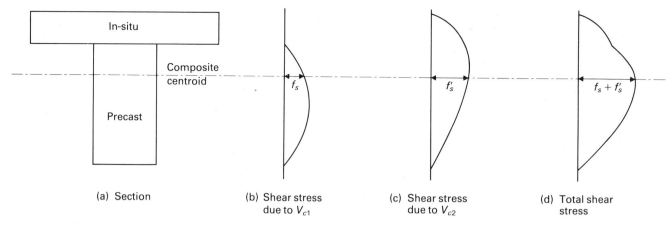

**Fig. 8.6(a)–(d)** Shear in composite beam and slab section

## Shear

It is not necessary to consider interface shear at the ultimate limit state because the interface shear criteria discussed in Chapter 4, for the serviceability limit state, are intended to ensure adequate strength at the ultimate limit state in addition to full composite action at the serviceability limit state.

Thus it is necessary to consider only vertical shear at the ultimate limit state. The fact that interface shear is checked at the serviceability limit state and vertical shear at the ultimate limit state causes a minor problem in the organisation of the calculations and introduces the possibility of errors being made. It is understood that in the proposed amendments to CP 110, which are being drafted at the time of writing, the interface shear criteria are different to those in the Code, and the calculations will be carried out at the ultimate limit state.

The design of a precast element to resist vertical shear can be carried out in accordance with the methods described in Chapter 6. However, the design of a composite section to resist vertical shear is more complicated and there does not seem to be an established method. The latter fact reflects the lack of appropriate test data.

The appropriate Code clause merely states that the design rules for prestressed and reinforced concrete should be applied and, when in-situ concrete is placed between precast prestressed units, the principal tensile stress in the prestressed units should not anywhere exceed $0.24 \sqrt{f_{cu}}$ (see Chapter 6). It is thus best to consider the problem from first principles. In the following, it will be assumed that the precast units are prestressed and thus the problem is one of determining the shear capacity of a prestressed-reinforced composite section when it is flexurally uncracked ($V_{co}$) and also when it is flexurally cracked ($V_{cr}$). The terminology and notation are the same as those of Chapter 6. It is emphasised that the suggested procedures are tentative and that test data are required.

There are two general cases to consider:

1. Precast prestressed beams with an in-situ reinforced concrete top slab to form a composite beam and slab bridge.
2. Precast prestressed beams with in-situ concrete placed between and over the beams to form a composite slab.

*Composite beam and slab* It is suggested by Reynolds, Clarke and Taylor [161] that $V_{co}$ for a composite section should be determined on the basis of a limiting principal tensile stress of $0.24 \sqrt{f_{cu}}$ at the centroid of the composite section. This approach is thus similar to that for prestressed concrete.

The shear force ($V_{c1}$) due to self weight and construction loads produces a shear stress distribution in the precast member as shown in Fig. 8.6(b). The shear stress in the *precast* member at the level of the *composite* centroid is $f_s$.

The additional shear force ($V_{c2}$) which acts on the composite section produces a shear stress distribution in the composite section as shown in Fig. 8.6(c). The shear stress at the level of the composite centroid, due to $V_{c2}$, is $f'_s$.

The total shear stress distribution is shown in Fig. 8.6(d) and the shear stress at the level of the composite centroid is ($f_s + f'_s$).

Let the compressive stress at the level of the composite centroid due to self weight and construction loads plus 0.8 of the prestress be $f'_{cp}$. (The factor of 0.8 is explained in Chapter 6.) Then the major principal stress at the composite centroid is given by

$$f_1 = -f'_{cp}/2 + \sqrt{(f'_{cp}/2)^2 + (f_s + f'_s)^2}$$

This stress should not exceed the limiting value of $f_t = 0.24 \sqrt{f_{cu}}$. Hence, on substituting $f_1 = f_t$, and rearranging

$$f'_s = \sqrt{f_t^2 + f'_{cp}f_t} - f_s$$

But $f'_s = V_{c2} A \bar{y}/Ib$, where $I$, $b$ and $A\bar{y}$ refer to the composite section.

Hence

$$V_{c2} = \frac{Ib}{A\bar{y}} \left[ \sqrt{f_t^2 + f'_{cp}f_t} - f_s \right] \tag{8.7}$$

Finally, the total shear capacity ($V_{co}$) is given by

$$V_{co} = V_{c1} + V_{c2} \tag{8.8}$$

The above calculation would be carried out at the junction of the flange and web of the composite section if the centroid were to occur in the flange (see Chapter 6).

The above approach, suggested by Reynolds, Clarke and Taylor, seems reasonable except, possibly, when the section is subjected to a hogging bending moment. In such a situation the reinforced concrete in-situ flange might be cracked, and it could be argued that the in-situ concrete should then be ignored. However, as explained in Chapter 6, a significant amount of shear can be transmitted by dowel action of reinforcement, which would be present in the flange, and by aggregate interlock across the cracks. It thus seems reasonable to include the in-situ flange as part of a homogeneous section, whether or not the section is subjected to hogging bending.

When calculating $V_{cr}$, which is the shear capacity of the member cracked in flexure, it is reasonable, if the member is subjected to sagging bending, to apply the prestressed concrete clauses to the *composite* section because the in-situ flange would be in compression. However, when the section is subjected to hogging bending, it would be conservative to ignore the cracked in-situ flange and to carry out the calculations by applying the prestressed concrete clauses to the precast section alone.

When the prestressed concrete clauses are applied to the *composite* section, the author would suggest that the following amendments be made.

1. Classes 1 and 2:
   (a) Replace equation (6.12) with

   $$M_t = M_b(1 - y_b I_c/y_c I_b) +$$
   $$(0.37 \sqrt{f_{cub}} + 0.8 f_{pt}) I_c/y_c \qquad (8.8)$$

   where the subscripts $b$ and $c$ refer to the precast beam and the composite sections respectively and $M_b$ is the moment acting on the precast beam alone. Equation (8.8) is derived as follows. The design value of the compressive stress at the bottom fibre due to prestress and the moment acting on the precast section alone is

   $$0.8 f_{pt} - M_b y_b/I_b$$

   The additional stress to cause cracking is (see Chapter 6)

   $$0.37 \sqrt{f_{cub}} + 0.8 f_{pt} - M_b y_b/I_b$$

   Thus the additional moment, applied to the composite section to cause cracking, is

   $$M_a = (0.37 \sqrt{f_{cub}} + 0.8 f_{pt} - M_b y_b/I_b)I_c/y_c$$

   The total moment is the cracking moment:

   $$M_t = M_a + M_b$$

   which on simplification gives equation (8.8)
   (b) Replace the term $(d \sqrt{f_{cu}})$ in equation (6.11) with $(d_b \sqrt{f_{cub}} + d_i \sqrt{f_{cui}})$, where the subscripts $b$ and $i$ refer to the precast beam and in-situ concrete respectively, and $d_b$ and $d_i$ are defined in Fig. 8.7(a). In this context $d_b$ is measured to the centroid of all of the tendons.
2. Class 3:
   (a) Calculate $M_o$ from

   $$M_o = M_b (1 - y_b I_c/y_c I_b) + 0.8 f_{pt} I_c/y_c \qquad (8.9)$$

   which can be derived in a similar manner to equation (8.8).
   (b) Replace the term $(v_c d)$ in equation (6.16) with

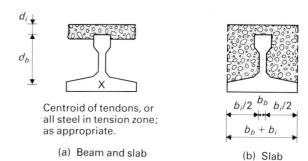

Centroid of tendons, or all steel in tension zone; as appropriate.

(a) Beam and slab         (b) Slab

**Fig. 8.7(a),(b)** Composite sections

$(v_{cb}d_b + v_{ci}d_i)$, where $v_{cb}$ and $v_{ci}$ are the nominal shear stresses appropriate to $f_{cub}$ and $f_{cui}$ respectively. $d_b$ is measured to the centroid of all of the steel in the tension zone.

Finally, it is suggested that the maximum allowable shear force should be calculated from

$$V_u = 0.75b(d_b \sqrt{f_{cub}} + d_i \sqrt{f_{cui}}) \qquad (8.10)$$

$d_b$ is measured to the centroid of all of the steel in the tension zone.

The above suggested approach is slightly different to that of BE 2/73.

*Composite slab* In order to comply strictly with the Code when calculating $V_{co}$ for the composite section, the principal tensile stress at each point in the precast beams should be checked. However, the author would suggest that it is adequate to check only the principal tensile stress at the centroid of the composite section. When carrying out the $V_{co}$ calculation, no consideration is given to whether the in-situ concrete *between* the beams is cracked. This is because the adjacent prestressed concrete restrains the in-situ concrete and controls the cracking [113]. This effect is discussed in Chapter 4 in connection with the allowable flexural tensile stresses in the in-situ concrete. Provided that the latter stresses are not exceeded, the in-situ and precast concretes should act compositely.

It could be argued that, when subjected to hogging bending, the in-situ concrete above the beams should be ignored. However, the author would suggest that it be included for the same reasons put forward for including it in beam and slab composite construction.

The general Code approach differs from the approach of BE 2/73, in which areas of plain in-situ concrete which develop principal tensile stresses in excess of the limiting value are ignored.

When calculating $V_{cr}$, the in-situ concrete could be flexurally cracked before the precast concrete cracks. It is not clear how to calculate $V_{cr}$ in such a situation although the author feels that the restraint to the in-situ concrete provided by the precast beams should enable one to apply the prestressed concrete clauses to the entire *composite* section. However, in view of the lack of test data, the author would suggest either of the following two conservative approaches.

1. Ignore all of the in-situ concrete and apply the prestressed concrete clauses to the precast beams alone, as proposed by Reynolds, Clarke and Taylor [161].

2. Apply the *reinforced* concrete clauses to the entire composite section.

Finally, it is suggested that the maximum allowable shear force for a section $(b_b + b_i)$ wide should be calculated from

$$V_u = 0.75d(b_b \sqrt{f_{cub}} + b_i \sqrt{f_{cui}}) \qquad (8.11)$$

where $b_b$ and $b_i$ are defined in Fig. 8.7(b).

The above suggested approach is different to that of BE 2/73 in which a modified form of equation (6.11) is adopted for $V_{cr}$.

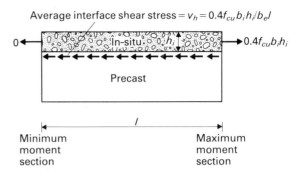

Average interface shear stress $= v_h = 0.4f_{cu}b_ih_i/b_el$

$0.4f_{cu}b_ih_i$

Fig. 8.8 Interface shear at ultimate limit state

## Serviceability limit state

### General

It is mentioned in Chapter 4 that the stresses which have to be checked in a composite member are the compressive and tensile stresses in the precast concrete, the compressive and tensile stresses in the in-situ concrete and the shear stress at the interface between the two concretes.

For the usual case of a prestressed precast unit acting compositely with in-situ reinforced concrete, the stress calculations are complicated by the fact that different load levels have to be adopted for checking the various stresses. This is because, as discussed in Chapter 4, different $\gamma_{f3}$ values have to be applied when carrying out the various stress calculations. It is explained in Chapter 4 that the value of $\gamma_{f3}$ implied by the Code is unity for all stress calculation under any load except for HA and HB loading. For the latter loadings, $\gamma_{f3}$ is unity for all stress calculations except for the compressive and tensile stresses in the prestressed concrete, when it takes the values of 0.83 and 0.91 for HA and HB loading respectively. It is emphasised that the actual values to be adopted for the various stress calculations are not stated in the Code. The values quoted above can be deduced from the assumption that it was the drafters' intention that, in general, $\gamma_{f3}$ should be unity at the serviceability limit state. However, there is an important implication of this intention, which may have been overlooked by the drafters. The tensile stresses in the in-situ concrete have to be checked under a design load of 1.2 HA or 1.1 HB (because $\gamma_{fL}$ is 1.2 and 1.1 respectively and $\gamma_{f3} = 1.0$) as compared with 1.0 HA or 1.0 HB when designing in accordance with BE 2/73, despite the allowable tensile stresses in the Code and BE 2/73 being identical. This suggests that, perhaps, the drafters intended the Code design load to be 1.0 HA or 1.0 HB and, hence, the $\gamma_{f3}$ values to be 0.83 and 0.91 respectively. This argument also throws some doubt on the actual intended values of $\gamma_{f3}$ to be adopted when checking interface shear stresses.

In conclusion, although it is not entirely clear what value of $\gamma_{f3}$ should be adopted for HA and HB loading, and it could be argued that $(\gamma_{fL} \gamma_{f3})$ should always be taken to be unity, the following values of $\gamma_{f3}$ will be assumed.

1. HA – 1.0 except for stresses in prestressed concrete when it is 0.83.

2. HB – 1.0 except for stresses in prestressed concrete when it is 0.91.

### Compressive and tensile stresses

Compressive and tensile stresses in either the precast or in-situ concrete should not exceed the values discussed in Chapter 4. Such stresses can be calculated by applying elastic theory to the precast section or to the composite section as appropriate. The difference between the elastic moduli of the two concretes should be allowed for if their strengths differ by more than one grade.

It is emphasised that the allowable flexural tensile stresses of Table 4.4 for in-situ concrete are applicable only when the in-situ concrete is in direct contact with precast prestressed concrete. If the adjacent precast unit is not prestressed then the flexural cracks in the in-situ concrete should be controlled by applying equation (7.4).

The allowable flexural tensile stresses in the in-situ concrete are interpreted differently in the Code and BE 2/73. In the Code it is explicitly stated that they are stresses at the contact surface; whereas in BE 2/73 they are applicable to all of the in-situ concrete, but those parts of the latter in which the allowable stress is exceeded are not included in the composite section.

In neither the Code nor BE 2/73 is it necessary to calculate the flexural tensile stresses in any in-situ concrete which is not considered, for the purposes of stress calculations, to be part of the composite section.

### Interface shear stresses

In terms of limit state design, it is necessary to check interface shear stresses for two reasons.

1. It is necessary to ensure that, at the serviceability limit state, the two concrete components act compositely. Since shear stress can only be transmitted across the interface after the in-situ concrete has hardened, the loads considered when calculating the interface shear stress at the serviceability limit state should consist only of those loads applied after the concrete has hardened. Thus the self weight of the precast unit and the in-situ concrete should be considered in propped but not in unpropped construction. In addition, at the serviceability limit state it is reasonable to calculate the interface shear stress by using elastic theory; hence

$$v_h = VS_c/Ib_e \qquad (8.12)$$

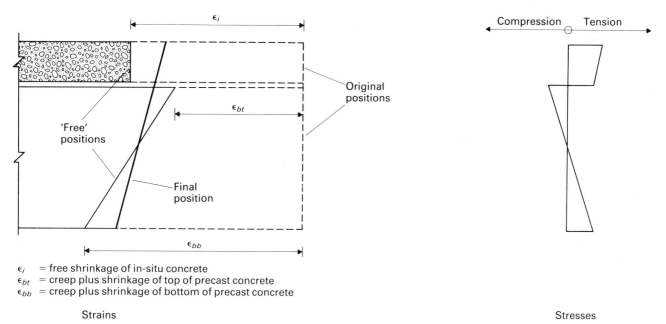

$\epsilon_i$ = free shrinkage of in-situ concrete
$\epsilon_{bt}$ = creep plus shrinkage of top of precast concrete
$\epsilon_{bb}$ = creep plus shrinkage of bottom of precast concrete

Strains                                         Stresses

**Fig. 8.9** Differential shrinkage plus creep

where $v_h$ = horizontal interface shear stress
  $V$ = shear force at point considered
  $S_c$ = first moment of area, about the neutral axis of the transformed composite section, of the concrete to one side of the interface
  $I$ = second moment of area of the transformed composite section
  $b_e$ = width of interface.

2. It is necessary to ensure adequate horizontal shear strength at the ultimate limit state. The shear force per unit length which has to be transmitted across the interface is a function of the normal forces acting in the in-situ concrete at the ultimate limit state. The latter forces result from the *total design load* at the ultimate limit state. If a constant flexural stress of $0.4f_{cu}$ (see Chapter 5) is assumed at the point of maximum moment, then the maximum normal force is $0.4f_{cu}\,b_i\,h_i$ where $b_i$ is the effective breadth of in-situ concrete above the interface and $h_i$ is the depth of in-situ concrete or the depth to the neutral axis if the latter lies within the in-situ concrete. It is conservative to assume that the normal force is zero at the point of minimum moment, which will be considered to be distance $l$ from the point of maximum moment. Hence an interface shear force of $0.4f_{cu}\,b_i\,h_i$ must be transmitted over a distance $l$ (see Fig. 8.8); thus the *average* interface shear stress is

$$v_h = 0.4\,f_{cu}\,b_i\,h_i/b_e\,l \qquad (8.13)$$

A Technical Report of the Fédération Internationale de la Précontrainte [213] suggests that the average shear stress should be distributed over the length $l$ in proportion to the vertical shear force diagram.

It can be seen from the above that, in order to be thorough, two calculations should be carried out:

1. An elastic calculation at the serviceability limit state which considers only those loads which are applied after the in-situ concrete hardens.

2. A plastic calculation at the ultimate limit state which considers the total design load at the ultimate limit state.

However, the Code does not require these two calculations to be carried out: instead a single elastic calculation is carried out at the serviceability limit state which considers *the total design load at the serviceability limit state*. This is achieved by taking $V$ in equation (8.12) to be the shear force due to the total design load at the serviceability limit state. This procedure, though illogical, is intended to ensure that both the correct serviceability and ultimate limit state criteria can be satisfied by means of a single calculation.

Finally, as mentioned previously in this chapter, the interface shear clauses in CP 110, which are essentially identical to those in the Code, are being redrafted at the time of writing. It is understood that the new clauses will require the calculation to be carried out only at the ultimate limit state by considering the total design load at this limit state. This procedure, if adopted, would be more logical than the existing procedure.

## Differential shrinkage

When in-situ concrete is cast on an older precast unit, much of the movement of the latter due to creep and shrinkage will already have taken place, whereas none of the shrinkage of the in-situ concrete will have occurred. Hence, at any time after casting the in-situ concrete, there will be a tendency for the in-situ concrete to shorten relative to the precast unit. Since the in-situ concrete acts compositely with the precast unit, the latter restrains the movement of the former but is itself strained as shown in Fig. 8.9. Hence, stresses are developed in both the in-situ and precast concretes as shown in Fig. 8.9. It is possible to calculate the stresses from considerations of equilibrium, and the necessary equations are given by Kajfasz, Somerville and Rowe [113].

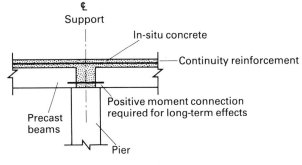

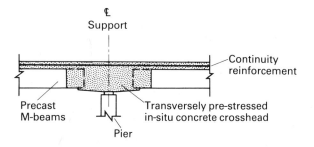

(a) Beams supported on pier

(b) Beams embedded in crosshead

**Fig. 8.10(a),(b)** Continuity in composite construction

It is emphasised that the above calculations need to be carried out only at the serviceability limit state since the stresses arise from restrained deformations and can thus be ignored at the ultimate limit state. The explanation of this is given in Chapter 13 in connection with a discussion of thermal stresses.

The Code does not give values of $\gamma_{fL}$ and $\gamma_{f3}$ to be used when assessing the effects of differential shrinkage at the serviceability limit state; but it would seem to be reasonable to use 1.0 for each.

The most difficult part of a differential shrinkage calculation is the assessment of the shrinkage strains of the two concretes, and the creep strains of the precast unit. These strains depend upon many variables and, if data from tests on the concretes and precast units are not available, estimated values have to be used. For beam and slab bridges in a normal environment the Code gives a value of $100 \times 10^{-6}$ for the differential shrinkage strain, which is defined as the difference between the shrinkage strain of the in-situ concrete and the average shrinkage plus creep strain of the precast unit. This value was based upon the results of tests on composite T-beams reported by Kajfasz, Somerville and Rowe [113]. The test results indicated differential shrinkage strains which varied greatly, but the value quoted in the Code is a reasonable value to adopt for design purposes.

Although it is not stated in the Code, it was intended that the current practice [6, 113] of ignoring differential shrinkage effects in composite slabs, consisting of pretensioned beams with solid in-situ concrete infill, be continued.

Finally, the stresses induced by the restraint to differential shrinkage are relieved by creep and the Code gives a reduction factor of 0.43. The derivation of this factor is discussed later in this chapter.

## Continuity

### Introduction

A multi-span bridge formed of precast beams can be made continuous by providing an in-situ concrete diaphragm at each support as shown in Fig. 8.10(a). An alternative form of connection in which the ends of the precast beams are not supported directly on piers but instead are embedded in a transversely prestressed in-situ concrete crosshead, with some tendons passing through the ends of the beams, has been described by Pritchard [214] (see Fig. 8.10 (b)).

A bridge formed by either of the above methods is statically determinate for dead load but statically indeterminate for live load; and thus the in-situ concrete diaphragm or crosshead has to be designed to resist the hogging moments which will occur at the supports. The design rules for reinforced concrete can be applied to the diaphragm, but consideration should be given to the following points.

### Moment redistribution

Tests have been carried out on half-scale models of continuous girders composed of precast I-sections with an in-situ concrete flange and support diaphragm, as in Fig. 8.10(a), at the Portland Cement Association in America [215, 216]. It was found that, at collapse, moment redistributions causing a reduction of support moment of about 30% could be achieved. It thus seems to be reasonable to redistribute moments in composite bridges provided that the Code upper limit of 30% for reinforced concrete is not exceeded.

### Flexural strength

The Code permits the effect of any compressive stresses due to prestress in the ends of the precast units to be ignored when calculating the ultimate flexural strength of connections such as those in Fig. 8.10.

This recommendation is based upon the results of tests carried out by Kaar, Kriz and Hognestad [215]. They carried out tests on continuous girders with three levels of prestress (zero, $0.42 f_{cyl}$ and $0.64 f_{cyl}$) and three percentages of continuity reinforcement (0.83, 1.66 and 2.49). It was found to be safe to ignore the precompression in the precast concrete except for the specimens with 2.49% reinforcement. In addition, it was found that the difference between the flexural strengths calculated by, first, ignoring and, second, including the precompression was negligible except for the highest level of prestress. Kaar, Kriz and Hognestad thus proposed that the precompression be ignored provided that the reinforcement does not exceed 1.5%, and the stress due to prestress does not exceed $0.4 f_{cyl}$ (i.e. about $0.32 f_{cu}$). Although the Code does not quote these criteria, they will generally be met in practice.

In addition to the above tests, good agreement between calculated and observed flexural strengths of continuous connections involving inverted T-beams with added in-situ concrete has been reported by Beckett [217].

## Shear strength

The shear strength of continuity connections of the type shown in Fig. 8.10(a) has been investigated by Mattock and Kaar [218], who tested fifteen half-scale models. The reinforced concrete connections contained no shear reinforcement, but none of the specimens failed in shear in the connection. The actual failures were as follows: thirteen by shear in the precast beams, one by flexure of a precast beam and one by interface shear.

When designing, to resist shear, the end of a precast beam which is to form part of a continuous composite bridge, it should be remembered that the end of the beam will be subjected to hogging bending. Thus flexural cracks could form at the top of the beam; consequently this region should be given consideration in the shear design.

Sturrock [219] tested models of continuity joints which simulated the type shown in Fig. 8.10(b). The tests showed that no difficulty should be experienced in reinforcing the crosshead to ensure that a flexural failure, rather than a shear failure in either the joint or a precast beam away from the joint, would occur.

## Serviceability limit state

Crack widths and stresses in the reinforced concrete diaphragm can be checked by the method discussed for reinforced concrete in Chapter 7.

Since the section in the vicinity of the ends of the precast beams is to be designed as reinforced concrete, it is almost certain that tensile stresses will be developed at the tops of the precast beams. If the beams are prestressed, the Code implies that these stresses should not exceed the allowable stresses, for the appropriate class of prestressed member, given in Chapter 4. This means that no tensile stresses are permitted in a Class 1 member: this seems rather severe in view of the fact that any cracks in the in-situ concrete will be remote from the tendons. Consequently, the CP 110 handbook [112] suggests that the ends of prestressed units, when used as shown in Fig. 8.10, should always be considered as Class 3 and, hence, cracking permitted at the serviceability limit state.

## Shrinkage and creep

The deflection of a simply supported composite beam changes with time because of the effects of differential shrinkage and of creep due to self weight and prestress. Hence, the ends of a simply supported beam rotate as a function of time, as shown in Fig. 8.11(a), and, since there is no restraint to the rotation, no bending moments are developed in the beam. But, in the case of a beam made continuous by providing an in-situ concrete diaphragm, as shown in Fig. 8.10, the diaphragm restrains the rotation at the end of the beam and bending moments are developed as shown in Fig. 8.11(b).

A positive rotation occurs at the end of the beam because of creep due to prestress and thus a sagging moment is developed at the connection. The sagging moment is relieved by the fact that negative rotations occur as a result of both differential shrinkage and creep due to self weight, which cause hogging moments to be

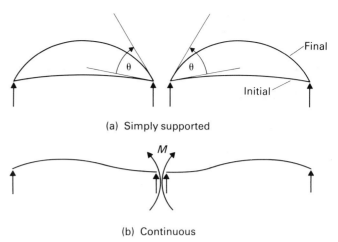

**(a)** Simply supported

**(b)** Continuous

**Fig. 8.11** Long-term effects

developed at the connection. Nevertheless, a net sagging moment at the connection will generally be developed. Since this sagging moment exists when no imposed loading is on the bridge, bottom reinforcement, as shown in Fig. 8.10(a), is often necessary, together with the usual top reinforcement needed to resist the hogging moment under imposed loading. The provision of bottom reinforcement has been considered by Mattock [220].

In order to examine the influence of creep and shrinkage on connection behaviour, Mattock [220] tested two continuous composite beams for a period of two years. One beam was provided with bottom reinforcement in the diaphragm and could thus transmit a significant sagging moment, whereas the other beam had no bottom reinforcement. The latter beam cracked at the bottom of the diaphragm after about one year and the behaviour of the beam under its design load was adversely affected.

Mattock found that the observed variation due to creep and shrinkage of the centre support reactions could be predicted by the 'rate of creep' approach [198]. He thus suggested that this approach should be used to predict the support moment, which is of particular interest in design. The 'rate of creep' approach, which is adopted in the Code, assumes that under variable stress the rate of creep at any time is independent of the stress history. Hence, if the ratio of creep strain to elastic strain at time $t$ is $\beta$ and the stress at time $t$ is $f$, then the increment of creep strain ($\delta\varepsilon_c$) in time $\delta t$ is given by

$$\delta\varepsilon_c = (f/E)\delta\beta \tag{8.14}$$

where $E$ is the elastic modulus. In the continuity problem under consideration, it is more convenient to work in terms of moment ($M$) and curvature ($\psi$); and thus, by analogy with stress and strain

$$\delta\psi = (M/EI)\delta\beta \tag{8.15}$$

The effects of creep and shrinkage are discussed in detail in reference [220]. In the following analyses, they are treated less rigorously but with sufficient detail to illustrate the derivation of the relevant formulae in the Code.

*Creep due to prestress* The general method is illustrated here by considering the two span continuous beam shown in Fig. 8.12. The beam has constant flexural stiffness and

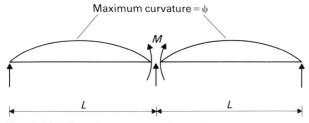

Maximum curvature $= \psi$

$M$

$L$ $L$

**Fig. 8.12** Effect of creep on continuous beam

equal spans. The prestressing force is $P$ and the maximum eccentricity in each span is $e$. Hence the maximum prestressing moment in each span is $Pe$. At time $t$, let the curvatures associated with these moments be $\psi$ and the restraint moment at the internal support be $M$.

In time $\delta t$, the curvature $\psi$ will increase due to creep by $\delta\psi$, where

$$\delta\psi = \psi\delta\beta = (Pe/EI)\delta\beta$$

If the spans were freely supported, the change in end rotation of span 1 at the internal support would be

$$k\delta\psi = k(Pe/EI)\delta\beta \qquad (8.16)$$

and of span 2 would be

$$-k\delta\psi = -k(Pe/EI)\delta\beta \qquad (8.17)$$

where $k$ depends upon the tendon profile. For a straight profile $k = L$, and for a parabolic profile $k = 2L/3$.

The rotation due to $M$ if the spans were simply supported would be $\pm 2ML/3EI$ with the negative and positive signs being taken for spans 1 and 2 respectively. In time $\delta t$, these rotations would change due to creep by

$$(-2ML/3EI)\,\delta\beta \qquad (8.18)$$

and

$$(+2ML/3EI)\delta\beta \qquad (8.19)$$

Also in time $\delta t$, the restraint moment would change by $\delta M$. The rotation, due to this change, at the ends of spans 1 and 2 respectively would be

$$(-2\delta ML/3EI) \qquad (8.20)$$

and

$$(+2\delta ML/3EI) \qquad (8.21)$$

Since the two spans are joined at the support, their net changes of rotation must be equal. The net change of rotation for span 1 is obtained by summing expression (8.16), (8.18) and (8.20); and, for span 2, by summing expressions (8.17), (8.19) and (8.21). If the two net changes of rotation are equated and the resulting equation rearranged, the following differential equation is obtained

$$\frac{dM}{d\beta} + M = \frac{3kPe}{2}$$

The right hand side of this equation is the *sagging* restraint moment which would result, in the absence of creep, if the beam were cast and prestressed as a monolithic continuous beam. This moment will be designated $M_p$. Hence

$$\frac{dM}{d\beta} + M = M_p$$

The solution of this equation with the boundary condition

that, when $t = 0$, $\beta = 0$, $M = 0$ is

$$M = M_p[1 - \exp(-\beta)]$$

In the Code, the expression $[1 - \exp(-\beta)]$ is designated $\phi_1$; thus

$$M = M_p\phi_1 \qquad (8.22)$$

Although equation (8.22) has been derived for a two-span beam, it is completely general and is applicable to any span arrangement, provided that the appropriate value of $M_p$ is used. Hence, for any continuous beam, the restraint moments at any time can be obtained by calculating the restraint moments which would occur if the beam were cast and prestressed as a monolithic continuous beam and by then multiplying these moments by the creep factor $\phi_1$.

The above analysis implies that the prestress moment $(Pe)$ should be calculated by considering the prestressing forces applied to the entire composite section. Hence the eccentricity to be used should be that of the prestressing force relative to the *centroid of the composite section*. This is explained as follows. Consider the composite beam shown in Fig. 8.13. The eccentricities of the prestressing force $P$ are $e_b$ and $e_c$ with respect to the precast beam and composite centroids respectively. In time $t$ after casting the in-situ concrete, creep of the precast concrete will cause the axial strain of the precast beam to change by $(P/A_b)\beta$ and its curvature to change by $(Pe_b/E_bI_b)\,\beta$, where the subscript $b$ refers to the precast beam. The in-situ concrete is initially unstressed and thus does not creep. In order to maintain compatibility between the two concretes it is necessary to apply an axial force of $P\beta$ and a moment of $Pe_b\beta$ to the *precast* beam. However, since the *composite* section must be in a state of internal equilibrium, it is now necessary to apply a cancelling force of $P\beta$ and a cancelling moment of $Pe_b\beta$ to the *composite* section. The net moment applied to the *composite* section and which produces a curvature of the *composite* section is thus

$$(Pe_b\beta) + (P\beta)(e_c - e_b) = Pe_c\beta$$

Hence, the eccentricity used should be that relative to the centroid of the composite section. A more rigorous proof is given in reference [220].

The creep factor $\beta$ is dependent upon a great number of variables. Appendix C of Part 4 of the Code gives data for the assessment of the following effects on $\beta$ : relative humidity, age at loading, cement content, water–cement ratio, thickness of member and time under load. Unfortunately, the basic data required to assess these effects are not generally known at the design stage. For design purposes, one is interested in $\beta_{cc}$, which is the value towards which $\beta$ would eventually tend. In the absence of more precise data, Mattock [220] suggested that $\beta_{cc}$ should be taken as 2. This value implies that the creep factor $\phi_1$ to be applied to the restraint moment due to creep is 0.87. In practice the value of $\beta_{cc}$ is likely to be between 1.5 and 2.5, and the adoption of the average value of 2 for $\beta_{cc}$ implies a maximum error of 10% in the value of $\phi_1$.

If $\beta$ is calculated from the data in Appendix C of the Code, it should be remembered that its value should be based on the increase in creep strain from the time that the beam is made continuous by casting the in-situ concrete, and not from the time of prestressing.

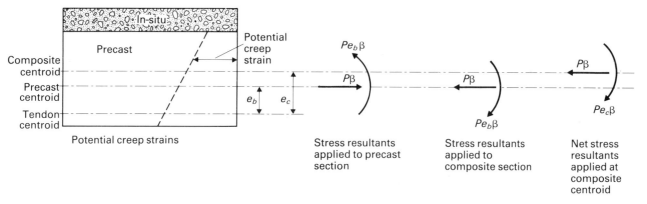

Fig. 8.13 Effect of creep on composite section

*Creep due to dead load*  By an analysis similar to that given above, it can be shown that the restraint moment due to creep under dead load is given by

$$M = M_d \phi_1 \qquad (8.23)$$

$M_d$ is the *hogging* restraint moment which would occur if the beam were cast as a monolithic continuous beam; i.e. the restraint moment due to the combined dead loads of the precast beam and in-situ concrete applied to the composite section. For the two-span beam considered in the last section, $M_d$ is a hogging moment of magnitude $(wL^2/8)$ where $w$ is the total (precast plus in-situ) dead load per unit length. $\phi_1$ can again be taken as 0.87.

*Shrinkage*  Before considering the effects of shrinkage on the restraint moments of a continuous beam, the relief of shrinkage stresses, in general, due to creep is examined. In the analysis which follows, it is assumed that the relationships between creep strain per unit stress (specific creep) and time, and between shrinkage strain ($\varepsilon_s$) and time, have the similar forms shown in Fig. 8.14, so that,

$$\varepsilon_s = K(\varepsilon_c/f)$$

where $K$ is a constant. Hence, using equation (8.14)

$$\varepsilon_s = K\beta/E \text{ and } \delta\varepsilon_s = K\delta\beta/E \qquad (8.24)$$

Consider a piece of concrete which is restrained against shrinkage so that a tensile shrinkage stress is developed. At time $t$, let this stress be $f$. In time $\delta t$, let the increase of shrinkage be $\delta\varepsilon_s$ and the change of the shrinkage stress be $\delta f$. Also in time $\delta t$, there would be a potential creep strain of $(f/E)\delta\beta$. Thus the net potential change of strain, which results in a change of shrinkage stress, is

$$\delta\varepsilon_s - (f/E)\delta\beta$$

Hence, the change of stress is

$$\delta f = E\delta\varepsilon_s - f\delta\beta$$

or

$$\frac{df}{d\beta} + f = E\frac{d\varepsilon_s}{d\beta}$$

Hence, using equation (8.24)

$$\frac{df}{d\beta} + f = K$$

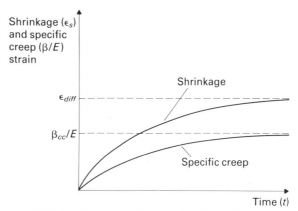

Fig. 8.14 Specific creep–time and shrinkage–time curves

The solution of this equation with the boundary condition that, when $t = 0$, $\beta = 0$, $f = 0$ is

$$f = K[1 - \exp(-\beta)] = K\phi_1$$

Hence, using equation (8.24)

$$f = E\varepsilon_s \frac{\phi_1}{\beta}$$

In the Code, the expression $\phi_1/\beta$ is designated $\phi$; thus

$$f = E\varepsilon_s\phi \qquad (8.25)$$

Hence, the shrinkage stress, at any time $t$, is the shrinkage stress ($E\varepsilon_s$) which would occur in the absence of creep multiplied by the creep factor $\phi$. If the limiting value ($\beta_{cc}$) of $\beta$ is again taken as 2, then $\phi = 0.43$. This factor is referred to earlier in this chapter in connection with the relief of differential shrinkage stresses.

*Differential shrinkage*  The general method is again illustrated by considering a two-span continuous beam which is symmetrical about the internal support. Due to the differential shrinkage between the precast beam and the in-situ flange, at any time $t$, there will be a constant curvature ($\psi$) imposed throughout the length of the beam, and the change ($\delta\psi$) of curvature in time $\delta t$ can be calculated by considering Fig. 8.15. In time $\delta t$, the differential shrinkage strain will change by $\delta\varepsilon_s$. If it is required that the precast beam and the flange stay the same length, it is necessary to apply to the centroid of the flange a tensile force of

$$\delta F = \delta\varepsilon_s E_{cf} A_{cf}$$

113

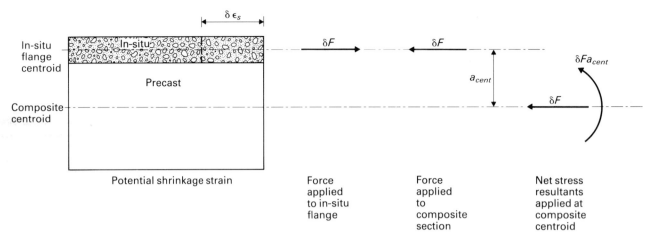

**Fig. 8.15** Differential shrinkage

where $A_{cf}$ and $E_{cf}$ are the area and elastic modulus respectively of the flange concrete. Since the *composite* section must be in a state of internal equilibrium, it is now necessary to apply a cancelling compressive force of $\delta F$ to the *composite* section. This force has an eccentricity of $a_{cent}$ with respect to the centroid of the composite section. Hence a moment $\delta F a_{cent}$ is effectively applied to the composite section and this moment induces the curvature $\psi$. Thus

$$\delta\psi = \delta F a_{cent}/EI = \delta\varepsilon_s E_{cf} A_{cf} a_{cent}/EI$$

where the $EI$ value is appropriate to the composite section. If at time $t$, the restraint moment at the internal support is $M$ and it changes by $\delta M$ in time $\delta t$, then an analysis, identical to that presented earlier in this chapter for creep due to prestress, results in the following differential equation

$$\frac{dM}{d\beta} + M = \frac{3}{2} E_{cf} A_{cf} a_{cent} \frac{d\varepsilon_s}{d\beta}$$

Using equation (8.24)

$$\frac{dM}{d\beta} + M = \frac{3}{2} E_{cf} A_{cf} a_{cent} \frac{K}{E}$$

The solution of this equation with the boundary condition that, when $t = 0$, $\beta = 0$, $\varepsilon_s = 0$, $M = 0$ is

$$M = \frac{3}{2} E_{cf} A_{cf} a_{cent} \frac{K}{E}\left[1 - \exp(-\beta)\right]$$

Using equation (8.24)

$$M = \frac{3}{2} (\varepsilon_s E_{cf} A_{cf} a_{cent}) \frac{[1 - \exp(-\beta)]}{\beta} \tag{8.26}$$

or

$$M = M_{cs}\phi \tag{8.27}$$

where $M_{cs}$ is the *hogging* restraint moment which would occur in the absence of creep. Although equation (8.26) has been derived for a two-span beam, equation (8.27) is completely general and is applicable to any span arrangement. The Code again assumes a value of 0.43 for $\phi$. For design purposes, one is interested in $\varepsilon_{diff}$, which is the value towards which $\varepsilon_s$ would eventually tend. Appendix C of the Code gives data which enable the effects on

shrinkage of the following to be assessed: relative humidity, cement content, water–cement ratio, thickness of member and time. Hence, the differential shrinkage strain can be assessed: a typical value would be about $200 \times 10^{-6}$. This value is much greater than the recommended value of $100 \times 10^{-6}$ quoted earlier in this chapter, when differential shrinkage calculations for beam and slab bridges were discussed. This is because the latter value allows for the creep strains of the precast beam whereas, when carrying out the restraint moment calculations, the creep and shrinkage effects are treated independently (compare Figs. 8.9 and 8.15).

Finally, it should be noted that the Code states that $M_{cs}$ can be taken as

$$M_{cs} = \varepsilon_{diff} E_{cf} A_{cf} a_{cent} \phi \tag{8.28}$$

for an internal support. This is approximately correct for a large number of spans (5 or more) but it will underestimate the restraint moment for beams with fewer spans. In general the value of $M_{cs}$ calculated from equation (8.28) needs to be multiplied by a constant which depends upon the span arrangement. Equation (8.26) shows that the constant is 3/2 for a two-span beam with equal span lengths. Appropriate values for other numbers of equal length spans are:

1. Three spans: 1.2 for both internal supports.
2. Four spans: 1.29 for first internal supports, 0.86 for centre support.
3. Five or more spans: 1.27 for first internal supports, 1.0 for all other supports.

Values for unequal spans would have to be calculated from first principles.

*Combined effects of creep and differential shrinkage*   The net *sagging* restraint moment due to creep under prestress and dead load and due to differential shrinkage can be obtained by summing equations (8.22), (8.23) and (8.27) with due account being taken of sign. Hence

$$M = (M_p - M_d)\phi_1 - M_{cs}\phi \tag{8.29}$$

Examples of these calculations are given in [113], [216] and [219]. The calculations need to be carried out only at the serviceability limit state since the restraint moments are

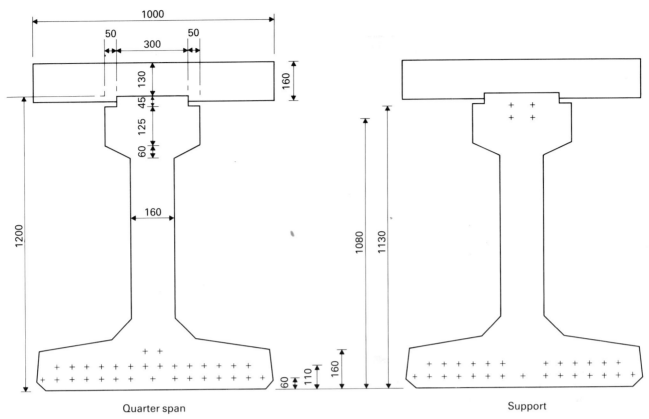

**Fig. 8.16** Composite beam cross-sections

due to imposed deformations and can be ignored at the ultimate limit state.

*Code notation*  The notation adopted in the Code for the creep factors is confusing. Appendix C of Part 4 of the Code adopts the symbol φ for the creep factor which, in this chapter, has been given the symbol β (or $\beta_{cc}$ after a long period of time). However, the main body of Part 4 of the Code uses $\beta_{cc}$, φ and $φ_1$ in the same sense as they are used in this chapter. Thus care should be exercised when assessing a creep factor from Appendix C of the Code for use in composite construction calculations.

The notation adopted for the creep factors is also referred to in Chapter 7 in connection with the calculation of long-term curvatures and deflections.

## Example – Shear in composite construction

A bridge deck consists of pretensioned precast standard M8 beams at 1 m centres acting compositely with a 160 mm thick in-situ concrete top slab. The characteristic strengths of the shear reinforcement to be designed and of the precast and in-situ concretes are 250 N/mm², 50 N/mm² and 30 N/mm² respectively. Four tendons are deflected at the quarter points and the tendon patterns at mid-span and at a support are shown in Fig. 8.16. The total tendon force after all losses have occurred is 3450 kN. The span is 25 m, the overall beam length 25.5 m and the nominal values per beam of the critical shear forces and moments at the support and at quarter-span for load combination 1 are given in Table 8.1.

Design reinforcement for both vertical and interface shear at the two sections.

**Table 8.1**  Nominal values of stress resultants

| Load | Support | | Quarter span | |
|---|---|---|---|---|
| | Shear force (kN) | Moment (kNm) | Shear force (kN) | Moment (kNm) |
| Self weight | 163 | 0 | 81 | 763 |
| Parapet | 27 | 0 | 14 | 132 |
| Surfacing | 29 | 0 | 15 | 135 |
| HB + associated HA | 332 | 0 | 196 | 1333 |

## Section properties

The modular ratio for the in-situ concrete is $\sqrt{30/50} = 0.775$. Reference [36] gives section properties for the precast and composite sections; these are summarised in the upper part of Table 8.2. The composite values are based upon a modular ratio of 0.8 which is only slightly different to the correct value of 0.775.

**Table 8.2**  Section properties

| Property | Precast | Composite |
|---|---|---|
| Area (mm²) | 393450 | – |
| Height of centroid above bottom fibre (mm) | 454 | 642 |
| Second moment of area (mm⁴) | $65.19 \times 10^9$ | $124.55 \times 10^9$ |
| First moment of area above composite centroid (mm³) | $44.4 \times 10^6$ | $116.0 \times 10^6$ |
| First moment of area above interface (mm³) | | $71.6 \times 10^6$ |

The first moments of area, about the composite centroid, of the sections above the composite centroid and above the interface are also required. These have been calculated and are given in the lower part of Table 8.2.

## Vertical shear

### At support

Centroid of tendons from soffit

$$= (15 \times 60 + 12 \times 110 + 2 \times 1080 + 2 \times 1130)/31$$
$$= 214 \text{ mm}$$

There is no applied moment acting, thus the stress at the composite centroid is due only to the prestress and is

$$f_{cp} = (3.45 \times 10^6/393\,450) -$$
$$(3.45 \times 10^6)\,(454 - 214)\,(642 - 454)/(65.19 \times 10^9)$$
$$= 6.38 \text{ N/mm}^2$$

Allowable principal tensile stress (see Chapter 6) = $f_t$ = $0.24\sqrt{50}$ = 1.70 N/mm². Design shear force at the ultimate limit state acting on the precast section alone is nominal value $x\gamma_{fL}\, x\gamma_{f3}$

$$V_{c1} = 163 \times 1.2 \times 1.15 = 225 \text{ kN}$$

Inclination of the four deflected tendons
= arctan (970/6500) = 8.49°
Vertical component of inclined prestress
= (4/31) (3.45) (sin 8.49)10³ = 66 kN
Net design shear force on precast section
= 225 − (0.8) (66) = 172 kN = $V$ where 0.8 is the partial safety factor applied to the prestress (see Chapter 6). Shear stress at composite centroid is

$$f_s = V(A\bar{y})/Ib = \frac{(172 \times 10^3)\,(44.4 \times 10^6)}{(65.19 \times 10^9)\,(160)} = 0.73 \text{ N/mm}^2$$

Additional shear force ($V_{c2}$) which can be carried by the composite section before the principal tensile stress at the composite centroid reaches 1.7 N/mm² is, from equation (8.7),

$$V_{c2} = \frac{(124.55 \times 10^9)\,(160)}{116 \times 10^6} \times$$
$$\left(\sqrt{(1.7)^2 + (0.8)\,(6.38)\,(1.7)} - 0.73\right)10^{-3}$$
$$= 459 \text{ kN}$$

$$V_{c0} = V_{c1} + V_{c2} = 225 + 459 = 684 \text{ kN}$$

It is not necessary to consider the section cracked in flexure at the support and thus the ultimate shear resistance of the concrete alone is

$$V_c = V_{c0} = 684 \text{ kN}$$

Design shear force is

$$V = 225 + (27 \times 1.2 \times 1.15) +$$
$$(29 \times 1.75 \times 1.15) + (332 \times 1.3 \times 1.1)$$
$$= 795 \text{ kN}$$

$V > V_c$, thus links are required such that

$$\frac{A_{sv}}{s_v} = \frac{V - V_c}{0.87\,f_{yv}d_t}$$

$d_t$ = distance from extreme compression fibre of composite section to the centroid of the lowest tendon
= 1270 mm

$$\frac{A_{sv}}{s_v} = \frac{(795 - 684)10^6}{(0.87)(250)(1270)} = 402 \text{ mm}^2/\text{m}$$

$V \not> 1.8\,V_c$ (Clause 7.3.4.3), thus maximum link spacing is lesser of $0.75d_t$ = 952 mm and 4 × 160 = 640 mm.
10 mm links (2 legs) at 390 mm centres give 403 mm²/m.
Finally, in the above calculations, the full value of the prestress has been taken at the support, although the support would lie within the transmission zone. BE 2/73 requires a reduced value of prestress to be adopted but the Code does not state that this should be done. However, in practice, one would normally estimate the build-up of prestress within the transmission zone, and calculate the shear capacity in accordance with the estimated prestress.

### At quarter span

Centroid of tendons from soffit

$$= (15 \times 60 + 14 \times 110 + 2 \times 160)/31 = 89 \text{ mm}$$

Design moment at ultimate limit state acting on precast section alone

$$= 763 \times 1.2 \times 1.15 = 1053 \text{ kNm}$$

Stress at composite centroid due to the moment

$$= 1053 \times 10^6\,(642 - 454)/(65.19 \times 10^9)$$
$$= 3.04 \text{ N/mm}^2 \text{ (tension)}$$

Stress at composite centroid due to prestress

$$= (3.45 \times 10^6/393\,450) -$$
$$(3.45 \times 10^6)\,(454 - 89)\,(642 - 454)/(65.19 \times 10^9)$$
$$= 5.14 \text{ N/mm}^2$$

Total stress at composite centroid to be used in equation (8.7) is

$$f'_{cp} = (0.8)\,(5.14) - 3.04 = 1.07 \text{ N/mm}^2$$

Design shear force at the ultimate limit state acting on the precast section alone is

$$V_{c1} = 81 \times 1.2 \times 1.15 = 112 \text{ kN}$$

Shear stress at composite centroid is

$$f_s = (112 \times 10^3)(44.4 \times 10^6)/(65.19 \times 10^9)\,(160)$$
$$= 0.48 \text{ N/mm}^2$$

Additional shear force ($V_{c2}$) which can be carried by the composite section before the principal tensile stress at the composite centroid reaches 1.70 N/mm² is

$$V_{c2} = \frac{(124.55 \times 10^9)\,(160)}{116 \times 10^6} \times$$
$$\left(\sqrt{(1.7)^2 + (1.07)\,(1.7)} - 0.48\right)10^{-3}$$
$$= 290 \text{ kN}$$

$$V_{c0} = V_{c1} + V_{c2} = 112 + 290 = 402 \text{ kN}$$

The section must now be considered to be cracked in flexure. Stress at extreme tension fibre due to prestress is

$f_{pt} = (3.45 \times 10^6 / 393\,450) +$
$$(3.45 \times 10^6)(454 - 89)(454)/(65.19 \times 10^9)$$
$$= 17.54 \text{ N/mm}^2$$

The cracking moment is given by equation (8.8)

$M_t = (1053 \times 10^6) \times$
$$(1 - 454 \times 124.55 \times 10^9/642 \times 65.19 \times 10^9) +$$
$$(0.37 \sqrt{50} + 0.8 \times 17.54) \, 124.55 \times 10^9/642$$
$$= 2860 \times 10^6 \text{ Nmm} = 2860 \text{ kNm}$$

Design moment at the ultimate limit state is

$M = 1053 + (132 \times 1.2 \times 1.15) +$
$$(135 \times 1.75 \times 1.15) + (1333 \times 1.3 \times 1.1)$$
$$= 3413 \text{ kNm}$$

Design shear force at the ultimate limit state is

$V = 112 + (14 \times 1.2 \times 1.15) +$
$$(15 \times 1.75 \times 1.15) + (196 \times 1.3 \times 1.1)$$
$$= 442 \text{ kN}$$

From the modified form of equation (6.11), suggested in this chapter

$V_{cr} = (0.037)(160)(1111 \sqrt{50} + 130 \sqrt{30})10^{-3} +$
$$442 \, (2860/3413) = 421 \text{ kN}$$

Ultimate shear resistance ($V_c$) is the lesser of $V_{c0}$ and $V_{cr}$ and thus

$$V_c = V_{c0} = 402 \text{ kN}$$

$V > V_c$, thus links are required such that

$$\frac{A_{sv}}{s_v} = \frac{(442 - 402)10^6}{0.87 \times 250 \times 1270} = 145 \text{ mm}^2/\text{m}$$

However, minimum links must be provided such that

$$\frac{A_{sv}}{s_v} \left( \frac{0.87 \, f_{yv}}{b} \right) = 0.4 \text{ N/mm}^2$$

or

$$\frac{A_{sv}}{s_v} = \frac{(0.4)(160)10^3}{0.87 \times 250} = 294 \text{ mm}^2/\text{m}$$

$V \ngtr 1.8V_c$, thus the maximum link spacing is the same as at the support (= 640 mm)
10 mm diameter links (2 legs) at 500 mm give 314 mm²/m

### Maximum allowable shear force

At support, distance of centroid of tendons in tension zone from soffit is
$(15 \times 60 + 12 \times 110)/27 = 82$ mm.
Thus, in equation (8.10)

$$d_i = 130 \text{ mm}$$

$d_b = 1200 - 82 = 1118$ mm
$V_u = (0.75)(160)(1118 \sqrt{50} + 130 \sqrt{30})10^{-3}$
$$= 1034 \text{ kN}$$

Allowing for vertical component of inclined prestress, because uncracked

$$V_u = 1034 + (0.8)(66) = 1087 \text{ kN}$$

Maximum design shear force = 795 kN, thus section is adequate.

### Interface shear

*At support*

Total design shear force at serviceability limit state is

$(163 \times 1.0 \times 1.0) + (27 \times 1.0 \times 1.0) +$
$$(29 \times 1.2 \times 1.0) + (332 \times 1.1 \times 1.0)$$
$$= 590 \text{ kN}$$

From equation (8.12), interface shear stress is

$$v_h = \frac{(590 \times 10^3)(71.6 \times 10^6)}{(124.55 \times 10^9)(300)} = 1.16 \text{ N/mm}^2$$

Type 1 surface is not permitted for beam and slab bridge construction. The allowable shear stress for Type 2 surface is 0.38 N/mm² (see Table 4.5) and this cannot be increased by providing links in excess of the required minimum of 0.15%.

The allowable shear stress for Type 3 surface is 1.25 N/mm² (see Table 4.5); and thus the minimum amount of steel of 0.15% is all that is required, but the laitence must be removed from the top surface of the beam. Thus provide

$$(0.15)(300)(1000)/100 = 450 \text{ mm}^2/\text{m}$$

This exceeds that provided for vertical shear. Thus use 10 mm links (2 legs) at 350 mm centres which give 449 mm²/m

*At quarter span*

Total design shear force at serviceability limit state is

$(81 \times 1.0 \times 1.0) + (14 \times 1.0 \times 1.0) +$
$$(15 \times 1.2 \times 1.0) + (196 \times 1.1 \times 1.0)$$
$$= 329 \text{ kN}$$

$$v_h = 0.63 \text{ N/mm}^2$$

Again only Type 3 surface can be used and 10 mm links at 350 mm centres would be required. This amount of reinforcement exceeds that required for vertical shear.

*Chapter 9*

# Substructures and foundations

## Introduction

The Code does not give design rules which are specifically concerned with bridge substructures. Instead, design rules, which are based upon those of CP 110, are given for columns, walls (both reinforced and plain) and bases. In addition, design rules for pile caps, which did not originate in CP 110, are given. The CP 110 clauses were derived for buildings and, thus, the column and wall clauses in the Code are also more relevant to buildings than to bridges. In view of this, the approach that is adopted in this chapter is, first, to give the background to the Code clauses and, then, to discuss them in connection with bridge piers, columns, abutments and wing walls. However, these structural elements are treated in general terms only, and, for a full description of the various types of substructure and of their applications, the reader is referred to [221].

The author anticipates that the greatest differences in section sizes and reinforcement areas, between designs carried out in accordance with the current documents and with the Code, will be noticed in the design of substructures and foundations. The design of the latter will also take longer because, as explained later in this chapter, more analyses are required for a design in accordance with the Code because of the requirement to check stresses and crack widths at the serviceability limit state in addition to strength at the ultimate limit state. In CP 110, from which the substructure clauses of the Code were derived, it is only necessary to check the strength at the ultimate limit state, since compliance with the serviceability limit state criteria is assured by applying deemed to satisfy clauses. The column and wall clauses of CP 110 were derived with this approach in mind. The fact that the CP 110 clauses have been adopted in the Code without, apparently, allowing for the Code requirement that stresses and crack widths at the serviceability limit state should be checked, has led to a complicated design procedure. This complication is mainly due to the fact that it has not yet been established which limit state will govern the design under a particular set of load effects. Presumably, as experience in using the Code is gained, it will be possible to indicate the most likely critical limit state for a particular situation.

## Columns

### General

#### Definition

A column is not defined in the Code; but a wall is defined as having an aspect ratio, on plan, greater than 4. Thus a column can be considered as a member with an aspect ratio not greater than 4.

#### Effective height

The Code gives a table of effective heights ($l_e$), in terms of the clear height ($l_o$), which is intended only to be a guide for various end conditions. The table, which is here summarised in Table 9.1, is based upon a similar table in CP 110 which, in turn, was based upon a table in CP 114. The effective heights have been derived mainly with framed buildings in mind and do not cover, specifically, the types of column which occur in bridge construction. Indeed, in view of the variety of different types of articulation which can occur in bridges, it would be difficult to produce a table covering all situations. It would thus appear necessary to consider each particular case individually by examining the likely buckling mode. In doing this, consideration should be given to the way that movement can be accommodated by the bearings, the flexibility of the column base (and soil in which it is founded), and whether the articulation of the bridge is such that the columns are effectively braced or can sway. Some of these aspects are discussed by Lee [222].

**Table 9.1** Effective heights of columns

| Column type | $l_e/l_o$ |
|---|---|
| Braced, restrained in direction at both ends | 0.75 |
| Braced, partially restrained in direction at one or both ends | 0.75–1.0 |
| Unbraced or partially braced, restrained in direction at one end, partially restrained at other | 1.0–2.0 |
| Cantilever | 2.0–2.5 |

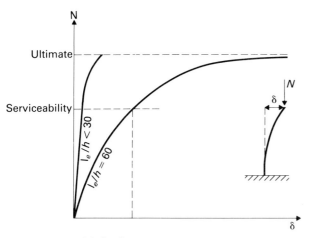

**Fig. 9.1** Lateral deflection

Reference is made in Table 9.1 to braced and unbraced columns: the Code states that a column is braced, in a particular plane, if lateral stability to the *structure as a whole* is provided in that plane. This can be achieved by designing bracing or bearings to resist all lateral forces.

### Slenderness limits

A column is considered to be short, and thus the effects of its lateral deflection can be ignored, if the slenderness ratios appropriate to each principal axis are both less than 12. The slenderness ratio appropriate to a particular axis is defined as the effective height in respect to that axis divided by the overall depth in respect to that axis. The overall depth should be used irrespective of the cross-sectional shape of the column.

If the slenderness ratios are not less than 12, the column is defined as slender and lateral deflection has to be considered by using the additional moment concept which is explained later. The limiting slenderness ratio is taken as 12 because work carried out by Cranston [223] indicated that buckling is rarely a significant design consideration for slenderness ratios less than 12. This work formed the basis of the CP 110 clauses for slender columns.

It is possible for slender columns to buckle by combined lateral bending and twisting: Marshall [224] reviewed all of the available relevant test data and concluded that lateral torsional buckling will not influence collapse provided that, for simply supported ends,

$$l_o \leqslant 500 \, b^2/h$$

where $h$ is the depth in the plane under consideration and $b$ is the width. Cranston [223] suggested that this limit should be reduced, for design purposes, to

$$l_o \leqslant 250 \, b^2/h \qquad (9.1)$$

Cranston also suggested the following limit for columns for which one end is not restrained against twisting, and this limit has been adopted in the Code for cantilever columns.

$$l_o \leqslant 100 \, b^2/h \qquad (9.2)$$

In addition, the Code requires that $l_o$ should not exceed 60 times the minimum column thickness. This limit is stipulated because Cranston's study did not include col-

umns having a greater slenderness ratio. This limit on $l_o$ is, generally, more onerous than that implied by equation (9.1) and, thus, the latter equation does not appear in the Code.

It should be noted that, for unbraced columns (which frequently occur in bridges), excessive lateral deflections can occur at the serviceability limit state for large slenderness ratios. This is illustrated in Fig. 9.1. An analysis, which allows for lateral deflections, is not required at the serviceability limit state and, consequently, CP 110 suggests a slenderness ratio limit of 30 for unbraced columns. This limit does not appear in the Code, but it would seem prudent to apply a similar limit to bridge columns, unless it is intended to consider, by a non-linear analysis, lateral deflections at the serviceability limit state.

In the above discussion of slenderness limits, it is implicitly assumed that the column has a constant cross-section throughout its length. However, many columns used for bridges are tapered: for such columns, data given by Timoshenko [225] indicate that, generally, it is conservative to calculate the slenderness ratio using the average depth of column.

## Ultimate limit state

### Short column

*Axial load* Since lateral deflections can be neglected in a short column, collapse of an axially loaded column occurs when all of the material attains the ultimate concrete compressive strain of 0.0035 (see Chapter 4). The design stress–strain curve for the ultimate limit state (see Fig. 4.4) indicates that, at a strain of 0.0035, the compressive stress in the concrete is $0.45 \, f_{cu}$. Similarly Fig. 4.4 indicates that, at this strain, the compressive stress in the reinforcement is the design stress which can vary from $0.718 \, f_y$ to $0.784 \, f_y$ (see Chapter 4). The Code adopts an average value of $0.75 \, f_y$ for the steel stress. It is not clear why the Code adopts the average value for columns, but a minimum value of $0.72 \, f_y$ for beams (see Chapter 5).

If the areas of concrete and steel are $A_c$ and $A_{sc}$ respectively, then the axial strength of the column is

$$N = 0.45 \, f_{cu}A_c + 0.75 \, f_y A_{sc} \qquad (9.3)$$

The Code recognises the fact that some eccentricity of load will occur in practice and, thus, the Code requires a minimum eccentricity of 5% of the section depth to be adopted. It is not necessary to calculate the moment due to this eccentricity, since allowance for it is made by reducing the ultimate strength, obtained from equation (9.3), by about 10%. This reduction leads to the Code formula for an axially loaded column:

$$N = 0.4 \, f_{cu}A_c + 0.67 \, f_y A_{sc} \qquad (9.4)$$

*Axial load plus uniaxial bending* When a bending moment is present, three possible methods of design are given in the Code:

1. For symmetrically reinforced rectangular or circular columns, the design charts of Parts 2 and 3 of CP 110

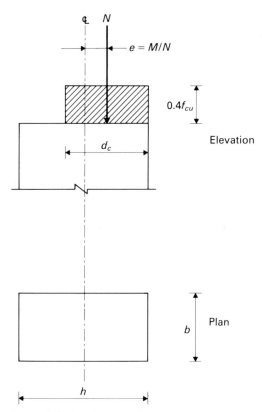

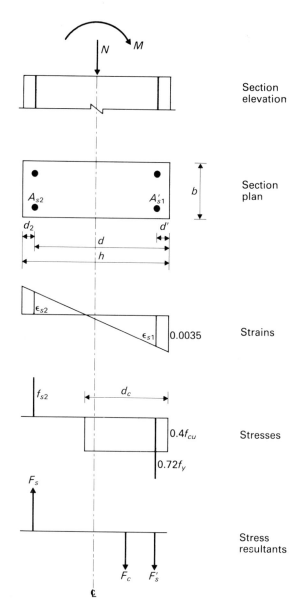

**Fig. 9.2** Plain column section

**Fig. 9.3(a),(b)** Reinforced column section

[128, 130] may be used. These charts were prepared using the rectangular – parabolic stress–strain curve for concrete and the tri-linear stress–strain curve for reinforcement discussed in Chapter 4. Allen [203] gives useful advice on the use of the design charts.

2. A strain compatibility approach (see Chapter 5) can be adopted for any cross-section. An area of reinforcement is first proposed and then the neutral axis depth is guessed. Since the extreme fibre compressive strain is 0.0035, the strains at all levels are then defined. Hence, the stresses in the various steel layers can be determined from the stress–strain curve. The axial load and bending moment that can be resisted by the column can then be determined. These values can be compared with the design values and, if deficient, the area of reinforcement and/or neutral axis depth modified. The procedure is obviously tedious and is best performed by computer.

3. The Code gives formulae for the design of *rectangular* columns only. The formulae, which are described in the next section, require a 'trial and error' design method which can be tedious. An example of their use is given by Allen [203]. Although the Code formulae are for rectangular sections only, similar formulae could be derived for other cross-sections.

In conclusion, it can be seen that a computer, or a set of design charts, is required for the efficient design of columns subjected to an axial load and a bending moment.

*Code formulae* Consider an *unreinforced* section at collapse under the action of an axial load ($N$) and a bending moment ($M$). If the depth of concrete in compression is $d_c$, as shown in Fig. 9.2, then by using the simplified rectan-

gular stress block for concrete with a constant stress of $0.4f_{cu}$:

Eccentricity $(e) = M/N = (h/2) - (d_c/2)$
$\therefore d_c = h - 2e$           (9.5)
For equilibrium
$N = 0.4 f_{cu} b d_c = 0.4 f_{cu} b (h - 2e)$    (9.6)

Hence, only nominal reinforcement is required if the axial load does not exceed the value of $N$ given by equation (9.6). However, it should be noted from this equation that, when $e > h/2$, $N$ is negative. Hence, equation (9.6) should not be used for $e > h/2$; however, the Code specifies the more conservative limit of $e = h/2 - d'$, where $d'$ is the depth from the surface to the reinforcement in the more highly compressed face.

When $N$ exceeds the value given by equation (9.6), it is necessary to design reinforcement. At failure of a reinforced concrete column, the strains, stresses and stress resultants are as shown in Fig. 9.3. It should be noted that the Code now takes the design stress of yielding compression reinforcement to be its conservative value of $0.72 f_y$.

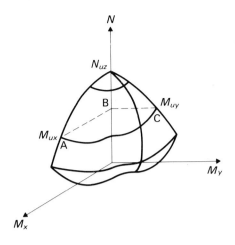

(a) Biaxial interaction diagram

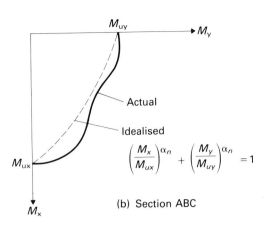

$$\left(\frac{M_x}{M_{ux}}\right)^{\alpha_n} + \left(\frac{M_y}{M_{uy}}\right)^{\alpha_n} = 1$$

(b) Section ABC

**Fig. 9.4** Biaxial bending interaction diagram

For equilibrium

$$N = 0.4 f_{cu}bd_c + 0.72 f_y A'_{s1} + f_{s2}A_{s2} \qquad (9.7)$$

and, by taking moments about the column centre line,

$$M = 0.2 f_{cu}bd_c (h - d_c) + 0.72 f_y A'_{s1} (h/2 - d') \\ - f_{s2}A_{s2} (h/2 - d_2) \qquad (9.8)$$

These equations are difficult to apply because the depth $(d_c)$ of concrete in compression and the stress $(f_{s2})$ in the reinforcement in the tension (or less highly compressed) face are unknown. The design procedure is thus to assume values of $d_c$ and $f_{s2}$, then calculate $A'_{s1}$ and $A_{s2}$ from equation (9.7), and check that the value of $M$ calculated from equation (9.8) is not less than the actual design moment. If $M$ is less than the actual design moment, the assumed values of $d_c$ and $f_{s2}$ should be modified and the procedure repeated. Guidance on applying this procedure is given by Allen [203]. However, the Code does not allow $d_c$ to be taken to be less than $2d'$. From Fig. 9.3 it can be shown that, at this limit, the strain in the more highly compressed reinforcement is 0.00175, which is less than the yield strain of 0.002 (see Chapter 4). However, serious errors in the required quantities of reinforcement should not arise by assuming that the stress is always 0.72 $f_y$, as in Fig. 9.3.

When the eccentricity is large $(e > h/2 - d_2)$ and thus the reinforcement in one face is in tension, the Code permits a simplified design method to be used in which the axial load is ignored at first and the section designed as a beam. The required design moment is obtained by taking moments about the tension reinforcement. Thus, from Fig. 9.3.

$$M + N(d - h/2) = F_c (d - d_c/2) + F'_s (d - d') \qquad (9.9)$$

The right-hand side of this equation is the ultimate moment of resistance $(M_u)$ of the section when considered as a beam. Hence, the section can be designed, as a beam, to resist the increased moment $(M_a)$ given by the left-hand side of equation (9.9), i.e.

$$M_a = M + N(d - h/2) \qquad (9.10)$$

Now, $d - h/2 = h/2 - d_2$, and the Code gives equation (9.10) in the form

$$M_a = M + N(h/2 - d_2) \qquad (9.11)$$

If force equilibrium is considered,
$$N = F_c + F'_s - F_s$$
or
$$F_s = (F_c + F'_s) - N \qquad (9.12)$$
In the absence of the axial force $N$,
$$F_s = F_c + F'_s = F_b$$
where $F_b$ is the tension steel force required for the section considered as a beam. Hence, from equation (9.12), when the axial force is present

$$F_s = F_b - N$$

The area of tension reinforcement is obtained by dividing $F_s$ by the tensile design stress of 0.87 $f_y$; hence

$$A_s = A_b - N/0.87 f_y \qquad (9.13)$$

where $A_s$ and $A_b$ are the required areas of tension reinforcement for the column section and a beam respectively.

Hence, the section can be designed by, first, designing it as a beam to resist the moment $M_a$ from equation (9.11) and then reducing the area of tension reinforcement by $(N/0.87 f_y)$.

*Axial load plus biaxial bending*  If a column of known dimensions and steel area is analysed rigorously, it is possible to construct an interaction diagram which relates failure values of axial load $(N)$ and moments $(M_x, M_y)$, about the major and minor axes respectively. Such a diagram is shown in Fig. 9.4(a), where $N_{uz}$ represents the full axial load carrying capacity given by equation (9.3). A section, parallel to the $M_x M_y$ plane, through the diagram for a particular value of $N/N_{uz}$ would have the shape shown by Fig. 9.4(b), where, assuming an axial load $N$, $M_{ux}$ and $M_{uy}$ are the maximum moment capacities for bending about the major and minor axes respectively.

The shape of the diagram in Fig. 9.4(b) varies according to the value of $N/N_{uz}$ but can be represented approximately by

$$(M_x/M_{ux})^{\alpha_n} + (M_y/M_{uy})^{\alpha_n} = 1 \qquad (9.14)$$

$\alpha_n$ is a function of $N/N_{uz}$. Appropriate values are tabulated in the Code.

When designing a column subjected to biaxial bending, it is first necessary to assume a reinforcement area. Values

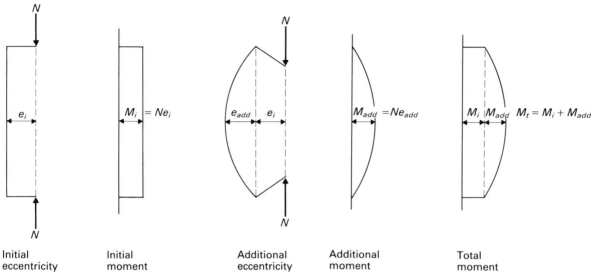

Initial eccentricity    Initial moment    Additional eccentricity    Additional moment    Total moment

**Fig. 9.5** Additional moment

of $N_{uz}$, $M_{ux}$ and $M_{uy}$ can then be calculated from first principles or obtained from the design charts for uniaxial bending. It is then necessary to check that the left-hand side of equation (9.14) does not exceed unity.

### Slender columns

*General approach* When an eccentric load is applied to any column, lateral deflections occur. These deflections are small for a short column and can be ignored, but they can be significant in the design of slender columns. The deflections and their effects are illustrated in Fig. 9.5, where it can be seen that the lateral deflections increase the eccentricity of the load and thus produce a moment ($M_{add}$) which is *additional* to the primary (or initial) moment ($M_i$). Hence, the total design moment ($M_t$) is given by

$$M_t = M_i + M_{add} \tag{9.15}$$
where
$$M_i = Ne_i \tag{9.16}$$
$$M_{add} = Ne_{add} \tag{9.17}$$

$e_i$ and $e_{add}$ are the initial and additional eccentricities respectively.

Since the section design is carried out at the ultimate limit state, it is necessary to assess the additional eccentricity at collapse. The additional eccentricity is the lateral deflection, and the latter can be determined if the distribution of curvature along the length of the column can be calculated. The distribution of curvature for a column subjected to an axial load and end moments is shown in Fig. 9.6, where $\psi_u$ is the maximum curvature at the centre of the column at collapse. The actual distribution of curvature depends upon the column cross-section, the extent of cracking and of plasticity in the concrete and reinforcement. However, it can be seen, from Fig. 9.6, that it is unconservative to assume a triangular distribution, and conservative to assume a rectangular distribution. For these two distributions the central deflections are given by, respectively:

$$e_{add} = l_e^2 \psi_u / 12$$
$$e_{add} = l_e^2 \psi_u / 8$$

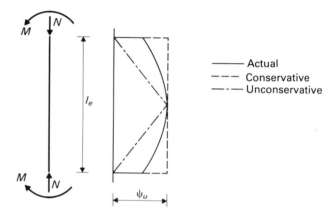

        ——— Actual
        – – – Conservative
        –·–·– Unconservative

**Fig. 9.6** Curvature distributions

Thus, Cranston [223] suggested that, for design purposes, a reasonable value to adopt would be:

$$e_{add} = l_e^2 \psi_u / 10 \tag{9.18}$$

$\psi_u$ can be determined if the strain distribution at collapse can be assessed. It is thus necessary to consider the mode of collapse. Unless the slenderness ratio is large, it is unlikely that a reinforced concrete column will fail due to instability, prior to material failure taking place [223]. Hence, instability is ignored initially and, for a balanced section in which the concrete crushes and the tension steel yields simultaneously, the strain distribution is as shown in Fig. 9.7.

The additional moment concept used in the Code is based upon that of the C.E.B. [226] in which the short-term concrete crushing strain ($\varepsilon_u$) is taken as 0.0030. In order to allow for long-term effects under service conditions, the latter strain has to be multiplied by a creep factor which Cranston [223] suggests should be conservatively taken as 1.25. Hence, $\varepsilon_u = 0.00375$. The strain ($\varepsilon_s$) in the reinforcement is that appropriate to the design stress at the ultimate limit state. Since the characteristic strength of the reinforcement is unlikely to exceed 460 N/mm², $\varepsilon_s$ can be conservatively taken as $0.87 \times 460/200 \times 10^3 = 0.002$. Hence, the curvature is given by

$$\psi_u = (0.00375 + 0.002)/d = 0.00575/d \tag{9.19}$$

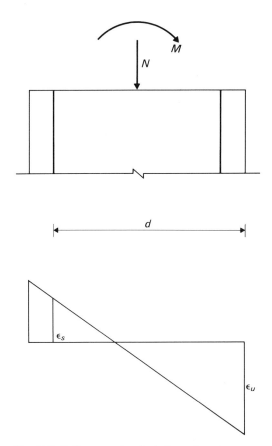

Fig. 9.7 Collapse strains for balanced section

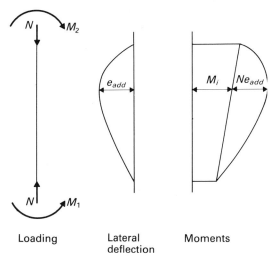

Fig. 9.8 Effect of unequal end moments

It is now necessary to consider the possibility of an instability failure as opposed to a material failure. In such a situation, the strains are less than their ultimate values and, hence, the curvature is less than that given by equation (9.19). The C.E.B. Code [226] allows for this by reducing the curvature obtained from equation (9.19) by the following empirical amount

$l_e/50\ 000h^2$

It also assumed that $d \simeq h$, so that the curvature is obtained finally as

$$\psi_u = (0.00575 - l_e/50\ 000h)/h \tag{9.20}$$

If this curvature is substituted into equation (9.18), then the following expression for the lateral deflection (or additional eccentricity) is obtained

$$e_{add} = (h/1750)(l_e/h)^2\ (1 - 0.0035\ l_e/h) \tag{9.21}$$

It should be noted that for a slenderness ratio of 12 (i.e. just slender), $e_{add} = 0.08h$, but for the maximum permitted slenderness ratio of 60, $e_{add} = 1.63h$. Hence, for very slender columns, the additional eccentricity and, hence, the additional moment can be very significant in design terms.

*Minor axis bending*    If $h$ is taken as the depth with respect to minor axis bending, then the additional eccentricity is given by equation (9.21). Hence, the total design moment, which is obtained from equations (9.15), (9.17) and (9.21), is given in the Code as

$$M_t = M_i + (Nh/1750)\ (l_e/h)^2\ (1 - 0.0035\ l_e/h) \tag{9.22}$$

$l_e$ is taken as the greater of the effective heights with respect to the major and minor axes. It should be noted that $M_i$ should not be taken to be less than $0.05\ Nh$, in order to allow for the nominal minimum initial eccentricity of $0.05h$.

The column should then be designed, by any one of the methods discussed earlier for short columns, to resist the axial load $N$ and the total moment $M_t$.

In Fig. 9.5, the maximum additional moment occurs at the same location as the maximum initial moment. If these maxima do not coincide, equation (9.15) is obviously conservative. Such a situation occurs when the moments at the ends of the column are different, as shown in Fig. 9.8.

To be precise, one should determine the position where the maximum total moment occurs and then calculate the latter moment. However, in order to simplify the calculation for a *braced* column, Cranston [223] has suggested that the initial moment, where the total moment is a maximum, may be taken as

$$M_i = 0.4\ M_1 + 0.6\ M_2 \tag{9.23}$$

but

$$M_i \not< 0.4\ M_2$$

where $M_1$ and $M_2$ are the smaller and larger of the initial end moments respectively. For a column bent in double curvature, $M_1$ is taken to be negative.

It is possible for the resulting total moment $(M_t)$ to be less than $M_2$. In such a situation, it is obviously necessary to design to resist $M_2$ and thus $M_t$ should never be taken to be less than $M_2$.

For an *unbraced* column, the Code requires the total moment to be taken as the sum of the additional moment and the maximum initial moment. This can be very conservative for certain bridge columns which are effectively fixed at both the base and the top, but which can sway under lateral load or imposed deformations (e.g. temperature movement).

*Major axis bending*    A column which is loaded eccentrically with respect to its major axis can fail due to large additional moments developing about the minor axis. This is because the slenderness ratio with respect to the minor axis is greater than that with respect to the major axis.

123

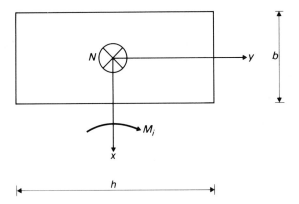

**Fig. 9.9** Major axis bending

Hence, with reference to Fig. 9.9 the column should be designed for biaxial bending to resist the following moments

$$M_{tx} = M_i + (Nh/1750)(l_{ex}/h)^2(1 - 0.0035\,l_{ex}/h) \qquad (9.24)$$

$$M_{ty} = (Nb/1750)(l_{ey}/b)^2(1 - 0.0035\,l_{ey}/b) \qquad (9.25)$$

where $M_{tx}$ and $M_{ty}$ are the total moments about the major $(x)$ axis and minor $(y)$ axis respectively, and $l_{ex}$ and $l_{ey}$ are the effective heights with respect to these axes.

Cranston [223] has shown that, for a braced column with $h > 3b$, it is conservative to design the column solely for bending about the major axis, but the slenderness ratio should then be calculated with respect to the minor axis. Hence, in such situations, the Code permits the column to be designed to resist the axial load $N$ and the following total moment about the major axis.

$$M_t = M_i + (Nh/1750)(l_e/b)^2(1 - 0.0035\,l_e/b) \qquad (9.26)$$

where $l_e$ is the greater of $l_{ex}$ and $l_{ey}$.

*Biaxial bending*  When subjected to biaxial bending, a column should be designed to resist the axial load $N$ and moments $(M_x = M_{tx}, M_y = M_{ty})$ such that equation (9.14) is satisfied. The total moments about the major and minor axes respectively are

$$M_{tx} = M_{ix} + (Nh/1750)(l_{ex}/h)^2(1 - 0.0035\,l_{ex}/h) \qquad (9.27)$$

$$M_{ty} = M_{iy} + (Nb/1750)(l_{ey}/b)^2(1 - 0.0035\,l_{ey}/b) \qquad (9.28)$$

where $M_{ix}$ and $M_{iy}$ are the initial moments with respect to the major and minor axes respectively.

## Serviceability limit state

### General

The design of columns in accordance with the Code is complicated by the fact that it is necessary to check stresses and crack widths at the serviceability limit state in addition to carrying out strength calculations at the ultimate limit state.

### Stresses

At the serviceability limit state, the compressive stress in the concrete has to be limited to $0.5\,f_{cu}$ and the reinforcement stresses to $0.8\,f_y$. Thus, for an axially loaded col-

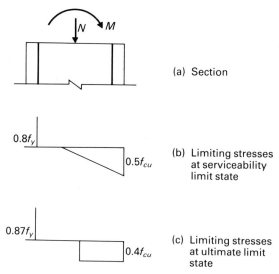

**Fig. 9.10(a)–(c)** Stress comparison

umn, the design resistance at the serviceability limit state is, usually,

$$N_s = 0.5\,f_{cu}A_c + (0.5\,f_{cu}E_s/E_c)A_{sc}$$

This value generally exceeds the design resistance at the ultimate limit state, as given by equation (9.4). Since the design load at the ultimate limit state exceeds that at the serviceability limit state, it can be seen that ultimate will generally be the critical limit state when the loading is predominantly axial.

When the loading is eccentric to the extent that one face is in tension, the stress conditions at the ultimate and serviceability limit states will be as shown in Fig. 9.10. Since the average concrete stress at the serviceability limit state $(0.25\,f_{cu})$ is much less than that at the ultimate limit state $(0.4\,f_{cu})$, it is likely that, with regard to concrete stress, the serviceability limit state will be critical.

### Crack widths

The Code considers that if a column is designed for an ultimate axial load in excess of $0.2\,f_{cu}A_c$, it is unlikely that flexural cracks will occur. For smaller axial loads, it is necessary to check crack widths by considering the column to be a beam and by applying equation (7.4). From Table 4.7, it can be seen that, since a column could be subjected to salt spray, the allowable design crack width could be as small as 0.1 mm. Hence, equation (7.4) implies a maximum steel strain of about $1000 \times 10^{-6}$, or a stress of about 200 N/mm². For high yield steel, this stress is equivalent to about $0.48\,f_y$. Hence, when crack control is considered, the reinforcement stress in Fig. 9.10(b) is limited to much less than $0.8\,f_y$, and thus crack control could be the critical design criterion for columns with a large eccentricity of load.

## Summary

Ultimate is likely to be the critical limit state for a column which is either axially loaded or has a small eccentricity of load. However, due to the fact that either the limiting compressive stress or crack width could be the critical design criterion, serviceability is likely to be the critical limit state for a column with a large eccentricity of load. It appears that, in order to simplify design, studies should be carried out with a view to establishing guidelines for identifying the critical limit state in a particular situation.

# Reinforced concrete walls

## General

### Definitions

Retaining walls, wing walls and similar structures which, primarily, are subjected to bending should be considered as slabs and designed in accordance with the methods of Chapters 5 and 7. The following discussion is concerned with walls subjected to significant axial loads.

In terms of the Code, a reinforced concrete wall is a vertical load-bearing member with an aspect ratio, on plan, greater than 4; the reinforcement is assumed to contribute to the strength, and has an area of at least 0.4% of the cross-sectional area of the wall. This definition thus covers reinforced concrete abutments. The limiting value of 0.4% is greater than that specified in CP 114 because tests have shown that the presence of reinforcement in walls reduces the in-situ strength of the concrete [227]. Hence, under axial loading, a plain concrete wall can be *stronger* than a wall with a small percentage of reinforcement.

### Slenderness

The slenderness ratio is the ratio of the effective height to the thickness of the wall.

A short wall has a slenderness ratio less than 12. Walls with greater slenderness ratios are considered to be slender.

In general, the slenderness ratio of a braced wall should not exceed 40, but, if the area of reinforcement exceeds 1%, the slenderness ratio limit may be increased to 45. These values are more severe than those for columns because walls are thinner than columns, and thus deflections are more likely to lead to problems. If lateral stability is not provided to the structure as a whole, then a wall is considered to be unbraced and its slenderness ratio should not exceed 30. This rule ensures that deflections will not be excessive.

The above slenderness limits were obtained from CP 110 and were thus derived with shear walls and in-fill panels in framed structures in mind. They are thus not necessarily applicable to the types of wall which are used in bridge construction. However, the slenderness ratios should not result in any further design restrictions compared with existing practice.

## Analysis

The Code requires that forces and moments in reinforced concrete walls should be determined by elastic analysis.

When considering bending perpendicular to an axis in the plane of a wall, a nominal minimum eccentricity of $0.05h$ should be assumed. Thus a wall should be designed for a moment per unit length of at least $0.05\,n_w h$ where $n_w$ is the maximum load per unit length.

## Ultimate limit state

### Short walls

*Axial load*  An axially loaded wall should be designed in accordance with equation (9.4).

*Eccentric loads*  If the load is eccentric such that it produces bending about an axis in the plane of a wall, the wall should be designed on a unit length basis to resist the combined effects of the axial load per unit length and the bending moment per unit length. The design could be carried out either by considering the section of wall as an eccentrically loaded column of unit width or by using the 'sandwich' approach, described in Chapter 5, for designing against combined bending and in-plane forces.

If the load is also eccentric in the plane of the wall, an elastic analysis should be carried out, in the plane of the wall, to determine the distribution of the in-plane forces per unit length of the wall. The Code states that this analysis may be carried out assuming no tension in the concrete. In fact, any distribution of tension and compression, which is in equilibrium with the applied loads could be adopted at the ultimate limit state since, as explained in Chapter 2, a safe lower bound design would result.

Each section along the length of the wall should then be designed to resist the combined effects of the moment per unit length at right angles to the wall and the compression, or tension, per unit length of the wall. The design could be carried out by considering each section of the wall as an eccentrically loaded column or tension member of unit width, or by using the 'sandwich' approach.

### Slender walls

The forces and moments acting on a slender wall should be determined by the same methods previously described for short walls. The portion of wall, subject to the highest intensity of axial load, should then be designed as a slender column of unit width.

## Serviceability limit state

### Stresses

The comments made previously regarding stress calculations for columns are also appropriate to walls.

### Crack widths

Walls should be considered as slabs for the purposes of crack control calculations, and the details of the Code requirements are discussed in Chapter 7.

# Plain concrete walls

## General

A plain concrete wall or abutment is defined as a vertical load-bearing member with an aspect ratio, on plan, greater than 4; any reinforcement is *not* assumed to contribute to the strength.

If the aspect ratio is less than 4, the member should be considered as a plain concrete column. The following design rules for walls can also be applied to columns, but, as indicated later, certain design stresses need modification.

The definitions of 'short', 'slender', 'braced' and 'unbraced', which are given earlier in this chapter for reinforced concrete walls, are also applicable to plain concrete walls.

The clauses, concerned with slenderness and lateral support of plain walls, were taken directly from CP 110 which in turn were based upon those in CP 111 [228]. In order to preclude failure by buckling the slenderness ratio of a plain wall should not exceed 30 [229]. The effective heights given in the Code are summarised in Table 9.2.

**Table 9.2** Effective heights of plain walls

| Wall type | $l_e/l_o$ |
|-----------|-----------|
| Unbraced, laterally spanning structure at top | 1.5 |
| Unbraced, no laterally spanning structure at top | 2.0 |
| Braced against lateral movement and rotation | 0.75* or 2.0† |
| Braced against lateral movement only | 1.0* or 2.5† |

\* $l_o$ = distance between centres of support
† $l_o$ = distance between a support and a free edge

In order to be effective, a lateral support to a braced wall must be capable of transmitting to the structural elements, which provide lateral stability to the structure as a whole, the following forces:

1. The static reactions to the applied horizontal forces.
2. 2.5% of the total ultimate vertical load that the wall has to carry.

A lateral support could be a horizontal member (e.g., a deck) or a vertical member (e.g., other walls), and may be considered to provide rotational restraint if one of the following is satisfied:

1. The lateral support and the wall are detailed to provide bending restraint.
2. A deck has a bearing width of at least two-thirds of the wall thickness, or a deck is connected to the wall by means of a bearing which does not allow rotation to occur.
3. The wall supports, at the same level, a deck on each side of the wall.

## Forces

Members, which transmit load to a plain wall, may be considered simply supported in order to calculate the reaction which they transmit to the wall.

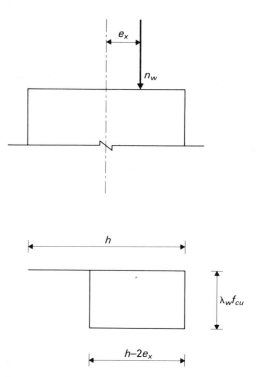

**Fig. 9.11** Eccentrically loaded short wall at collapse

If the load is eccentric in the plane of the wall, the eccentricity and distribution of load along the wall should be calculated from statics. When calculating the distribution of load (i.e., the axial load per unit length of wall), the concrete should be assumed to resist no tension.

If a number of walls resist a horizontal force in their plane, the distributions of load between the walls should be in proportion to their relative stiffnesses. The Code clause concerning horizontal loading refers to shear connection between walls and was originally written for CP 110 with shear walls in buildings in mind. However, the clause could be applied, for example, to connected semi-mass abutments.

When considering eccentricity at right angles to the plane of a wall, the Code states that the vertical load transmitted from a deck may be assumed to act at one-third the depth of the bearing area back from the loaded face. It appears from the CP 110 handbook [112] that this requirement was originally intended for floors or roofs of buildings bearing directly on a wall. However, the intention in the Code is, presumably, also to apply the requirement to decks which transmit load to a wall through a mechanical or rubber bearing.

## Ultimate limit state

### Axial load plus bending normal to wall

*Short braced wall* The effects of lateral deflections can be ignored in a short wall and thus failure is due solely to concrete crushing. The concrete is assumed to develop a constant compressive stress of $\lambda_w f_{cu}$ at collapse, where $\lambda_w$ is a coefficient to be discussed. The concrete stress distribution at collapse of an eccentrically loaded wall is as shown in Fig. 9.11. The centroid of the stress block must

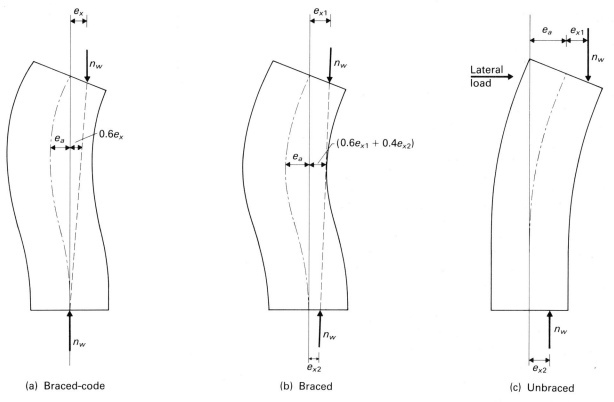

(a) Braced-code      (b) Braced      (c) Unbraced

**Fig. 9.12(a)–(c)** Lateral deflection of a slender wall

coincide with the line of action of the axial load per unit length of wall $(n_w)$, which is at an eccentricity of $e_x$. Hence the depth of concrete in compression is

$$2(h/2 - e_x) = h - 2e_x$$

Thus the maximum possible value of $n_w$ is given by

$$n_w = (h - 2e_x)\,\lambda_w f_{cu} \tag{9.29}$$

The coefficient $\lambda_w$ varies from 0.28 to 0.5. It is tabulated in the Code and depends upon the following:

1. Concrete strength. For concrete grades less than 25, lower values of $\lambda_w$ are adopted than for concrete grades 25 and above. This is because of the difficulty of controlling the quality of low grade concrete in a wall. Hence, essentially, a higher value of $\gamma_m$ is adopted for low grades than for high grades.

2. Ratio of clear height between supports to wall length. Tests reported by Seddon [229] have shown that the stress in a wall at failure increases as its height to length ratio decreases. This is because the base of the wall and the structural member(s) bearing on the wall restrain the wall against lengthwise expansion. Hence, a state of biaxial compression is induced in the wall which increases its apparent strength above its uniaxial value. The biaxial effect decreases with distance from the base or bearing member and, thus, the average stress, which can be developed in a wall, increases as the height to length ratio decreases. CP 111 permits an increase in allowable stress which varies linearly from 0%, at a height to length ratio of 1.5, to 20%, at a ratio of 0.5 or less. Similar increases have been adopted in the Code.

3. Ratio of wall length to thickness. It is not clear why the Code requires $\lambda_w$ to be reduced, when the ratio of wall length to thickness is less than 4 (i.e., when the wall becomes a column). The reduction coefficient varies linearly from 1.0 to 0.8 as the length to thickness ratio reduces from 4 to 1. The reason could be to ensure that the value of $\lambda_w$ does not exceed 0.4 when the aspect ratio is 1, because 0.4 is the value adopted for reinforced concrete columns and beams.

*Slender braced wall* At the base of a wall, the eccentricity of loading is assumed to be zero. Thus the eccentricity varies linearly from zero at the base to $e_x$ at the top. A slender wall deflects laterally under load in the same manner as a slender column. The lateral deflection increases the eccentricity of the load and the Code takes the net maximum eccentricity to be $(0.6 e_x + e_a)$, as shown in Fig. 9.12(a). The additional eccentricity $(e_a)$ is taken, empirically, to be $l_e^2/2500h$, where $l_e$ and $h$ are the effective height and thickness of the wall respectively. It should be noted that the Code mistakenly gives the additional eccentricity as $l_e/2500h$. If $e_x$ in equation (9.29) is replaced by $(0.6 e_x + e_a)$, the following equation is obtained for the ultimate strength of a slender braced wall:

$$n_w = (h - 1.2 e_x - 2 e_a)\,\lambda_w f_{cu} \tag{9.30}$$

The above assumption of zero eccentricity at the base of a braced wall is based upon considerations of walls in buildings [112]. In the case, for example, of an abutment an eccentricity could exist at the bottom of the wall as shown in Fig. 9.12(b). If the eccentricities at the top and bottom are $e_{x1}$ and $e_{x2}$ respectively, the author would sug-

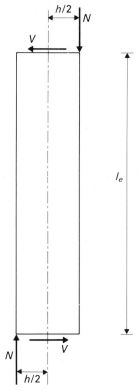

**Fig. 9.13** Shear normal to wall

gest that, by analogy with equation (9.23), the maximum net eccentricity should be taken as the greater of

$(0.4 \, e_{x1} + 0.6 \, e_{x2} + e_a)$
and
$(0.6 \, e_{x1} + 0.4 \, e_{x2} + e_a)$

The appropriate net eccentricity should then be substituted for $e_x$ in equation 9.29.

*Slender unbraced wall* The lateral deflection of a slender unbraced wall is shown in Fig. 9.12(c). The net eccentricities, from the wall centre line, at the top and bottom of the wall are $e_{x1}$ and $(e_{x2} + e_a)$ respectively. The Code requires every section of the wall to be capable of resisting the load at each of these eccentricities. Hence, by replacing $e_x$ in equation (9.29) by each of these eccentricities, the ultimate strength of a slender unbraced wall is the lesser of:

$$n_w = (h - 2e_{x1}) \, \lambda_w f_{cu} \qquad (9.31)$$
and
$$n_w = (h - 2e_{x2} - 2e_a) \, \lambda_w f_{cu} \qquad (9.32)$$

### Shear

In general the total shear force in a horizontal plane should not exceed one-quarter of the associated vertical load. The reason for this requirement is not clear but, since the requirement was taken from CP 110, it was intended presumably for shear walls bearing on a footing or a floor. Thus it appears that the design criterion was taken to be shear friction with a coefficient of friction of 0.25.

A shear force at right angles to a wall arises from a change in bending moment down the wall. The maximum moments at the ends of a wall occur when the load is at its greatest eccentricity of $h/2$. The maximum change of moment over the length of the wall occurs when the eccentricities at each end are of opposite sign, as shown in Fig. 9.13, and is given by $N(h/2 + h/2) = Nh$. Hence, the constant shear force throughout the length of the wall is

$$V = Nh/l_e$$

In order that $V$ does not exceed $0.25N$, it is necessary that $l_e/h$ should exceed 4. In fact, the Code states that it is not necessary to consider shear forces normal to the wall if $l_e/h$ exceeds 6. The Code is thus conservative in this respect.

When considering shear forces in the plane of the wall, it is necessary to check that the total shear force does not exceed 0.25 of the associated total vertical load, and that the average shear stress does not exceed 0.45 N/mm$^2$ for concrete of grade 25 or above, or 0.3 N/mm$^2$ for lower grades of concrete. The reason for assigning these allowable stresses is not apparent.

### Bearing

The bearing stress under a localised load should not exceed the limiting value given by equation (8.4).

## Serviceability limit state

### Deflection

The Code states that excessive deflections will not occur in a cantilever wall if its height-to-length ratio does not exceed 10. The basis of this criterion is not apparent, but the CP 110 handbook [112] adds that the ratio can be increased to 15 if tension does not develop in the wall under lateral loading.

### Crack control

It is necessary to control cracking due to both applied loading (flexural cracks) and the effects of shrinkage and temperature.

*Flexural cracking* Reinforcement, specifically to control flexural cracking, only has to be provided when tension occurs over at least 10% of the length of a wall, when subjected to bending in the plane of the wall. In such situations, at least 0.25% of high yield steel or 0.3% of mild steel should be provided in the area of wall in tension: the spacing should not exceed 300 mm. These percentages are identical to those discussed in the next section when considering the control of cracking due to shrinkage and temperature effects. The spacing of 300 mm is in accordance with the maximum spacing discussed in Chapter 7.

*Shrinkage and temperature effects* In order to control cracking due to the restraint of shrinkage and temperature movements, at least 0.25% of high yield steel or 0.3% of mild steel should be provided both horizontally and vertically. These percentages are identical to those for water-retaining structures in CP 2007 [230], but it should be noted that they are much less than those given in the new standard for water retaining structures (BS 5337) [231], and are also much less than those suggested by Hughes

[185]. The author would thus suggest that the values of 0.25% and 0.3% should be used with caution.

# Bridge piers and columns

Earlier in this chapter, the Code clauses concerned with columns and reinforced walls are presented and brief mention made of their application. In the following discussion the design of bridge piers and columns in accordance with the Code is considered briefly and compared with present practice.

## Effective heights

The Code clauses concerning effective heights are intended, primarily, for buildings and are not necessarily applicable to bridge piers and columns. However, this criticism is equally applicable to the effective heights given in the existing design document (CP 114). Thus, there is no difference in the assessment of effective heights in accordance with the Code and with CP 114.

## Slender columns and piers

CP 114 defines a slender column as one with a slenderness ratio in excess of 15, whereas the Code critical slenderness ratio is 12. This means that some columns, which could be considered to be short at present, would have to be considered as slender when designed in accordance with the Code.

CP 114 allows for slenderness by applying a reduction factor to the calculated permissible load for a short column. The reduction factor is a function of the slenderness ratio. This approach is simple, but does not reflect the true behaviour of a slender column at collapse. Thus the reduction factor approach has not been adopted in the Code: instead, the additional moment concept, which is described earlier in this chapter, is used. Use of the latter concept requires more lengthy calculations, and thus the design of slender columns, in accordance with the Code, will take longer than their design in accordance with CP 114.

## Design procedure

In accordance with CP 114, only one calculation has to be undertaken – the permissible load has to be checked under working load conditions. However, in accordance with the Code, three calculations, each under a different load condition, have to be carried out. These calculations are concerned with strength at the ultimate limit state, stresses at the serviceability limit state and, if appropriate, crack width at the serviceability limit state. Hence, the design procedure will be much longer for a column designed in accordance with the Code.

A further problem arises when applying the Code: it is not clear in advance which of the three design calculations will be critical. However, it is likely that ultimate will be the critical limit state for a column, which is either axially loaded or is subjected to a relatively small moment. For columns subjected to a large moment, either the limiting concrete compressive stress at the serviceability limit state, or the limiting crack width at the serviceability limit state could be critical. If the latter criterion is critical then it may be necessary to specify columns with greater cross-sectional areas than are adopted at present. This is because a very large amount of reinforcement would be required to control the cracks. For a column size currently adopted, the required amount of reinforcement may exceed the maximum amount permitted by the Code. This possibility is increased by the fact (see Chapter 10) that the maximum amount of reinforcement permitted in a vertically cast column is 6% in the Code as compared with 8% in CP 114.

# Bridge abutments and wing walls

The design of abutments and wing walls in accordance with the Code is very different to their design to current practice. A major difference is the number of analyses which need to be carried out. At present a single analysis covers all aspects of design but, in accordance with the Code, five analyses, each under a different design load, have to be carried out for the following five design aspects:

1. Strength at the ultimate limit state.
2. Stresses at the serviceability limit state.
3. Crack widths at the serviceability limit state; but deemed to satisfy rules for bar spacing are appropriate in some situations (see Chapter 7).
4. Overturning. The Code requires the least restoring moment due to unfactored nominal loads to exceed the greatest overturning moment due to the design loads (given by the effects of the nominal loads multiplied by their appropriate $\gamma_{fL}$ values at the ultimate limit state).
5. Factor of safety against sliding and soil pressures due to unfactored nominal loads in accordance with CP 2004 [92].

A further important difference in design procedures occurs when considering the effects of applied deformations described in the Code and in the present documents. In the latter, all design aspects are considered under working load conditions, and thus the effects of applied deformations (creep, shrinkage and temperature) need to be considered for all aspects of design. However, as explained in Chapter 13, the effects of applied deformations can be ignored under collapse conditions. Thus Part 4 of the Code permits creep, shrinkage and temperature effects to be ignored at the ultimate limit state. The implication of this is that less main reinforcement would be required in an abutment designed to the Code than one designed to the existing documents.

Although the effects of applied deformations can be

ignored at the ultimate limit state, they have to be considered at the serviceability limit state. The effects of applied deformations thus contribute to the stresses at the serviceability limit state. Since less reinforcement would be present in an abutment designed to the Code than one designed to the existing documents, the stresses at the serviceability limit state would be greater in the former abutment. However, it is unlikely that they would exceed the Code limiting stresses of $0.8 f_y$ and $0.5 f_{cu}$ for reinforcement and concrete respectively.

The main bar spacings will generally be greater for abutments designed in accordance with the Code than for those designed in accordance with the present documents. This is because the Code maximum spacing of 150 mm (see Chapter 7) will generally be appropriate for abutments, whereas spacings of about 100 mm are often necessary at present. The Code should thus lead to less congestion of main reinforcement.

Finally, the Code does permit the use of plastic methods of analysis, and the design of abutments and wing walls is an area where plastic methods could usefully be applied. In particular, Lindsell [232] has tested a model abutment with cantilever wing walls, and has shown that yield line theory gives reasonable estimates of the loads at collapse of the abutment and of the wing walls. An alternative plastic method of design is the Hillerborg strip method, which is applied to an abutment in Example 9.2 at the end of this chapter.

# Foundations

## General

A foundation should be checked for sliding and soil bearing pressure in accordance with the principles of CP 2004 [92]. The latter document is written in terms of working stress design and thus unfactored nominal loads should be used when checking sliding and soil bearing pressure. However, when carrying out the structural design of a foundation, the design loads appropriate to the various limit states should be adopted. Hence, more calculations have to be carried out when designing foundations in accordance with the Code than for those designed in accordance with the current documents.

In the absence of a more accurate method, the Code permits the usual assumption of a linear distribution of bearing stress under a foundation.

## Footings

### Ultimate limit state

*Flexure* The critical section for bending is taken at the face of the column or wall as shown in Fig. 9.14(a). Reinforcement should be designed for the total moment at this critical section and, except for the reinforcement parallel to the shorter side of a rectangular footing, it should be

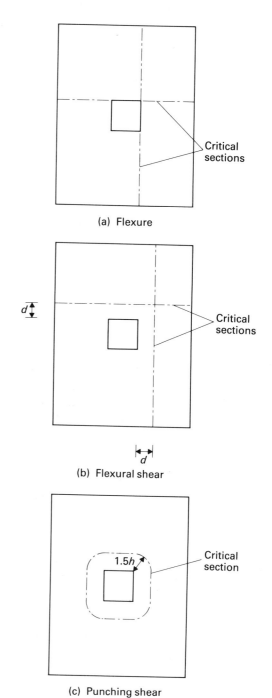

(a) Flexure

(b) Flexural shear

(c) Punching shear

**Fig. 9.14(a)–(c)** Critical sections for footing

spread uniformly across the base. Reinforcement parallel to the shorter side should be distributed as shown in Fig. 9.15. The latter requirement is empirical and was based upon a similar requirement in the ACI Code [168]. The Code is thus more precise than CP 114 with regard to the distribution of reinforcement.

*Flexural shear* The total shear on a section, at a distance equal to the effective depth from the face of the column or wall (see Fig. 9.14(b)), should be checked in accordance with the method given in Chapter 6 for flexural shear in beams. These requirements, when allowance is made for the different design loads, are very similar to those of BE 1/73.

It is worth mentioning that the Code clause is identical to that in CP 110, except that the latter document requires

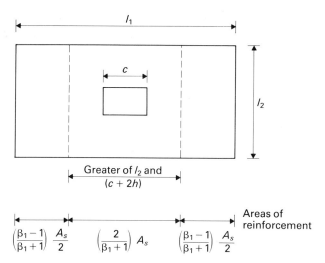

**Fig. 9.15** Distribution of reinforcement in rectangular footing

$h$ = overall slab depth

$A_s$ = total area of reinforcement parallel to shorter side

$\beta_1 = \dfrac{l_1}{\text{Central band width}}$

Areas of reinforcement:

$\left(\dfrac{\beta_1 - 1}{\beta_1 + 1}\right)\dfrac{A_s}{2}$  $\left(\dfrac{2}{\beta_1 + 1}\right)A_s$  $\left(\dfrac{\beta_1 - 1}{\beta_1 + 1}\right)\dfrac{A_s}{2}$

the critical section to be at 1½ times the effective depth from the face. This critical section was adopted in CP 110 because a critical section, at a distance equal to the effective depth from the loaded face, would have resulted in much deeper foundations than those previously required in accordance with CP 114.

*Punching shear*  The critical perimeter and design method discussed for slabs in Chapter 6 should be used for footings. The perimeter is shown in Fig. 9.14(c).

### Serviceability limit state

*Stresses*  It is necessary to restrict the stresses to the limiting values of $0.8f_y$ and $0.5f_{cu}$ in the reinforcement and concrete respectively.

*Crack widths*  As discussed in Chapter 7, footings should be treated as slabs when considering crack control.

## Piles

The Code does not give specific design rules for piles. However, once the forces acting on a pile have been assessed, the pile can be designed as a column in accordance with CP 2004 and the Code.

## Pile caps

### Ultimate limit state

The reinforcement in a pile cap may be designed either by bending theory or truss analogy. The shear strength then has to be checked.

*Bending theory*  When applying the bending theory [233], the pile cap is considered to act as a wide beam in each direction. The total bending moment at any section

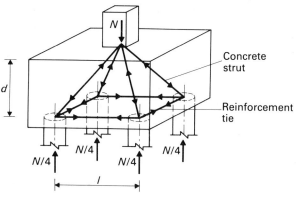

**Fig. 9.16** Truss analogy for pile cap

can be obtained at that section, and the total amount of reinforcement at the section determined from simple bending theory as described in Chapter 5. Such a design method is not correct because a pile cap acts as a deep, rather than a shallow, beam; however, the method has been shown by tests to result in adequate designs [234]. This is probably because most pile caps fail in shear and the method of design of the main reinforcement is, largely, irrelevant [234]. The total amount of reinforcement calculated at a section should be uniformly distributed across the section.

*Truss analogy*  The truss analogy assumes a strut and tie system within the cap, and is in the spirit of a lower bound method of design. The strut and tie system for a four-pile cap is shown in Fig. 9.16. Formulae for determining the forces in the ties for various arrangements of piles are given by Allen [203] and Yan [235]. It can be seen from Fig. 9.16 that, because of the assumed structural action, the reinforcement, calculated from the tie forces, should be concentrated in strips over the piles. However, since it is considered good practice to have some reinforcement throughout the cap, the Code requires 80% of the reinforcement to be concentrated in strips joining the piles and the remainder to be uniformly distributed throughout the cap.

Tests carried out by Clarke [234] have demonstrated the adequacy of the truss analogy.

*Flexural shear*  The Code requires flexural shear to be checked across the full width of a cap at a section at the face of the column, as shown in Fig. 9.17(a). It should be noted that the critical section is not intended to coincide with the actual failure plane, but is chosen merely because it is convenient for design purposes.

The question now arises as to what allowable design shear stress should be used in association with the above critical section. Tests carried out by Clarke [234] have indicated that the basic design stresses given in Table 6.1 should be used, except for those parts of the critical section which are crossed by flexural reinforcement which is fully anchored by passing over a pile. For the latter parts of the critical section, the basic design shear stresses should be enhanced to allow for the increased shear resistance due to the short shear span (see Chapter 6). The enhancement factor $(2d/a_v)$ where $d$ is the effective depth and $a_v$, is the shear span which, in the present context, is

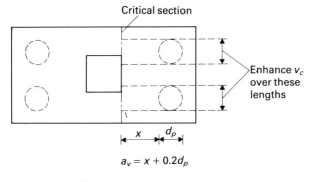

Critical section

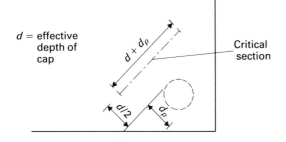

Enhance $v_c$ over these lengths

$a_v = x + 0.2d_p$

(a) Flexural shear

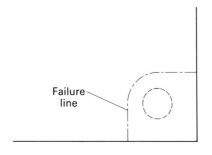

$d$ = effective depth of cap

Critical section

(b) Punching shear—Code

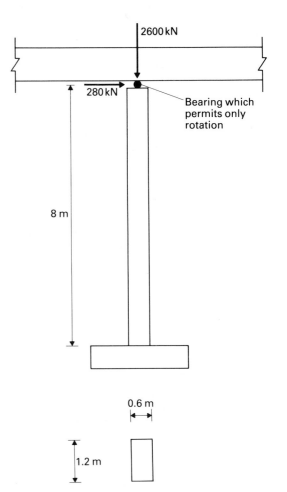

2600 kN

280 kN

Bearing which permits only rotation

8 m

0.6 m

1.2 m

**Fig. 9.18** Bridge column

Failure line

(c) Punching shear—actual

**Fig. 9.17(a)–(c)** Shear in pile caps

resulting shear force at the critical section will be only marginally different.

*Punching shear* Clarke [234] suggests that punching of the column through the cap need only be considered if the pile spacing exceeds four times the pile diameter, which is unlikely; thus the Code only requires punching of a pile through the cap to be considered.

The critical section given in the Code for punching of a corner pile is extremely difficult to interpret and originates from the 1970 CEB recommendations [226]. The relevant diagram in the latter document shows that the correct interpretation is as shown in Fig. 9.17(b). The Code does not state what value of allowable design shear stress should be used with the critical section. In view of this, the author would suggest using the value from Table 6.1 which is appropriate to the average of the two areas of reinforcement which pass over the pile. This suggestion is not based upon considerations of the Code section of Fig. 9.17(b) but of the section which would actually occur as shown in Fig. 9.17(c). The basic shear stress, obtained from Table 6.1 should then be enhanced by $(2d/a_v)$, where $a_v$ should be taken as the distance from the pile to the critical section (i.e. $d/2$). Thus, in all cases in which failure could occur along the Code critical section, the enhancement factor would be equal to 4.

It is understood that, in a proposed amendment to CP 110, punching shear is checked by limiting the shear stress calculated on the perimeter of the *column* to

taken as the distance between the face of the column and the nearer edge of the piles, viewed in elevation, plus 20% of the pile diameter. The Code states that the reason for adding 20% of the pile diameter is to allow for driving tolerances. However, Clarke [234] has suggested the same additional distance in order to allow for the fact that the piles are circular, rather than rectangular, and thus the 'average' shear span is somewhat greater than the clear distance between pile and critical section. The value of 20% of the pile diameter, chosen by Clarke [234], is similar to the absolute value of 150 mm suggested by Whittle and Beattie [233] to allow for dimensional errors. However, the result is that the allowable design stress varies along the critical section, as shown in Fig. 9.17(a), and the total shear capacity at the section should be obtained by summing the shear capacities of the component parts of the section.

It is understood that, in a proposed amendment to CP 110, the critical section for flexural shear is located at 20% of the diameter of the pile inside the face of the pile. Thus the critical section is at the distance $a_v$, defined in Fig 9.17(a), from the face of the column. This critical section is more logical than that defined in the Code, but the

$0.8 \sqrt{f_{cu}}$. This limiting shear stress is very similar to the maximum nominal shear stress of $0.75 \sqrt{f_{cu}}$ which is specified in the Code (see Chapter 6).

### Serviceability limit state

Realistic values of stresses and crack widths in a pile cap, at the serviceability limit state, could only be assessed by carrying out a proper analysis; such an analysis would probably need to be non-linear to allow for cracking. Since it is difficult to imagine serviceability problems arising in a cap which has been properly designed and detailed at the ultimate limit state, a sophisticated analysis at the serviceability limit state cannot be justified. Thus, the author would suggest ignoring the serviceability limit state criteria for pile caps.

# Examples

## 9.1 Slender column

A reinforced concrete column is shown in Fig. 9.18. The loads indicated are design loads at the ultimate limit state.

Design reinforcement for the column, at the ultimate limit state, if the characteristic strengths of the reinforcement and concrete are 425 N/mm² and 40 N/mm² respectively. Assume that the articulation of the deck is (a) such that sidesway is prevented and (b) such that sidesway can occur.

### No sidesway

With sidesway prevented, the column can be considered to be braced. Consider the column to be partially restrained in direction at both ends, and, from Table 9.1, take the effective height to be the same as the actual height, i.e. $l_e$ = 8 m.

Slenderness ratio = 8/0.6 = 13.3. This exceeds 12, thus the column is slender.

Assume a minimum eccentricity of $0.05h = 0.03$ m for the vertical load.

Initial moment at top of column = $M_1 = 0$
Initial moment at bottom of column = $M_2$
$\qquad = 280 \times 8 + 2600 \times 0.03 = 2318$ kNm

Since the column is braced, the initial moment to be added to the additional moment is, from equation (9.23),
$M_i = (0.4)(0) + (0.6)(2318) = 1391$ kNm
From equation (9.22), the total moment is
$M_t =$
$\qquad 1391 + (2600 \times 0.6/1750)(13.3)^2(1 - 0.0035 \times 13.3)$
$\qquad = 1391 + 150 = 1541$ kNm

However, this moment is less than $M_2$, thus design to resist
$M_t = M_2 = 2318$ kNm.
$M_t/bh^2 = 2318 \times 10^6/(1200 \times 600^2) = 5.37$ N/mm²
$N/bh = 2600 \times 10^3/(1200 \times 600) = 3.61$ N/mm²
Assume 40 mm bars with 40 mm cover in each face, so that $d/h = 540/600 = 0.9$.

Hence use Design Chart 84 of CP 110: Part 2 [128]: from which
$100\, A_{sc}/bh = 2.8$
$\therefore A_{sc} = 2.8 \times 600 \times 1200/100 = 20160$ mm²
Use 16 No. 40 mm bars (20160 mm²) with 8 bars in each face.

### Sidesway

Since sidesway can occur, consider the column to be a cantilever with effective height equal to twice the actual height, i.e. $l_e$ = 16 m.
Slenderness ratio = 16/0.6 = 26.7
The initial and additional moments are both maximum at the base. Hence,
$M_i = 2318$ kNm and, from equation (9.22), the total moment is
$M_t =$
$\qquad 2318 + (2600 \times 0.6/1750)(26.7)^2\,(1 - 0.0035 \times 26.7)$
$\qquad = 2318 + 576 = 2894$ kNm
$M_t/bh^2 = 2894 \times 10^6/(1200 \times 600^2) = 6.70$ N/mm²
$N/bh = 3.61$ N/mm² (as before)
From Design Chart 84 of CP 110: Part 2,
$100\, A_{sc}/bh = 3.6$
$\therefore A_{sc} = 3.6 \times 600 \times 1200/100 = 25\,920$ mm²
Use 22 No. 40 mm bars (27 720 mm²) with 11 bars in each face.

It should be noted that the above designs have been carried out only at the ultimate limit state. In an actual design, it would be necessary to check the stresses and crack widths at the serviceability limit state by carrying out elastic analyses of the sections.

## 9.2 Hillerborg strip method applied to an abutment

A reinforced concrete abutment is 7 m high and 12 m wide. At each end of the abutment there is a wing wall which is structurally attached to the abutment.

The lateral loads acting on the abutment are the earth pressure, which varies from zero at the top to $5H$ kN/m² at a depth $H$; HA surcharge, the nominal value of which is 10 kN/m² (see Chapter 3); and the HA braking load, which acts at the top of the abutment and may be taken to have a nominal value of 30 kN/m width of abutment.

The Hillerborg strip method will be used to obtain a lower bound moment field for the abutment.

At the ultimate limit state, the design loads (nominal load $\times \gamma_{fL} \times \gamma_{f3}$) are:

Earth pressure = $5H \times 1.5 \times 1.15 = 8.625H$ kN/m²
HA surcharge = $10 \times 1.5 \times 1.10 = 16.5$ kN/m²
HA braking = $30 \times 1.25 \times 1.10 = 41.25$ kN/m

The wing walls and abutment base are considered to provide fixity to the abutment, which will thus be designed as if it were fixed on three sides and free on the fourth. The load distribution is chosen to be as shown in Fig. 9.19 (see also Chapter 2). Thus at the top of the abutment all of the load is considered to be carried in the $y$ direction; at the centre of the base, all of the load is considered to be carried in the $x$ direction; in the bottom corners, the load is

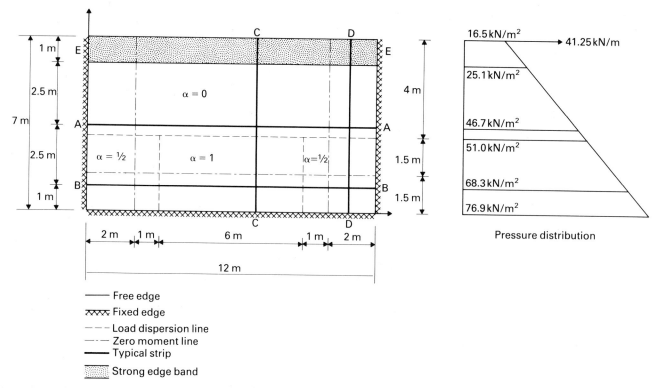

**Fig. 9.19** Abutment

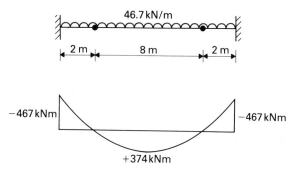

**Fig. 9.20** Strip AA

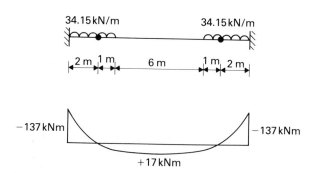

**Fig. 9.21** Strip BB

considered to be shared equally between the $x$ and $y$ directions. It is emphasised that any distribution of load could be chosen and that shown in Fig. 9.19 is merely one possibility.

In order that the resulting moments do not depart too much from the elastic moments, and thus serviceability problems do not arise, the zero moment lines shown in Fig. 9.19 are chosen.

Typical strips AA, BB, CC, DD and EE of unit width are now considered.

### Strip AA

The loading and bending moments are shown in Fig. 9.20.

### Strip BB

The loading and bending moments are shown in Fig. 9.21.

### Strip CC

Strip EE, which carries the braking load, earth pressure and surcharge at the top of the abutment, must also pro-

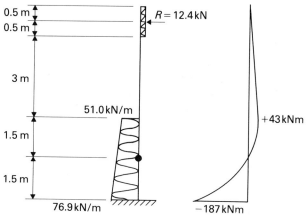

**Fig. 9.22** Strip CC

vide a reaction to strip CC. Hence, the loading and bendings moments for strip CC are shown in Fig. 9.22. The reaction $(R)$ can first be obtained by taking moments about the point of zero moment, and then the bending moment diagram can be calculated.

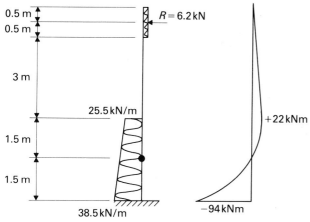

**Fig. 9.23** Strip DD

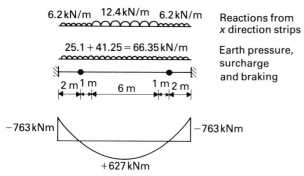

Reactions from
x direction strips

Earth pressure,
surcharge
and braking

**Fig. 9.24** Strip EE

### Strip DD

Strip DD is similar to strip CC and its loading and bending moments are shown in Fig. 9.23.

### Strip EE

Strip EE acts as a strong edge band (1 m wide) which not only supports the surcharge, earth pressure and braking loads but also supports the ends of typical strips CC and DD. Thus the loading and bending moments are as shown in Fig. 9.24. The loading is taken, conservatively, as that at a depth of 1 m.

## 9.3 Pile cap

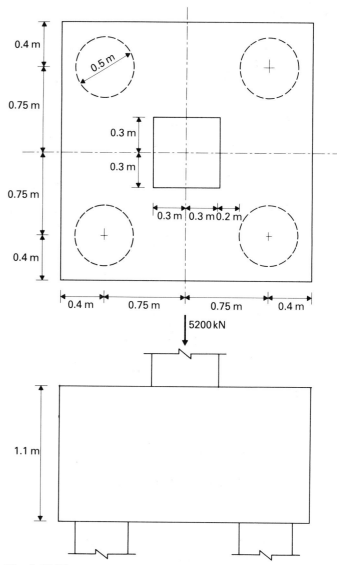

**Fig. 9.25** Pile cap

Design the four-pile cap shown in Fig. 9.25 if the characteristic strengths of the reinforcement and concrete are 425 N/mm² and 30 N/mm² respectively. The design load at the ultimate limit state is 5200 kN.

Load per pile = 5200/4 = 1300 kN
The cap will be designed by both bending theory and truss analogy methods.

### Bending theory

#### Bending

Total bending moment at column centre line
= 2 × 1300 × 0.75 = 1950 kNm
Assume effective depth = $d$ = 980 mm
From equation (5.7), the lever arm is

$$z = 0.5 \times 980 \left(1 + \sqrt{1 - \frac{5 \times 1950 \times 10^6}{30 \times 2300 \times 980^2}}\right)$$

$$= 943 \text{ mm}$$

But maximum allowable $z = 0.95d = 0.95 \times 980$
= 931 mm
Thus $z = 931$ mm
From equation (5.6), required reinforcement area is
$A_s = 1950 \times 10^6/(0.87 \times 425 \times 931) = 5665$ mm²
Use 19 No. 20 mm bars (5970 mm²)

#### Flexural shear

100 $A_s/bd$ = (100 × 5970)/(2300 × 980) = 0.26
From Table 5 of Code, allowable shear stress without shear reinforcement = $v_c$ = 0.36 N/mm². This stress may be enhanced by $(2d/a_v)$ for those parts of the critical section indicated in Fig. 9.17(a).
$a_v$ = 200 + 0.2 × 500 = 300 mm
Enhancement factor = 2 × 980/300 = 6.53
Enhanced $v_c$ = 6.53 × 0.36 = 2.35 N/mm²
Shear capacity of critical section
= [(2)(2.35)(500) + (0.36)(2300 – 2 × 500)] 980 × 10⁻³
= 2760 kN
Actual shear force = 2 × 1300 = 2600 kN < 2760 kN
∴ O.K.

*Punching shear*

The critical section shown in Fig. 9.17(b) would occur under the column in this example. Thus take critical section at corner of column, as shown in Fig. 9.25, at $[(0.75 - 0.3)\sqrt{2} - 0.25] = 0.386$ m from pile. The latter value will be assumed for $a_v$.

From Fig. 9.17(b), length of perimeter is $980 + 500$
$$= 1480 \text{ mm}$$

As for flexural shear, $v_c = 0.36$ N/mm$^2$

Enhancement factor $= (2 \times 980/386) = 5.08$

Enhanced $v_c = 5.08 \times 0.36 = 1.83$ N/mm$^2$

Shear capacity of critical section
$$= 1.83 \times 1480 \times 980 \times 10^{-3} = 2650 \text{ kN}$$

Actual shear force $= 1300$ kN $< 2650$ kN $\therefore$ O.K.

### Truss analogy

*Truss*

For equilibrium, the force in each of the reinforcement ties of Fig. 9.16 is $Nl/8d$.
$$= (5200 \times 1500)/(8 \times 980) = 995 \text{ kN}$$

Required reinforcement area is
$A_s = 995 \times 10^3/(0.87 \times 425) = 2690$ mm$^2$

Since there at two ties in each direction, the total reinforcement area in each direction is $2 \times 2690 = 5380$ mm$^2$. It can be seen that the truss theory requires less reinforcement than the bending theory, and this is generally the case.

80% of the tie reinforcement should be provided over the piles, i.e.

$0.8 \times 2690 = 2150$ mm$^2$

Use 7 No. 20 mm bars (2200 mm$^2$) over the piles.

The remaining 20% (540 mm$^2$) should be placed between the cap centre line and the piles, i.e. $2 \times 540 = 1080$ mm$^2$ should be placed between the piles. Use 4 No. 20 mm bars (1260 mm$^2$) between the piles.

*Flexural shear*

Over a pile, $100 A_s/bd = (100 \times 2200)/(500 \times 980)$
$$= 0.45.$$

From Table 5 of Code, $v_c = 0.51$ N/mm$^2$

Enhancement factor $= 6.53$ (as for bending theory)

Enhanced $v_c = 6.53 \times 0.51 = 3.33$ N/mm$^2$

Between piles, $100 A_s/bd = (100 \times 1260)/(1000 \times 980)$
$$= 0.13.$$

From Table 5 of Code, $v_c = 0.35$ N/mm$^2$

Assume no reinforcement outside piles, thus
$$v_c = 0.35 \text{ N/mm}^2$$

Shear capacity of critical section
$$= [(2)(3.33)(500) + (0.35)(2300 - 2 \times 500)] 980 \times 10^{-3}$$
$$= 3710 \text{ kN}$$

Actual shear force $= 2 \times 1300 = 2600$ kN $< 3710$ kN
$\therefore$ O.K.

*Punching shear*

Length of critical section $= 1480$ mm (as for bending theory)

As for flexural shear, $v_c = 0.51$ N/mm$^2$

Enhancement factor $= 5.08$ (as for bending theory)

Enhanced $v_c = 5.08 \times 0.51 = 2.59$ N/mm$^2$

Shear capacity of critical section
$$= 2.59 \times 1480 \times 980 \times 10^{-3} = 3760 \text{ kN}$$

Actual shear force $= 1300$ kN $< 3760$ kN $\therefore$ O.K.

It can be seen from the above calculations that the truss theory design results in a greater shear capacity than does the bending theory design.

## Chapter 10

# Detailing

## Introduction

In this chapter, the Code clauses, concerned with considerations affecting design details for both reinforced and prestressed concrete, are discussed and compared with those in the existing design documents.

**Table 10.1** Nominal covers

| Conditions of exposure | Nominal cover (mm) for concrete grade | | | |
|---|---|---|---|---|
| | 25 | 30 | 40 | ≥50 |
| Moderate<br>Surfaces sheltered from severe rain and against freezing whilst saturated with water, e.g.<br>(1) surfaces protected by a waterproof membrane;<br>(2) internal surfaces whether subject to condensation or not;<br>(3) buried concrete and concrete continuously under water | 40 | 30 | 25 | 20 |
| Severe<br>(1) Soffits<br>(2) Surfaces exposed to driving rain, alternate wetting and drying, e.g. in contact with back-fill and to freezing whilst wet | 50 | 40 | 30 | 25 |
| Very severe<br>(1) Surfaces subject to the effects of de-icing salts or salt spray, e.g. roadside structures and marine structures | N/A | 50* | 40* | 25 |
| (2) Surfaces exposed to the action of sea water with abrasion or moorland water having a pH of 4.5 or less. | N/A | N/A | 60 | 50 |

* Only applicable if the concrete has entrained air (see text)

## Reinforced concrete

### Cover

The cover to a particular bar should be at least equal to the bar diameter, and is also dependent upon the exposure condition and the concrete grade as shown by Table 10.1. These values are very similar to those given in Amendment 1 to BE 1/73: however, there are two important differences.

First, the Code considers all soffits to be subjected to severe exposure conditions, whereas BE 1/73 distinguishes between sheltered soffits and exposed soffits. Thus, for the soffit of a slab between precast beams, the Code would require, for grade 30 concrete, a minimum cover of 40 mm, whereas BE 1/73 would require only 30 mm. Hence, top slabs in beam and slab construction may need to be thicker than they are at present.

Second, for roadside structures subjected to salt spray and constructed with grade 30 or 40 concrete, the Code requires the concrete to have entrained air. BE 1/73 does not have this requirement. The Code thus requires a dramatic change in current practice. The footnote to Table 10.1 appears in the Code with a reference to Part 7 of the Code. However, Part 7 refers only to the permitted variation in specified air content without giving the latter, but Clause 3.5.6 of Part 8 of the Code does specify air contents for various maximum aggregate sizes.

### Bar spacing

#### Minimum distance between bars

For ease of placing and compacting concrete, the Code relates the minimum distance between bars to the maximum aggregate size. The Code clauses were taken from CP 116 and are thus more detailed than those in CP 114, although they are very similar in implication.

In addition to rules for single bars and pairs of bars, the Code gives rules for bundled bars since the latter are allowable.

### Maximum spacing of bars in tension

In order to control crack widths to the values given in Table 4.7, the maximum spacing of bars has to be limited. The procedures for calculating maximum bar spacings are discussed in Chapter 7. The Code also stipulates that, in no circumstances, should the spacing exceed 300 mm. This was considered a reasonable maximum spacing to ensure that, in all reinforced concrete bridge members, the bars would be sufficiently close together for them to be assumed to form a 'smeared' layer of reinforcement, rather than act as individual bars.

## Minimum reinforcement areas

### Shrinkage and temperature reinforcement

In those parts of a structure where cracking could occur due to restraint to shrinkage or thermal movements, at least 0.3% of mild steel or 0.25% of high yield steel should be provided. These values are less than those suggested by Hughes [185] and the author would suggest that they be used with caution. The Code values originated in CP 2007.

### Beams and slabs

A minimum area of tension reinforcement is required in a beam or slab in order to ensure that the cracked strength of the section exceeds its uncracked strength; otherwise, any reinforcement would yield as soon as cracking occurred, and extremely wide cracks would result.

The cracking moment of a rectangular concrete beam is given by

$$M_t = f_t bh^2/6$$

where $f_t$ is the tensile strength of the concrete, and $b$ and $h$ are the breadth and overall depth respectively. If the beam is reinforced with an area of reinforcement $(A_s)$ at an effective depth $(d)$ and having a characteristic strength $(f_y)$, the ultimate moment of resistance is given by

$$M_u = f_y A_s z$$

where $z$ is the lever arm.
Since it is required that $M_u \geq M_t$, then

$$f_y A_s z \geq f_t bh^2/6$$

or

$$\frac{100A_s}{bd} \geq \frac{16.7 f_t}{(z/d)f_y} \left( \frac{h^2}{d} \right) \simeq 16.7 \frac{f_t}{f_y}$$

Beeby [119] has shown that $f_t \approx 0.556 \sqrt{f_{cu}}$: thus, for the maximum allowable value of $f_{cu}$ of 50 N/mm², $f_t = 3.9$ N/mm². Hence, for $f_y = 250$ N/mm² and 410 N/mm² respectively, the required minimum reinforcement percentages are 0.26 and 0.16. These values agree very well with the Code values of 0.25 and 0.15 respectively. The latter values cannot be compared directly with those in CP 114 because the CP 114 values are expressed as a percentage of the gross section, rather than the effective section. However, the Code will generally require greater minimum areas of reinforcement than does CP 114.

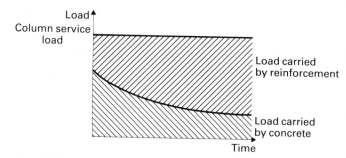

**Fig. 10.1** Load transfer in column under service load conditions

The above minimum reinforcement areas are given in the Code under the heading 'Minimum area of main reinforcement', but, since a minimum area of secondary reinforcement is not specified for solid slabs, it would seem prudent also to apply the Code values to secondary reinforcement.

### Voided slabs

Although a minimum area of secondary reinforcement in solid slabs is not specified, values are given for voided slabs. These values are discussed in detail in Chapter 7.

### Columns

Under long-term service load conditions, load is transferred from the concrete to the reinforcement as shown in Fig. 10.1. The load transfer occurs because the concrete creeps and shrinks. If the area of reinforcement is very small, there is a danger of the reinforcement yielding under service load conditions. In order to prevent yield, ACI Committee 105 [236] proposed a minimum reinforcement area of 1%. This value is adopted in the Code and is a little greater than that (0.8%) in CP 114. However, if a column is lightly loaded, the area of reinforcement is allowed to be less than 1% but not less than $(0.15 N/f_y)$, where $N$ is the ultimate axial load and $f_y$ is the characteristic strength of the reinforcement. This requirement is intended to cover a case where a column is made much larger than is necessary to carry the load.

In order to ensure the stability of a reinforcement cage prior to casting, the Code requires (as does CP 114) the main bar diameter to be at least 12 mm. In addition, the Code requires at least six main bars for circular columns and four bars for rectangular columns.

### Walls

It is explained in Chapter 9 that a reinforced concrete wall, which carries a significant axial load, should have at least 0.4% vertical reinforcement. This requirement is necessary because smaller amounts of reinforcement can result in a reinforced wall which is weaker than a plain concrete wall [227].

### Links

Links are generally present in a member for two reasons: to act as shear or torsion reinforcement, and to restrain main compression bars.

The minimum requirements for links to act as shear reinforcement are discussed in detail in Chapter 6. In the following, the requirements for a link to restrain a compression bar are discussed.

*Beams and columns*  The link diameter should be at least one-quarter of the diameter of the largest compression bar, and the links should be spaced at a distance which is not greater than twelve times the diameter of the smallest compression bar. These requirements are the same as those in CP 114, except that the latter also requires the link spacing in columns not to exceed the least lateral dimension of the column nor 300 mm, and the link diameter not to be less than 5 mm. The latter requirement is automatically satisfied by the fact that the smallest available bar has a diameter of 6 mm (although it is now difficult to obtain reinforcement of less than 8 mm diameter).

*Walls and slabs*  When the *designed* amount of compression reinforcement exceeds 1%, links have to be provided. The link diameter should not be less than 6 mm nor one-quarter of the diameter of the largest compression bar. In the direction of the compressive force, the link spacing should not exceed 16 times the diameter of the compression bar. In the cross-section of the member, the link spacing should not exceed twice the member thickness. These requirements were taken from the ACI Code [168] and are different to those of CP 114.

## Maximum steel areas

In order to ease the placing and compacting of concrete, the amount of reinforcement in a member must be restricted to a maximum value. The Code values are as follows.

### Beams and slabs

Neither the area of tension reinforcement nor that of compression reinforcement should exceed 4%. CP 114 requires only that the area of compression reinforcement should not exceed 4%.

### Columns

The amount of longitudinal reinforcement should not exceed 6% if vertically cast, 8% if horizontally cast nor 10% at laps. The CP 114 amount is always 8%. Hence, the Code is more restrictive with regard to vertically cast columns, and this fact, coupled with the small allowable design crack width, could result in larger columns – as discussed in Chapter 9.

### Walls

The area of vertical reinforcement should not exceed 4%. No limit is given in CP 114.

## Bond

### General

All bond calculations in accordance with the Code are carried out at the ultimate limit state.

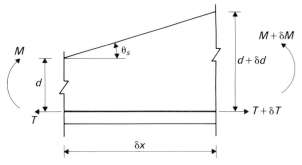

**Fig. 10.2** Local bond

The Code, like Amendment 1 to BE 1/73, recognises two types of deformed bars:

1. Type 1, which are, generally, square twisted.
2. Type 2, which have, generally, transverse ribs.

Type 2 bars have superior bond characteristics to type 1 bars. However, unless it is definitely known at the design stage which type of bar is to be used on site, it is necessary to assume type 1 for design purposes.

### Local bond

Consider a beam of variable depth subjected to moments which increase in the same direction as the depth increases, as shown in Fig. 10.2.
The tension steel force at any point $(x)$ is $T$, where

$$T = M/z$$

and $M$ and $z$ are the moment and lever arm respectively at $x$. The rate of change of $T$ is

$$\frac{dT}{dx} = \frac{z(dM/dx) - M(dz/dx)}{z^2}$$

But $dM/dx$ is the shear force $(V)$ at $x$ and the Code assumes $dz/dx \simeq \tan \theta_s$.
Hence

$$\frac{dT}{dx} = \frac{V - M \tan \theta_s/z}{z}$$

But $dT/dx$ is also equal to the bond force per unit length; which is $f_{bs} (\Sigma u_s)$, where $f_{bs}$ is the local bond stress and $(\Sigma u_s)$ is the sum of the perimeters of the tension reinforcement. Hence

$$f_{bs} (\Sigma u_s) = \frac{V - M \tan \theta_s/z}{z}$$

or

$$f_{bs} = \frac{V - M \tan \theta_s/z}{(\Sigma u_s)z} \tag{10.1}$$

The Code assumes that $z \simeq d$ (and adjusts the allowable values of $f_{bs}$ accordingly). If $M$ increases in the opposite direction to which $d$ increases, the negative sign in equation (10.1) becomes positive. Hence, the following Code equation is obtained

$$f_{bs} = \frac{V \pm M \tan \theta_s/d}{(\Sigma u_s)d} \tag{10.2}$$

In addition to the modification to allow for variable depth, this equation differs to that in CP 114 because the

CP 114 equation is written in terms of the lever arm rather than the effective depth.

The allowable local bond stresses at the ultimate limit state depend on bar type and concrete strength: they are given in Table 10.2. The bond stresses for plain, type 1 deformed and type 2 deformed are in the approximate ratio 1 : 1.25 : 1.5. It is understood that the tabulated values were obtained by considering the test data of Snowdon [237] and by scaling up the CP 114 values, for plain bars at working load conditions, to ultimate load conditions. Snowdon's tests on 150 mm lengths of various types of bar indicated that the bond stresses developed by plain, square twisted (type 1) and ribbed (type 2) bars were in the approximate ratio 1 : 1.3 : 3.5. Hence the Code ratio is reasonable for type 1 deformed bars but can be seen to be conservative for type 2 deformed bars. However, Snowdon found that the advantage of the latter bars over plain bars decreased with an increase in diameter, particularly with low strength concrete.

**Table 10.2** Ultimate local bond stresses

| Bar type | Local bond stress (N/mm²) for concrete grade | | | |
| | 20 | 25 | 30 | ≥ 40 |
| --- | --- | --- | --- | --- |
| Plain | 1.7 | 2.0 | 2.2 | 2.7 |
| Deformed Type 1 | 2.1 | 2.5 | 2.8 | 3.4 |
| Deformed Type 2 | 2.6 | 2.9 | 3.3 | 4.0 |

The Code local bond stresses are about 1.5 to 1.6 times those in BE 1/73 if overstress is ignored. However, it should be remembered that the Code bond stresses are intended for use at the ultimate limit state with a steel stress of $0.87 f_y$, whilst the BE 1/73 bond stresses are intended to be used at working load with a steel stress of about $0.56 f_y$. The ratio of these steel stresses is 1.55, and thus the result of carrying out local bond calculations in accordance with the Code and with BE 1/73 should be about the same.

The author understands that, in a proposed amendment to CP 110, local bond calculations are not required at all. If this proposal is adopted, then local bond calculations will, presumably, be omitted from the Code also.

### Anchorage bond

The conventional expression for the anchorage length ($L$) of a bar, which is required to develop a certain stress ($f_s$) can be obtained from any standard text on reinforced concrete, and is

$$L = (f_s/4f_{ba}) \, \phi \tag{10.3}$$

where $f_{ba}$ is the average anchorage bond stress and $\phi$ is the bar diameter.

Since bond calculations are carried out at the ultimate limit state, the reinforcement stress ($f_s$) is the design stress at the ultimate limit state and is $0.87 f_y$ for tension bars and $0.72 f_y$ for compression bars. However, if more than the required amount of reinforcement is provided, then the stress in the bar is less than the design stress and lower values of $f_s$, than those given above, may be used.

The allowable average anchorage bond stresses ($f_{ba}$) depend upon bar type, concrete strength and whether the bar is in tension or compression. Higher values are permitted for bars in compression because some force can be transmitted from the bars to the concrete by end bearing of the bar. The allowable anchorage bond stresses are given in Table 10.3. The stresses for plain, type 1 deformed and type 2 deformed bars are in the approximate ratio 1 : 1.4 : 1.8, and those for bars in compression are about 25% greater than those for bars in tension. It is understood that the values for bars in tension were obtained by considering the test data of Snowdon [237] and by scaling up the CP 114 values, for plain bars at working load conditions, to ultimate load conditions. Snowdon's tests indicated that the anchorage lengths for plain, square twisted (type 1) and ribbed (type 2) bars were in the approximate ratio 1 : 1.4 : 2. The Code ratio agrees very well with Snowdon's results. The increase of 25%, when bars are in compression, was taken from that implied in CP 114.

**Table 10.3** Ultimate anchorage bond stresses

| Bar type | Anchorage bond stress (N/mm²) for concrete grade | | | |
| | 20 | 25 | 30 | ≥ 40 |
| --- | --- | --- | --- | --- |
| Plain, in tension | 1.2 | 1.4 | 1.5 | 1.9 |
| Plain, in compression | 1.5 | 1.7 | 1.9 | 2.3 |
| Deformed, type 1, in tension | 1.7 | 1.9 | 2.2 | 2.6 |
| Deformed, type 1, in compression | 2.1 | 2.4 | 2.7 | 3.2 |
| Deformed, type 2, in tension | 2.2 | 2.5 | 2.8 | 3.3 |
| Deformed, type 2, in compression | 2.7 | 3.1 | 3.5 | 4.1 |

The Code average bond stresses for plain bars are about 1.5 those in BE 1/73, if overstress is ignored. However, since the ratio of steel stresses is 1.55 (see previous discussion of local bond stresses), anchorage lengths for plain bars will be about the same whether calculated in accordance with the Code or BE 1/73. But anchorage lengths for deformed bars will be shorter by about 13% for type 1 bars and 29% for type 2 bars. This is because BE 1/73 allows increases in bond stress, above the plain bar values, of only 25% for type 1 bars and 40% for type 2 bars. These percentages compare with 40% and 80%, respectively, in the Code.

### Bundled bars

The Code permits bars to be bundled into groups of two, three or four bars. The effective perimeter of a group of bars is obtained by calculating the sum of the perimeters of the individual bars and, then, by multiplying by a reduction factor of 0.8, 0.6, or 0.4 for groups of two, three or four bars respectively. The resulting perimeter, so calculated, is less than the actual exposed perimeter of the group of bars to allow for difficulties in compacting concrete around groups of bars in contact.

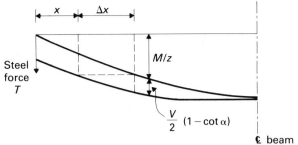

**Fig. 10.3** Steel force diagram

## Lap lengths

In general, as in CP 114, a lap length should be not less than the anchorage length calculated from equation (10.3). However, *for deformed bars in tension*, the lap length should be 25% greater than the anchorage length. This requirement is to allow for the stress concentrations which occur at each end of a lap, and which result in splitting of the concrete along the bars at a lower load than would occur for a single bar in a pull-out test [238]. Such splitting does not occur with plain bars, which fail in bond by pulling out of the concrete.

In addition to the above requirements, the Code requires the following minimum lap lengths to be provided for a bar of diameter $\phi$:

1.  Tension lap length $\triangleleft 25\,\phi + 150$ mm
2.  Compression lap length $\triangleleft 20\,\phi + 150$ mm

These minimum lengths are much more conservative than those in CP 114 for small diameter bars, and slightly less conservative for large diameter bars.

## Bar curtailment and anchorage

### General curtailment

As in CP 114, a bar should extend at least twelve diameters beyond the point at which it is no longer needed to carry load.

A bar should also be extended a minimum distance to allow for the fact that, in the presence of shear, a bar at a particular section has to carry a force greater than that calculated by dividing the bending moment ($M$) at the section by the lever arm ($z$). A rigorous analysis [239] of the truss of Fig. 6.4(a), rather than the simplified analysis of Chapter 6, shows that the total force, which has to be carried by the main tension reinforcement at a section where the moment and shear force are $M$ and $V$, respectively, is

$$T = M/z + (V/2)(\cot \theta - \cot \alpha)$$

The Code assumes $\theta = 45°$, thus

$$T = M/z + (V/2)(1 - \cot \alpha) \qquad (10.4)$$

In Fig. 10.3, the distributions of tension force due to bending ($M/z$) and total tension force ($T$) are plotted for a general case. It can be seen from Fig. 10.3 that the increase in steel force due to shear can be allowed for by designing the reinforcement at a section to resist only the moment at that section, and by extending the reinforce-

ment beyond that section by the distance $\triangle x$ in Fig. 10.3. The distance $\triangle x$ can be found by equating the total steel force at a section at $x$ to the steel force due to moment only at a section at $(x + \triangle x)$.

The maximum increase in steel force due to the shear force, and, hence, the maximum value of $\triangle x$ occurs when $\cot \alpha$ is zero (i.e. vertical stirrups) and equation (10.4) becomes

$$T = M/z + V/2 \qquad (10.5)$$

For a central point load ($2W$) on a beam of span $l$, the moment and shear force at $x$ are:

$$M_x = Wx$$
$$V_x = W$$

From equation (10.5)

$$T_x = Wx/z + W/2$$

The moment at $(x + \triangle x)$ is

$$M_{x+\triangle x} = W(x + \triangle x)$$

$\triangle x$ can be found from

$$T_x = M_{x+\triangle x}/z$$

Thus

$$Wx/z + W/2 = W(x + \triangle x)/z$$

From which

$$\triangle x = z/2$$

If this analysis is repeated for a uniformly distributed load, it can be shown that $\triangle x$ is, again, about $z/2$. Hence, if the longitudinal reinforcement is designed solely to resist the moment at a section, the reinforcement should be extended a distance $z/2$ beyond that section. However, the Code, conservatively, takes the extension length to be the effective depth.

### Curtailment in tension zones

*In addition* to the above general requirements, the Code requires any *one* of the following conditions to be met before a bar is curtailed in a tension zone.

1.  In order to control the crack width at the curtailment point, a bar should extend at least an anchorage length, calculated from equation (10.3) with $f_s = 0.87 f_y$, beyond the point at which it is no longer required to resist bending.
2.  Tests, such as those carried out by Ferguson and Matloob [240], have shown that the shear capacity of a section with curtailed bars can be up to 33% less than that of a similar section in which the bars are not curtailed. In order to be conservative, the Code requires the shear capacity at a section, where a bar is curtailed, to be greater than twice the actual shear force.
3.  In order to control the crack width at the curtailment point, at least double the amount of reinforcement required to resist the moment at that section should be provided. This requirement was taken from the ACI Code [168].

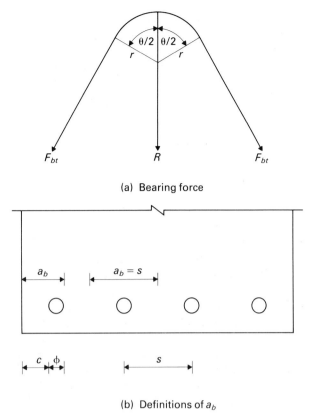

(a) Bearing force

(b) Definitions of $a_b$

**Fig. 10.4(a),(b)** Bearing force at bend

The above requirements are far more complicated than those of CP 114, unless one chooses to apply option 1 and continue a bar for a full anchorage length.

### Anchorage at a simply supported end

The Code requires *one* of the following conditions to be satisfied.

1. As in CP 114, a bar should extend for an anchorage length equivalent to twelve times the bar diameter; and no bend or hook should begin before the centre of the support.
2. If the support is wide and a bend or hook does not begin before $d/2$ from the face of the support (where $d$ is the effective depth of the member), a bar should extend from the face of a support for an anchorage length equivalent to $(d/2 + 12\phi)$, where $\phi$ is the bar diameter.
3. Provided that the local bond stress at the face of a support is less than half the value in Table 10.2, a straight length of bar should extend, beyond the centre line of the support, the greater of 30 mm or one-third of the support width. This clause was originally written, for CP 110, to cover small precast units [112], and it is not clear whether the clause is applicable to bridges.

### Bearing stresses inside bends

The bearing stress on the concrete inside a bend of a bar of diameter $\phi$, which is bent through an angle $\phi$ with a radius $r$, should be calculated by assuming the resultant force ($R$)

on the concrete is uniformly spread over the length of the bend. Hence, with reference to Fig. 10.4(a) the resultant force is

$$R = 2F_{bt} \sin(\theta/2)$$

where $F_{bt}$ is the tensile force in the bar at the ultimate limit state. The bearing area is

$$\phi[2r\sin(\theta/2)]$$

Thus the bearing stress $f_b$ is given by

$$f_b = 2F_{bt}\sin(\theta/2)/\phi[2r\sin(\theta/2)]$$
$$\therefore f_b = F_{bt}/r\phi \qquad (10.6)$$

which is the equation given in the Code.

The bearing stress calculated from equation (10.6) should not exceed the allowable value given by equation (8.4). In the present context, the bar diameter is the length of the loaded area and thus, in equation (8.4), $y_{po} = \phi/2$. Similarly the bar spacing ($a_b$) is the length of the resisting concrete block and thus, in equation (8.4), $y_o = a_b/2$. On substituting into equation (8.4), the following Code expression is obtained.

$$\frac{1.5 f_{cu}}{1 + 2\phi/a_b}$$

However, for a bar adjacent to a face of a member, as shown in Fig. 10.4(b), the length of the resisting concrete block is $(c + \phi + a_b/2)$, where $c$ is the side cover. Thus $y_o$ should be taken as $(c + \phi + a_b/2)/2$; but the Code, by redefining $a_b$ as $(c + \phi)$, implies that

$$y_o = (c + \phi)/2$$

The Code thus seems to be conservative in this situation. It appears that the Code requirements were based upon those of the CEB [226], which, in fact, defines $a_b$ as $(c + \phi/2)$ for a bar adjacent to a face of a member.

The Code definitions of $a_b$ are summarised in Fig. 10.4(b).

It is not necessary to carry out these bearing stress calculations if a bar is not assumed to be stressed beyond the bend. Hence, bearing stress calculations are not required for standard end hooks or bends.

## Prestressed concrete

The following points concerning detailing in prestressed concrete are intended to be additional to those discussed previously for reinforced concrete.

### Cover to tendons

#### Bonded tendons

As in BE 2/73, the Code requires the covers to bonded tendons to be the same as those to bars in reinforced concrete. Hence, the comments made earlier in this chapter regarding the reinforced concrete covers are relevant.

### Tendons in ducts.

Again, as in BE 2/73, the Code specifies a minimum cover of 50 mm to a duct. In addition a table, which is identical to that in BE 2/73, is provided (in Appendix D of Part 4 of the Code) for covers to curved tendons in ducts.

### External tendons

Part 4 of the Code refers to Clause 4.8.3 of Part 7 of the Code for the definition of an external tendon. However, the latter clause does not exist in Part 7, but exists in Part 8 of the Code as Clause 5.8.3. It defines an external tendon as one which 'after stressing and incorporation in the work, but before protection, is outside the structure'. This definition is essentially the same as that in BE 2/73.

As in BE 2/73, the Code requires that, when external tendons are to be protected by dense concrete, the cover to the tendons should be the same as if the tendons were internal. In addition, the protective concrete should be anchored, by reinforcement, to the prestressed member, and should be checked for cracking. The Code is not specific regarding how the latter check should be carried out: BE 2/73 refers to the reinforced concrete crack width formula of BE 1/73. When using the Code, the author would suggest that equation (7.4) for beams should be used.

## Spacing of tendons

### Bonded tendons

The minimum tendon spacing should comply with the minimum spacings specified for reinforcing bars. The latter spacings are similar to those of CP 116 and, as explained earlier in this chapter, are similar to those specified in BE 2/73. In addition, BE 2/73 requires compliance with the maximum spacings specified in CP 116, whereas the Code does not refer to maximum spacings.

### Tendons in ducts

The Code gives a number of requirements for the clear distance between ducts; these requirements are identical to those in BE 2/73.

## Transmission length in pre-tensioned members

In both the Code and BE 2/73, the transmission length is defined as the length required to transmit the initial prestressing force in the tendon to the concrete.

The transmission length depends upon a great number of variables (e.g. concrete compaction and strength, tendon type and size) and, if possible, should be determined from tests carried out under site or factory conditions, as appropriate. If such test data are not available, the Code gives recommended transmission lengths for wire and for strand. The Code implies values which are identical to those in BE 2/73.

The suggested transmission lengths for wires are based

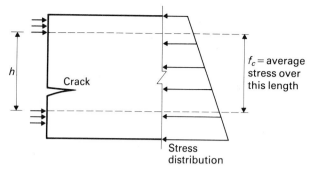

**Fig. 10.5** Splitting at end of prestressed member

on data from tests carried out in the laboratory and on site by Base [242]; and those for strands are based on data from tests carried out only in the laboratory by Base [243].

The CP 110 handbook [112] warns that the transmission lengths for strands, which were based upon laboratory data, could be exceeded on site; it also warns that they should not be used for compressed strands, for which transmission lengths can be nearly twice those given in the Code.

In members which have the tendons arranged vertically in widely spaced groups, the end section of the member acts like a deep beam when turned through 90° (see Fig. 10.5). This is due to the fact that, towards the end of the member, concentrated loads are applied by the tendons, and, at some distance from the end, the prestress is fully distributed over the section. Hence, the end face of the member is in tension and a crack can form, as shown in Fig. 10.5. Vertical links should be provided to control the crack, and Green [241] suggests that, by analogy with a deep beam, the required area of the vertical reinforcement ($A_s$) should be calculated from:

$$A_s = 0.2hf_cb_w/f_s \not< 0.04P_k/f_s \qquad (10.7)$$

where

$h$ = vertical clear distance between tendon groups

$b_w$ = width of web, or end block, at a distance $h$ from the end of the member

$f_c$ = average compressive stress *between* the tendons at a distance $h$ from the end of the member

$f_s$ = permissible reinforcement stress ($0.87f_y$)

$P_k$ = *total* initial prestressing force

In Fig. 10.5 and the above discussion, the prestressing forces have been considered to be applied at the end of the member, whereas, of course, they are transferred to the concrete over the transmission length, which is typically of the order of 400 mm. Thus the deep beam analogy tends to overestimate the tendency to crack, and equation (10.7) should be conservative.

## End blocks in post-tensioned members

### General

In a post-tensioned member, the prestressing forces are applied directly to the ends of the member by means of relatively small anchorages. The forces then spread out

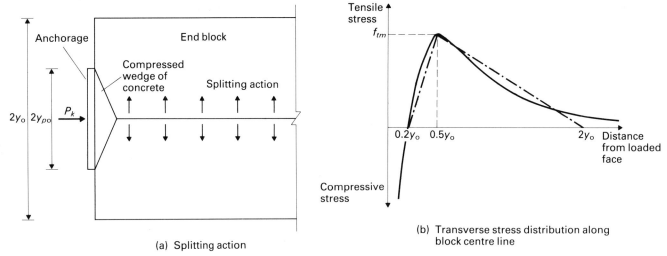

**Fig. 10.6(a),(b)** End block with symmetrical anchorage

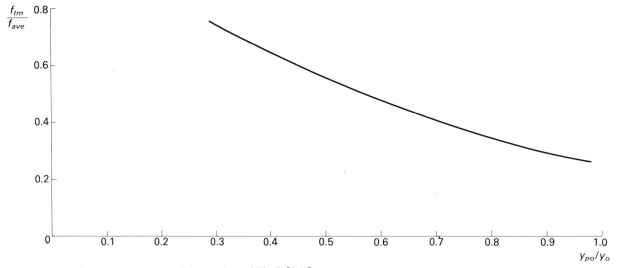

**Fig. 10.7** Maximum transverse tensile stress in end block [244]

over the cross-section of the member and, in this region of spread (the end block), high local stresses occur. In particular, large transverse bursting stresses occur: it is easiest to examine these stresses by considering an end block subjected to a single symmetrically placed prestressing force.

### Single anchorage

Single symmetrically placed anchorages have been studied both theoretically and experimentally [244]. Qualitatively, the structural behaviour consists of a cone of compressed concrete being driven into the end block and, thus, causing splitting of the end block as shown in Fig. 10.6(a). The splitting action causes transverse bursting stresses which are greatest across a horizontal or vertical section through the axis of the end block. The distribution of transverse stresses along such a section is of the form shown in Fig. 10.6(b), where it can be seen that compressive stresses exist near to the loaded face, but at a distance of about $0.2 \, y_o$ from the loaded face (where $2 \, y_o$ is the length of the side of the end block) the stress becomes tensile. The tensile stress reaches a maximum of $f_{tm}$ at about $0.5 \, y_o$ from the loaded face, and then decreases to nearly zero at about $2 \, y_o$ from the loaded face (i.e. at a distance equal to the

length of the side of the end block) [244]. It can be seen that it is reasonable to approximate the actual stress distribution to the triangular stress distribution shown in Fig. 10.6(b).

The maximum stress is mainly dependent upon the ratio of the length of the side of the loaded area ($2 \, y_{po}$) to that of the end block ($2 \, y_o$). The ratio of maximum transverse tensile stress ($f_{tm}$) to the average compressive stress, over the total cross-sectional area of the end block ($f_{ave}$), is plotted against $y_{po}/y_o$ in Fig. 10.7; the relationship was determined experimentally [244].

The bursting tensile stresses have to be resisted by reinforcement, and thus the total bursting tensile *force* to be resisted is of primary interest. The bursting tensile force can be obtained by integrating a number of stress diagrams, similar to that of Fig. 10.6(b). As one might expect, the bursting tensile force ($F_{bst}$) is mainly dependent upon $y_{po}/y_o$. The ratio of $F_{bst}$ to the maximum prestressing tendon force ($P_k$) is plotted against $y_{po}/y_o$ in Fig. 10.8. In this figure, the range of values obtained from various theories [244], values determined from tests [244] and the Code values are given. It can be seen that the experimental values exceed the theoretical values, and that the Code values have been *chosen* to lie between the theoretical and experimental values.

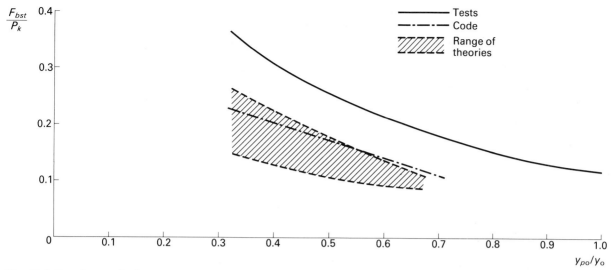

**Fig. 10.8** Bursting tensile force

The tendon force for use in determining the bursting force should be the greatest load that the tendon will carry during its life. This will be the jacking load for a bonded tendon, and the greater of the jacking load and the tendon load at the ultimate limit state for an unbonded tendon. The latter load may be assessed, as explained in Chapter 5, from Table 30 of the Code. However, the Code clause on end blocks refers to Tables 20 to 23 of the Code: these tables give characteristic strengths of tendons, and it is not clear whether the reference is an error, or whether it is intended that the load at the ultimate limit state should be taken as that equivalent to the characteristic strength of the tendon. It would seem more appropriate to use a load assessed from Table 30.

The bursting tensile force, calculated from the $F_{bst}/P_k$ ratios given in the Code and plotted in Fig. 10.8, should be resisted by reinforcement. From Fig. 10.6(b) it can be seen that this reinforcement should be distributed in a region extending from $0.2y_o$ to $2y_o$ from the loaded face of the end block. The reinforcement should be designed at the ultimate limit state and, thus, its design stress is $0.87f_y$. However, in order to control cracking, the reinforcement stress should be limited to a value corresponding to a strain of 0.001 (i.e. 200 N/mm²) if the cover to the reinforcement is less than 50 mm.

If the end block is rectangular, the value of $y_{po}/y_o$ is different in the two principal directions. Hence, $F_{bst}$ should be determined in each of the principal directions and reinforcement proportional accordingly. But, for detailing purposes, it is generally more convenient to use the greater area of reinforcement in both directions.

The above design method, in which all of the bursting tensile force is resisted by reinforcement, is the method given in the Code. However, the Code does permit the adoption of alternative design methods, in which some of the bursting tensile force is resisted by the concrete, to be adopted. One such method is that suggested by Zielinski and Rowe [244]; when using this method, the values of $F_{bst}/P_k$ given in the Code should not be used but, instead, the test values given in Fig. 10.8 should be used. This design procedure is, first for the appropriate value of $y_{po}/y_o$, to obtain the maximum tensile stress from Fig. 10.7

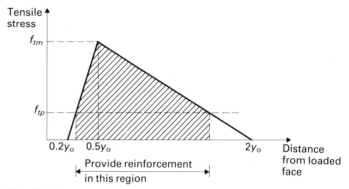

**Fig. 10.9** Stresses resisted by concrete end reinforcement

and the bursting tensile force from Fig. 10.8. The idealised triangular stress distribution diagram of Fig. 10.6(b) is then constructed and the permissible tensile stress in the concrete ($f_{tp}$) superimposed on the diagram as shown in Fig. 10.9. Only those areas where the stress exceeds the permissible tensile stress of the concrete need to be reinforced, as shown in Fig. 10.9. The bursting tensile force to be resisted by reinforcement ($F_s$) as a fraction of the total bursting tensile force ($F_{bst}$) is equal to the ratio of the area of the shaded part of the stress diagram to the total area; hence,

$$F_s = F_{bst} \left[ 1 - (f_{tp}/f_{tm})^2 \right] \tag{10.8}$$

The permissible steel stress used to calculate the required area of reinforcement is usually chosen to be 140 N/mm². The strain associated with this stress is generally too small to cause observable cracking of the concrete. Regarding the value to be taken for the permissible tensile stress in the concrete, BE 2/73 states that it should be the cylinder splitting strength of the concrete divided by 1.25, and this document also gives values of cylinder splitting strength for various grades of concrete.

The above two design methods lead to similar amounts of reinforcement because, although in the second method some of the bursting tensile force is resisted by the concrete, the total bursting tensile force to be resisted is greater (see Fig. 10.8).

BE 2/73 also permits the use of either of the two design methods.

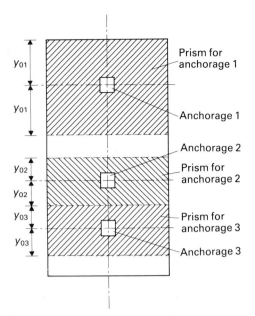

**Fig. 10.10** Multiple anchorages

## Multiple anchorages

Very often the total prestressing force is applied to the end of a member by a number of anchorages. Tests have been carried out on end blocks with multiple anchorages by Zielinski and Rowe [246], and the results indicate that each anchorage may be associated with a prism of concrete, which acts like an end block for the particular anchorage, as shown in Fig. 10.10. Each prism is symmetrically loaded by its anchorage, and its vertical dimension is the lesser of twice the distance from the centre line of its anchorage to the centre line of the nearer adjacent anchorage, and twice the distance from the centre line of its anchorage to the edge of the concrete.

The bursting tensile force and the required amount of reinforcement in each prism can be assessed by either of the methods for single anchorages described earlier.

In addition, the individual prisms should be tied together with reinforcement. No guidance is given in the Code on how to design such reinforcement, but a method has been suggested by Clarke [246], which is based on the French Code [247].

## Spalling tensile stresses

Tensile stresses occur at the loaded face of an end block between anchorages. These stresses arise for similar reasons to those, discussed previously in this chapter, in connection with transmission lengths in pre-tensioned members (see Fig. 10.5). Equation (10.7) may be used to design reinforcement to control the cracking caused by such stresses in post-tensioned members.

In addition, spalling tensile stresses occur at the loaded face of an end block, away from the axis of an anchorage, when the prestressing force is eccentric. The transverse

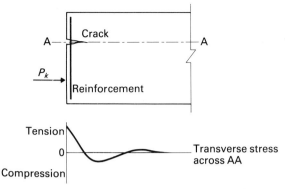

**Fig. 10.11** Spalling tensile stresses due to eccentric prestress

stress distribution along a line parallel to the axis of an eccentric prestressing force is as shown in Fig. 10.11 [248]. Figure 10.11 also shows the crack which can occur at the loaded face. Green [241] suggests that vertical reinforcement sufficient to resist a force of $0.04P_k$ should be placed as near as possible to the loaded face of the member to control the crack. It would seem prudent to provide a minimum amount of such reinforcement in all end blocks whether or not the prestressing force is highly eccentric.

Further guidance on resisting spalling tensile stresses, based upon the French Code [247], is given by Clarke [246].

## End block and beam of different shapes

Prestressed concrete beams are generally of a non-rectangular shape with flanges; however, the end blocks are often made rectangular. As one might expect, stress concentrations occur where the section changes from rectangular to non-rectangular, and, at the junction, a second region of transverse bursting tensile stress occurs. Tests [246] indicate that, at the junction of end block and beam, reinforcement should be provided to resist a bursting tensile force equal to 70% of that calculated within the end block.

## Summary

The steps in the design of a general end block can be summarised as follows.

1. Design reinforcement to prevent bursting of the individual prisms of concrete associated with each anchorage.
2. Design reinforcement to tie together the individual prisms.
3. Design reinforcement to resist the spalling tensile stresses between, and at a distance, from anchorages.
4. Design reinforcement to prevent bursting at the junction of a rectangular end block and a non-rectangular beam.

Clarke [246] gives a detailed worked example which illustrates the above design procedure.

## Chapter 11

# Lightweight aggregate concrete

## Introduction

The design recommendations which are discussed in previous chapters are intended only for concretes made from normal weight aggregates. All naturally occurring aggregates, with the exception of pumice, are of normal weight but, as such aggregates become scarce, it is likely that manufactured aggregates will become more popular than they are at present. The majority of manufactured aggregates (e.g., expanded shale and clay, foamed blast furnace slag and sintered pulverised fuel ash) are lightweight.

In addition to the above consideration of the future availability of natural aggregates, the use of lightweight aggregate concrete has obvious advantages where ground conditions are poor, and there is a need to reduce, as much as possible, dead loads and, thus, foundation loads.

Lightweight aggregate concrete has been used throughout the world for both reinforced and prestressed concrete construction, but it has been used far less in Great Britain than in some other countries. This is particularly true of bridge construction [280].

The first lightweight aggregate concrete road bridge built in Great Britain was the Redesdale Bridge, which was constructed by the Forestry Commission in Northumberland. This bridge has a single span of 16.8 m. It is constructed of prestressed precast inverted T-beams with in-situ concrete in-fill to form a composite slab. The beams were cast from a concrete composed of sintered pulverised fuel ash (Lytag) coarse aggregate and natural sand fine aggregate, with a wet density of 1890 kg/m³, to give a minimum strength at transfer of 35 N/mm² and a minimum 28-day strength of 48 N/mm². As explained later, greater losses of prestress occur with lightweight, as compared with normal weight, aggregate concrete; for this bridge, the losses were assumed to be 40% greater at transfer, and 30% greater finally.

The Transport and Road Research Laboratory carried out tests on beams identical to those used in the Redesdale Bridge and found that they behaved satisfactorily under both static and repeated loading [249].

Although the Redesdale Bridge was the first lightweight aggregate concrete road bridge in Great Britain, it does not form part of a public highway. The first lightweight aggregate concrete bridge to be built over a public highway, the Glasshouse Wood Footbridge at Kenilworth having a span of 31.5 m, was designed by the Warwickshire Sub-Unit of the Midland Road Construction Unit. This bridge was opened in 1974. Lytag was used for both coarse and fine aggregates: the 28-day strength was 45.7 N/mm² and the design air-dry density was 1700 kg/m³.

Examples are to be found in Staffordshire of composite slab motorway bridges with spans of about 11 m, which were constructed with normal weight aggregate precast concrete inverted T-beams with lightweight aggregate (Lytag) in-situ in-fill concrete. Lightweight aggregate concrete as in-fill for composite slabs has also been used elsewhere in Great Britain.

Recently Kerensky, Robinson and Smith [250] have reported the successful completion, in 1979, of the Friarton Bridge over the River Tay at Perth. This is a steel–concrete composite bridge consisting of a steel box girder with a composite lightweight aggregate concrete deck slab. Lytag was used for both coarse and fine aggregates, and the design strength and air-dry density were 30 N/mm² and 1680 kg/m³ respectively.

In the following, the structural properties of lightweight aggregate concrete, and how these are dealt with by the Code, are discussed. At present, the requirements for the structural use of lightweight aggregate concrete in highway structures are covered by BE 11 [251]. However, this document limits the use of lightweight aggregate concrete to in-filling (such as between inverted T-beams), and only gives data on density, modulus of elasticity and allowable tensile stresses for in-fill concrete.

## Durability

The durability of lightweight aggregate concrete can be very good, as was demonstrated by one of the early concrete ships – the *Selma*. This ship was constructed of a concrete with expanded shale aggregate and the reinforcement cover was only 10 mm. The reinforcement was still in excellent condition after forty years in service [96].

In non-marine environments, it is thought that, because of the greater porosity of lightweight aggregates, which permits the relatively easy diffusion of carbon dioxide through the concrete, carbonation of the concrete may occur to a greater depth when lightweight aggregates are used. The Code thus requires the cover to the reinforcement to be 10 mm greater than the appropriate value obtained from Table 10.1 for normal weight concrete. However, this requirement may be conservative because tests, carried out by Grimer [252], in which specimens of five different lightweight aggregates and one normal weight aggregate were exposed to a polluted atmosphere for six years, showed that the effect of the type of aggregate on the rate of penetration of the carbonation front was small in comparison with the effect of mix proportions.

# Strength

## Compressive strength

The minimum characteristic strengths permitted by the Code when using lightweight aggregate concrete are 15 N/mm², 30 N/mm² and 40 N/mm² for reinforced, post-tensioned and pre-tensioned construction respectively. It should be noted that BE 11 requires a minimum strength of 22.5 N/mm² for in-fill concrete. These strengths can be attained readily with lightweight aggregates [96], and details of mixes suitable for prestressed concrete have been given by Swamy *et al.* [253].

One important difference between concretes made with lightweight and normal weight aggregates is that the gain of strength with age may be different. In particular the gain in strength with certain lightweight aggregates may be very small for rich mixes [254].

## Tensile strength

The tensile strength of any concrete is greatly influenced by the moisture content of the concrete, because drying reduces the tensile strength. The flexural tensile strength tends to be reduced by drying more than the direct tensile strength.

Curing conditions affect the tensile strength of lightweight aggregate concrete more than normal weight aggregate concrete. Although the tensile strengths are similar for moist cured specimens, the tensile strength of lightweight aggregate concrete when cured in dry conditions can be up to 30% less than that of comparable normal weight concrete [96].

The relatively reduced tensile strength does not influence the design of reinforced concrete, but has to be allowed for in the design of prestressed concrete. No specific guidance is given in the Code, but the CP 110 handbook [112] suggests that all allowable tensile stresses, referred to in the prestressed concrete clauses for nor-

mal weight concrete, should be multiplied by 0.8. This value seems reasonable in view of the reduction in tensile strength referred to in the last paragraph. The implications of the factor of 0.8 are:

1. The allowable tensile stresses for Class 2 pre-tensioned and post-tensioned members are 0.36 $\sqrt{f_{cu}}$ and 0.29 $\sqrt{f_{cu}}$ respectively, instead of 0.45 $\sqrt{f_{cu}}$ and 0.36 $\sqrt{f_{cu}}$, respectively, for normal weight aggregate concrete (see Chapter 4).
2. The basic allowable hypothetical tensile stresses for Class 3 prestressed members, which are given in Table 4.6(a), should be multiplied by 0.8.
3. The allowable concrete flexural tensile stresses for in-situ concrete, when used in composite construction, which are given in Table 4.4, should be multiplied by 0.8. The allowable stresses, so obtained, are very close to those given in BE 11.
4. When calculating the shear strength of a prestressed member uncracked in flexure, the design tensile strength of the concrete ($f_t$) should be taken as 0.19$\sqrt{f_{cu}}$, instead of 0.24 $\sqrt{f_{cu}}$ for normal weight aggregate concrete (see Chapter 6).
5. When calculating the shear strength of a Class 1 or 2 prestressed member cracked in flexure, the design flexural tensile strength of the concrete ($f_r$) should be taken as 0.30 $\sqrt{f_{cu}}$, instead of 0.37 $\sqrt{f_{cu}}$ for normal weight aggregate concrete (see Chapter 6). Hence, equation (6.12), for the cracking moment, becomes

$$M_t = (0.3 \sqrt{f_{cu}} + 0.8 f_{pt})I/y \qquad (11.1)$$

## Shear strength

The shear cracks, which develop in members of lightweight aggregate concrete, frequently pass through the aggregate rather than around the aggregate, as occurs in members of normal weight aggregate concrete. Hence, the surfaces of a shear crack tend to be smoother for lightweight aggregate concrete, and less shear force can be transmitted by aggregate interlock across the crack (see Chapter 6). Since aggregate interlock can contribute 33% to 50% of the total shear capacity of a member [152], the shear strength of a lightweight aggregate concrete member can be appreciably less than that of a comparable normal weight concrete member.

Tests carried out by Hanson [255] and by Ivey and Buth [256] on beams without shear reinforcement have shown that, for a variety of lightweight aggregates, it is reasonable to calculate the shear strength of a lightweight aggregate concrete member by multiplying the shear strength of the comparable normal weight aggregate concrete member by the following factors.

1. 0.75 if both coarse and fine aggregates are lightweight.
2. 0.85 if the coarse aggregate is lightweight and the fine aggregate is natural sand.

In the Code an average value of 0.8 has been adopted for any lightweight aggregate concrete, and, by analogy, the same value has been adopted for torsional strength.

Hence:

1. The nominal allowable shear stresses for reinforced concrete ($v_c$), which are given in Table 6.1, should be multiplied by 0.8.

2. The maximum nominal flexural, or torsional, shear stress should be 0.6 $\sqrt{f_{cu}}$ (but not greater than 3.8 N/mm² and 4.6 N/mm² for reinforced and prestressed concrete respectively), instead of 0.75 $\sqrt{f_{cu}}$ (but not greater than 4.75 N/mm² and 5.8 N/mm² for reinforced and prestressed concrete respectively) for normal weight aggregate concrete (see Chapter 6). Only half of these values should be used for slabs (see Chapter 6).

3. When calculating the shear strength of a Class 1 or 2 prestressed member cracked in flexure, although it is not stated explicitly in the Code, the first term of the right-hand side of equation (6.11) should be taken as 0.03 $bd \sqrt{f_{cu}}$, instead of 0.037 $bd \sqrt{f_{cu}}$ as used for normal weight aggregate concrete.

4. The limiting torsional stress ($v_{tmin}$) above which torsion reinforcement has to be provided should be 0.054 $\sqrt{f_{cu}}$ (but not greater than 0.34 N/mm²), instead of 0.067 $\sqrt{f_{cu}}$ (but not greater than 0.42 N/mm²) as used for normal weight aggregate concrete (see Chapter 6).

5. Although not stated in the Code, it would seem prudent to multiply the basic limiting interface shear stresses for composite construction, which are given in Table 4.5, by 0.8.

## Bond strength

Shideler [257] has carried out comparative pull-out tests on deformed bars embedded in eight different types of lightweight aggregate concrete and one normal weight aggregate concrete. The average ultimate bond stresses developed with the lightweight aggregate concretes were, with the exception of foamed slag, at least 76% of those developed with the normal weight aggregate concrete. The Code reduction factor of 0.8 to be applied, for deformed bars, to the allowable bond stresses of Tables 10.2 and 10.3 thus seems reasonable.

Short and Kinniburgh [258] have reported the results of pull-out and 'bond beam' tests in which plain bars were embedded in three different types of lightweight aggregate concrete and in normal weight aggregate concrete. The average ultimate bond stresses developed with the lightweight aggregate concretes were 50% to 70% of those developed with the normal weight aggregate concrete. The Code reduction factor of 0.5 to be applied, for plain bars, to the allowable bond stresses of Tables 10.2 and 10.3 thus seems reasonable.

Shideler's tests with foamed slag aggregate indicated that the average bond stress could be as low as 66% for horizontal bars due to water gain forming voids in the concrete under the bars. The Code thus advises that allowable bond stresses should be reduced still further (than 20% and 50%) for horizontal reinforcement used with formed slag aggregate: appropriate reduction factors would seem to be 0.65 and 0.33 for deformed and plain bars respectively.

The lower bond stresses developed with lightweight aggregate concrete imply that transmission lengths of pretensioned tendons are greater than the values discussed in Chapter 10 for use with normal weight aggregate concrete. The Code gives no specific advice on transmission lengths for lightweight aggregate concrete, but the CP 110 handbook [112] suggests that they should be taken as 50% greater than those for normal weight aggregate concrete. This increase seems reasonable, as an upper limit, when compared with test data collated by Swamy [259].

## Bearing strength

It can be seen from Fig. 10.6(a), which illustrates an end block of a post-tensioned member, that a bearing failure is, essentially, a tensile splitting failure. Hence the allowable bearing stresses for lightweight aggregate concretes should reflect their reduced tensile strength discussed earlier in this chapter. Since the tensile strength of lightweight aggregate concrete can be up to 30% less than that of comparable normal weight aggregate concrete [96], the Code requires the limiting bearing stress for lightweight aggregate concrete to be two-thirds of that calculated from equation (8.4).

The Code implies that the above reduction should be applied only when considering bearing stresses inside bends of reinforcing bars, but it would seem prudent to apply the reduction to all bearing stress calculations involving lightweight aggregate concrete.

# Movements

## Thermal properties

Lightweight aggregate concrete has a cellular structure and, thus, its thermal conductivity can be as low as one-fifth of the typical value of 1.4 W/m°C for normal weight aggregate concrete [96]. The reduced thermal conductivity is of great benefit in buildings, because it provides good thermal insulation. However, for bridges, it implies that the differential temperature distributions are more severe than those discussed in Chapters 3 and 13 for normal weight aggregate concrete.

Although differential temperature distributions are more severe with lightweight aggregate concrete, their effects are mitigated by the fact that the coefficient of thermal expansion can be as low as $7 \times 10^{-6}/°C$ [96], as compared with approximately $12 \times 10^{-6}/°C$ for normal weight aggregate concrete. The lower coefficient of thermal expansion also means that overall thermal movements of a bridge are less when lightweight aggregate concrete is used. This fact, coupled with the lower elastic modulus of lightweight aggregate concrete (see next section), means that thermal stresses, which result from restrained thermal movements, are less than for normal weight aggregate concrete.

## Elastic modulus

The elastic modulus of lightweight aggregate concrete can range from 50% to 75% of that of normal weight aggregate concrete of the same strength [254]. The higher values are associated with foamed blast furnace slag aggregate and the lower values with expanded clay aggregate [254].

It is mentioned in Chapters 2 and 4 that the Code gives a table of short-term elastic moduli for normal weight aggregate concretes. These values are in good agreement with the following relationship, suggested by Teychenné, Parrot and Pomeroy [20] from considerations of test data.

$$E_c = 9.1 f_{cu}^{0.33}$$

where $E_c$ is the elastic modulus in kN/mm$^2$ and $f_{cu}$ is the characteristic strength in N/mm$^2$. The latter authors further suggested, from considerations of test data, that the elastic modulus of a lightweight aggregate concrete with a density of $D_c$ (kg/m$^3$) could be predicted from

$$E_c = 9.1 (D_c/2400)^2 f_{cu}^{0.33} \qquad (11.2)$$

Equation (11.2) was based upon data from sixty mixes covering four different lightweight aggregates with concrete densities in the range 1400 kg/m$^3$ to 2300 kg/m$^3$.

The Code states that the elastic modulus of a lightweight aggregate concrete, with a density in the above range, can be obtained by multiplying the elastic modulus of a normal weight aggregate concrete by $(D_c/2300)^2$. The resulting elastic modulus will thus agree closely with that predicted by equation (11.2). They will also be within 20% of those specified in BE 11.

The reduced elastic modulus of lightweight aggregate concrete has the following design implications.

1. Stresses arising from restrained shrinkage or thermal movements are less than for normal weight aggregate concrete.
2. Elastic losses in a prestressed member can be up to double those in normal weight aggregate concrete members.
3. Lateral deflections of columns are greater than for normal weight aggregate concrete. Hence stability problems are more likely to occur, and additional moments (see Chapter 9) are greater. These factors are allowed for in the Code by:
   (a) Defining a short column as one with a slenderness ratio of not greater than 10, as compared with the critical ratio of 12 for normal weight aggregate concrete columns (see Chapter 9);
   (b) Substituting the divisor of 1750, in the additional moment parts of equations (9.21) to (9.28), by a divisor of 1200. Hence, the additional moment is increased by nearly 50%. The requirement to reduce the divisor from 1750 to 1200 implies that the assumed extreme fibre concrete strain at failure for lightweight aggregate concrete is about 0.00633 as compared with 0.00375 for normal weight aggregate concrete (see Chapter 9). This increase is reasonable in view of the reduced elastic modulus and the greater creep of lightweight aggregate concrete (see next section).

## Creep

The data on creep of lightweight aggregate concrete, as compared with that of normal weight aggregate concrete, are conflicting. Although creep of lightweight aggregate concrete can be up to twice that of normal weight aggregate concrete [96], it has also been observed [260] that less creep may occur with structural lightweight aggregate concrete as compared with normal weight aggregate concrete.

As is true of all other concretes, creep of lightweight aggregate concrete depends upon a great number of factors, and it is desirable to obtain test data appropriate to the actual conditions under consideration. In lieu of such data, Spratt [96] suggests that creep of lightweight aggregate concrete should be assumed to be between 1.3 and 1.6 times that of normal weight concrete under the same conditions. In similar circumstances, the CP 110 handbook [112] suggests that the loss of prestress due to concrete creep should be assumed to be 1.6 times that calculated for normal weight aggregate concrete.

## Shrinkage

Great variations occur in the shrinkage values for lightweight aggregate concrete; values up to twice those for normal weight aggregate concrete have been reported [96]. In the absence of data pertaining to the actual conditions under consideration, Spratt [96] suggests the adoption of an unrestrained shrinkage strain of between 1.4 and 2.0 times that of normal weight aggregate concrete under the same conditions. In similar circumstances, the CP 110 handbook [112] suggests that the loss of prestress due to shrinkage should be assumed to be 1.6 times that calculated for normal weight aggregate concrete.

## Losses in prestressed concrete

From the previous considerations, it is apparent that total losses in prestressed lightweight aggregate concrete may be up to 50% greater than those in prestressed normal weight aggregate concrete. This is because of the smaller elastic modulus and the greater creep and shrinkage of lightweight aggregate concrete.

# Vibration and fatigue

## Introduction

In this chapter, the dynamic aspects of design are considered in terms of vibration and fatigue. Hence, reference is made to Parts 2, 4 and 10 of the Code.

## Vibration

### Design criterion

It is explained in Chapter 3 that it is not necessary to consider vibrations of highway bridges, because the stress increments due to the dynamic effects are within the allowance made for impact in the nominal highway loadings [107]. In addition, vibrations of railway bridges are allowed for by multiplying the nominal static standard railway loadings by a dynamic factor. Hence, specific vibration calculations only have to be carried out for footbridges and cycle track bridges.

It is explained in Chapter 4 that the appropriate design criterion for footbridges and cycle track bridges is that of discomfort to a user; this is quantified in the Code as a maximum vertical acceleration of $0.5 \sqrt{f_o}$ m/s², where $f_o$ is the fundamental natural frequency in Hertz of the unloaded bridge [107].

### Compliance

#### Introduction

The above design criterion is given in Appendix C of Part 2 of the Code, which also gives methods of ensuring compliance with the criterion. It should be noted that the criterion and the methods of compliance are the same as those in BE 1/77. The background to the compliance rules has been given by Blanchard, Davies and Smith [107]. They considered a pedestrian, with a static weight of 0.7 kN and a stride length of 0.9 m, walking in resonance

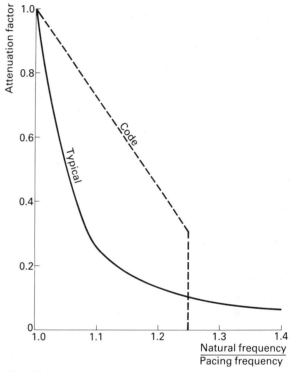

**Fig. 12.1** Attenuation factor

with the natural frequency of a footbridge. It was found that it was possible to excite a bridge in this way if the natural frequency did not exceed 4 Hz, since the latter value is a reasonable upper limit of applied pacing frequency (frequencies above 3 Hz representing running).

Thus, if the natural frequency exceeds 4 Hz, resonant vibrations do not occur; however, it is still necessary to calculate the amplitude of the non-resonant vibrations which do occur when a pedestrian strides at the maximum possible frequency of 4 Hz. Analyses were carried out to determine an attenuation factor defined as the ratio of the maximum acceleration when walking below the resonant frequency to that when walking at the resonant frequency. The results of such an analysis have the form shown in Fig. 12.1. It can be seen that the attenuation factor drops rapidly and, at a frequency ratio of 1.25 (for which the natural frequency is 5 Hz if the pacing frequency is 4 Hz), the attenuation factor is very small. It is thus considered

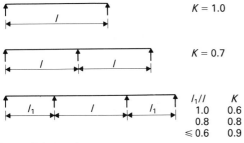

$K = 1.0$

$K = 0.7$

| $l_1/l$ | $K$ |
|---|---|
| 1.0 | 0.6 |
| 0.8 | 0.8 |
| $\leqslant 0.6$ | 0.9 |

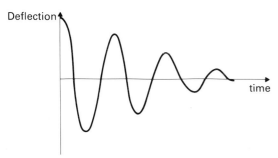

**Fig. 12.2** Configuration factor ($K$)

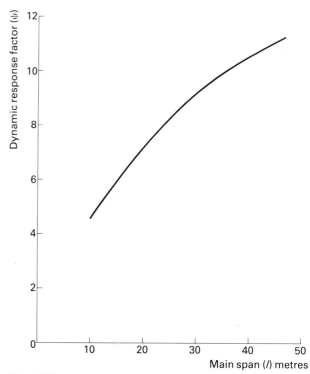

**Fig. 12.4** Dynamic response factor ($\psi$) for $\delta = 0.05$

Deflection

time

**Fig. 12.3** Decay due to damping

very difficult to excite bridges with natural frequencies greater than 5 Hz, and their vibration may be ignored. Hence the Code states that, if the natural frequency of the unloaded bridge exceeds 5 Hz, the vibration criterion is deemed to be satisfied.

If the natural frequency lies between 4 Hz and 5 Hz, Blanchard, Davies and Smith [107] suggest that the maximum bridge acceleration should first be calculated using the natural frequency. This maximum acceleration should then be multiplied by an attenuation factor which varies linearly from 1.0 at 4 Hz to 0.3 at 5 Hz, as shown in Fig. 12.1, to give the maximum acceleration due to a pacing frequency of 4 Hz. This approach has been adopted in the Code.

In the above discussion it is implied that it is necessary to consider only a single pedestrian crossing a bridge. This requirement was proposed by Blanchard, Davies and Smith [107] by considering some existing bridges: for each bridge the number of pedestrians required to produce a maximum acceleration just equal to the allowable value of $0.5 \sqrt{f_o}$ was calculated. It was concluded that, in order that the more sensitive of the existing bridges could be considered to be just acceptable to the Code vibration criterion, the applied loading should be limited to a single pedestrian.

### Calculation of natural frequency

The Code requires that the natural frequency of a concrete bridge should be determined by considering the uncracked section (neglecting the reinforcement), ignoring shear lag, but including the stiffness of parapets. The Code also requires the short-term elastic modulus of concrete to be used. It would seem appropriate to use the dynamic modulus, and Appendix B of Part 4 of the Code tabulates such moduli for various concrete strengths. The tabulated values are in good agreement with data presented by Neville [108].

Wills [261] has used the material and section properties, referred to in the last paragraph, to obtain good agreement

between predicted and observed natural frequencies of existing bridges.

The natural frequency of a bridge should be calculated by including superimposed dead load but excluding pedestrian live loading.

For bridges of constant cross-section and up to four spans, the fundamental natural frequency can be obtained easily by using, for example, tables presented by Gorman [262]. The fundamental natural frequency ($f_o$) is given by [262]

$$f_o = \frac{\beta^2}{2\pi L^2} \sqrt{\frac{EI}{\rho A}} \qquad (12.1)$$

where

$EI$ = flexural rigidity
$A$ = cross-sectional area
$\rho$ = density
$L$ = length of bridge
$\beta$ = parameter dependent on span arrangement and lengths, support conditions and the vibration mode.

For bridges of varying cross-section, it is necessary to use either a computer program, such as that adopted by Wills [261], or a simplified analysis based on a uniform cross-section. In the latter approach, a bridge of varying cross-section is replaced by a bridge having a constant cross-section with a mass per unit length and flexural rigidity equal to the weighted means of the actual masses per unit length and flexural rigidities of the bridge. Equation (12.1) can then be used. Wills [263] explains this procedure and shows that it leads to satisfactory estimates of the natural frequencies of bridges having cross-sections which vary significantly. Nine bridges were considered and the calculated frequencies were within 6% of the measured frequencies except for one bridge, which had an error of 15%.

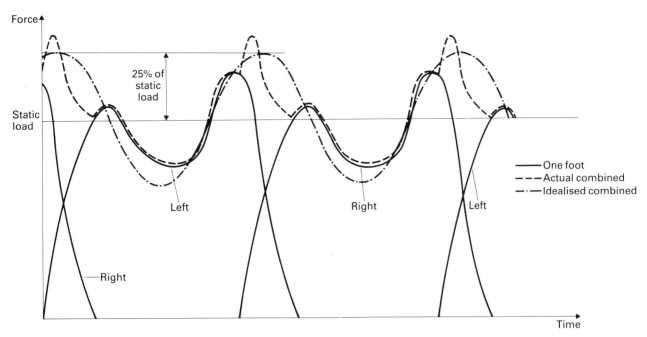

**Fig. 12.5** Moving pulsating force

## Calculation of acceleration

*Simplified method*   For a bridge of constant cross-section and up to three symmetric spans (as shown in Fig. 12.2), the Code gives the following formula for calculating the maximum vertical acceleration ($a$):

$$a = 4\pi^2 f^2_o\, y_s\, k\psi \qquad (12.2)$$

where

$f_o$  = fundamental natural frequency (Hz)
$y_s$  = static deflection (m)
$K$  = configuration factor
$\psi$  = dynamic response factor

If $f_o$ lies in the range 4 Hz to 5 Hz, the acceleration calculated from equation (12.2) should be reduced by applying the attenuation factor discussed earlier in this chapter.

Equation (12.2) was derived by Blanchard, Davies and Smith [107] and represents, in a simple form, the results of a study of a number of bridges with different span arrangements. For each bridge, a numerical solution to its governing equation of motion was obtained.

The static deflection ($y_s$) should be calculated, at the midpoint of the main span, for a vertical load of 0.7 kN, which represents a single pedestrian.

The configuration factor ($K$) depends upon the number and lengths of the spans; values are given in Fig. 12.2. Linear interpolation may be used for intermediate values of $l_1/l$ for three span bridges.

The dynamic response factor ($\psi$) depends upon the main span length and the damping characteristics of the bridge. If a bridge is excited, the amplitude of the vibration gradually decreases due to damping, as shown in Fig. 12.3. The damping is expressed in terms of the logarithmic decrement ($\delta$), which is the natural logarithm of the ratio of the amplitude in one cycle to the amplitude in the following cycle. The Code suggests that, in the absence of more precise data, the logarithmic decrement, of both reinforced

and prestressed concrete footbridges, should be assumed to be 0.05. In the past, a larger value has often been taken for reinforced concrete than for prestressed concrete, because it was felt that cracks in reinforced concrete dissipate energy and thus improve the damping. However, according to Tilly [264], this view is not supported by experimental data. Furthermore, when considering a bridge, it is the overall damping, due to energy dissipation at joints, etc, in addition to inherent material damping, which is of interest. Tests on existing concrete footbridges have indicated logarithmic decrements in the range 0.02 to 0.1, and thus the Code value of 0.05 seems reasonable [264].

The relationship between the dynamic response factor ($\psi$) and the main span length ($l$) is given graphically in the Code. The relationship is shown in Fig. 12.4 for the logarithmic decrement of 0.05 suggested in the Code.

*General method*   For bridges with non-uniform cross-sections, and/or unequal side spans, and/or more than three spans, the above simplified method of determining the maximum vertical acceleration is inappropriate. In such situations, it is necessary to analyse the bridge under the action of an applied moving and pulsating point load, which represents a pedestrian crossing the bridge. The amplitude of the point load was chosen so that, when applied to a simply supported single span bridge, it produces the same response as that produced by a pedestrian walking across the bridge [107]. In Fig. 12.5, the force–time relationship is shown for a single foot. The relationship obtained by combining consecutive single foot relationships is also shown. It can be seen that the combined effect can be represented by a sine wave with an amplitude of about 25% of the static single pedestrian load of 0.7 kN. In fact the Code takes the amplitude to be 180 N and, as discussed earlier in this chapter, the pacing frequency is taken to be equal to the natural frequency of the bridge ($f_o$). Hence the pulsating point load ($F$ in Newtons), given in the Code, is:

$$F = 180 \sin(2\pi f_o T) \qquad (12.3)$$

where $T$ is the time in seconds.

It is mentioned earlier in this Chapter that the assumed stride length is 0.9 m. Thus, if the pacing frequency is $f_o$Hz, the required velocity ($v_t$ in m/s) of the pulsating point load is given by:

$$v_t = 0.9 f_o \qquad (12.4)$$

Wills [261] discusses two methods of analysing a bridge under the above moving pulsating point load. One method requires a large amount of computer storage space and the other more approximate method requires much less storage space. Wills [261] shows that the approximate method is adequate for many footbridges.

## Forced vibrations

Up to now in this chapter, only those vibrations which result from normal pedestrian use of a footbridge have been considered. However, it is also necessary to consider the possibility of damage arising due to vandals deliberately causing resonant oscillations. It was not possible [107] to quantify a loading or a criterion for this action; thus the Code merely gives a warning that reversals of load effects can occur. However, the Code does suggest that, for prestressed concrete, the section should be provided with unstressed reinforcement capable of resisting a reverse moment of 10% of the static live load moment.

## Fatigue

### General

#### Code approach

Although Part 1 of the Code refers to fatigue under the heading 'ultimate limit state', it is the repeated application of *working loads* which cause deterioration to a stage where failure occurs. Alternatively, the working loads may cause minor fatigue damage, which could result in the bridge being considered unserviceable. Hence, fatigue calculations are carried out separately from the calculations to check compliance with the ultimate and serviceability limit state criteria: in Part 4 of the Code, fatigue is dealt with under 'Other considerations'. The *design* fatigue loading is specified in Part 10 of the Code and, since it is a design loading, partial safety factors ($\gamma_{fL}$ and $\gamma_{f3}$) do not have to be applied.

When determining the response of a bridge to fatigue loading, Part 1 of the Code requires the use of a linear elastic method with the elastic modulus of concrete equal to its short-term value.

With regard to concrete bridges, Part 4 of the Code requires only the fatigue strength (or life) of reinforcing bars to be assessed. Thus concrete and prestressing tendons do not have to be considered in fatigue calculations. Before presenting the Code requirements for reinforcing

bars, the implications of not considering concrete and pre-stressing tendons are discussed briefly in a simplistic manner.

#### Concrete

Concrete in compression can withstand, for 2 million cycles of repeated loading, a maximum stress of about 60% of the static strength if the minimum stress in a cycle is zero [265]. The maximum stress which can be tolerated increases as the minimum stress increases. Since Part 4 of the Code specifies a limiting compressive stress of $0.5 f_{cu}$ for concrete at the serviceability limit state (see Chapter 4), it is very unlikely that fatigue failure of concrete in compression would occur. It is thus reasonable for the Code not to require calculations for assessing the fatigue life of concrete in compression.

It should be noted that 2 million cycles of loading during the specified design life of 120 years are equivalent to about 50 applications of the full design load per day.

For concrete in tension, cracking occurs at a lower stress under repeated loading than under static loading. Cracking does not occur, in less than 2 million cycles of repeated loading, if the maximum tensile stress does not exceed about 60% of the static tensile strength, if the minimum stress is zero or compressive [265, 266]. The maximum tensile stress which can be tolerated without cracking increases if the minimum stress is tensile. The reduced resistance to cracking of concrete subjected to repeated loading has two implications.

1. Shear cracks may form at a lower load, with a possible decrease in shear strength.
2. Flexural cracks may form at a lower load, resulting in either cracks in Class 1 or 2 prestressed members, or cracks wider than the allowable values in reinforced concrete or Class 3 prestressed concrete members.

With the partial safety factors for loads and for material properties that have been adopted in the Code at the ultimate limit state, it is unlikely that principal tensile stresses under working load conditions would be great enough to cause fatigue shear cracks. Thus a bridge, designed to resist static shear in accordance with the Code, should exhibit adequate shear resistance when subjected to repeated loading.

The interface shear strength of composite members should also be adequate under repeated loading. Badoux and Hulsbos [279] have tested composite beams under 2 million cycles of loading. The test specimens were essentially identical to those, tested under static loading by Saemann and Washa [118], which are discussed in Chapter 4. It was found that, under repeated loading, the interface shear strength was reduced. However, the allowable interface shear stresses, which were proposed in [279] for repeated loading, exceed the allowable stresses given in the Code for static loading at the serviceability limit state (see Chapter 4). Since the latter stresses have to be checked under the full design load at the serviceability limit state (i.e. under dead plus imposed loads) it is very unlikely that interface shear fatigue failure would occur. The requirement to check interface shear stresses under the full design load is discussed fully in Chapter 8.

Flexural cracking under repeated loading is now considered.

*Reinforced concrete* Repeated loading causes cracks to form at a lower load than under static loading and, subsequently, the cracks are wider. However, the author would suggest that the breakdown of tension stiffening under repeated loading has a greater influence on crack widths than does the reduction in the load at which cracking occurs. Tension stiffening under repeated loading is discussed in Chapter 7, where the author suggests that, as an interim measure, tension stiffening under repeated loading should be taken as 50% of that under static loading.

*Class 1 prestressed concrete* Flexural tensile stresses are not permitted under service load conditions and thus repeated loading cannot cause cracking.

*Class 2 prestressed concrete* No flexural tensile stresses are permitted under dead plus superimposed dead loads (see Chapter 4), and thus the minimum flexural stress is always compressive. For repeated loading, the maximum tolerable tensile stress, appropriate to a compressive minimum stress, is about 60% of the static tensile strength. Since the flexural tensile stress permitted by the Code may be up to 80% of the static flexural tensile strength (see Chapter 4), it is possible that flexural fatigue cracking could occur in a Class 2 member designed in accordance with the Code. It is significant that, in CP 115, the allowable tensile stresses for repeated loading are about 65% of those for non-repeated loading; the latter stresses are very similar to the Code values. It would thus seem prudent to take about 65% of the Code values when considering repeated loading.

*Class 3 prestressed concrete* A Class 3 member is designed to be cracked under the serviceability limit state design load. Repeated loading may cause the cracks to be wider than under static loading. However, the permissible hypothetical tensile stresses in the Code (see Chapter 4) are conservative in comparision with test data [120, 122, 123], and thus excessive cracking under repeated loading should not occur.

*Prestressing tendons*

The allowable concrete flexural, compressive and tensile stresses specified in the Code imply that the stress range, under service load conditions, of a prestressing tendon in a Class 1 or 2 member cannot exceed about 10% of the ultimate static strength.

If it is assumed that the effective prestress in a tendon is about 45% to 60% of its ultimate strength, then test results indicate that it may be conservatively assumed that 2 million cycles of stress can be withstood by a tendon without failure, providing that the stress range does not exceed about 10% of the ultimate static strength [265]. It is thus unlikely that fatigue failure of a prestressing tendon, in a Class 1 or 2 member, would occur under service load conditions.

In a Class 3 member, designed for the maximum allowable hypothetical tensile stress of $0.25 f_{cu}$ (see Chapter 4), stress ranges in tendons could be up to 15% of the ultimate static strength of a tendon. Fatigue failure of tendons are thus possible in some Class 3 members.

However, it should be emphasised that, in the vast majority of Class 3 members, tendon stress ranges will be much less than the values quoted above and fatigue failures would then be unlikely. Nevertheless, it would seem prudent to ensure that, for all classes, tendon stress ranges do not exceed 10% of the ultimate static strength of the tendon.

It should be noted that the conservative tolerable stress range of 10% of the ultimate static strength, which is quoted in the last paragraph, is based upon tests on prestressing tendons in air. However, the work of Edwards [267, 268] has shown that tendon fatigue strength can be less when embedded in concrete than in air: for 7-wire strand, the stress range in concrete was about 8% of the ultimate static strength as compared with about 13% in air.

Fatigue failure of anchorages and couplers, rather than of a tendon, should also be considered, since they can withstand smaller stress ranges [265]. Particular attention should be given to the possibility of an anchorage fatigue failure when unbonded tendons are used, because stress changes in the tendon are transmitted directly to the anchorages.

## Code fatigue highway loading

A table in Part 10 of the Code gives the total number of commercial vehicles (above 15 kN unladen weight) per year which should be assumed to travel in each lane of various types of road. The number of vehicles varies from $0.5 \times 10^6$ for a single two-lane all purpose road to $2 \times 10^6$ for the slow lane of a dual three-lane motorway.

Part 10 of the Code also gives a load spectrum for commercial vehicles showing the proportions of vehicles having various gross weights (from 30 kN to 3680 kN) and various axle arrangements. The load spectrum depicts actual traffic data in terms of twenty-five typical commercial vehicle groups.

It is obviously tedious in design to have to consider a number of different axle arrangements, and thus it was decided to specify a standard fatigue vehicle. The intention was that each type of commercial vehicle in the load spectrum would be represented by a vehicle having the same gross weight as the actual vehicle, but having the axle arrangement of the standard fatigue vehicle.

The standard fatigue vehicle was chosen to give the same cumulative fatigue damage, for welded connections in steel bridges, as do the actual vehicles. However, Johnson and Buckby [24] have emphasised that equivalence of cumulative fatigue damage does not occur for shear connectors in composite (steel–concrete) construction. This is because fatigue damage in welded steel connections is proportional to the third or fourth power of the stress range whereas, for a shear connector, it is proportional to the eighth power. Fatigue damage of *unwelded* reinforcing bars is proportional to stress range to the power 9.5 (see

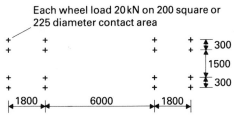

Each wheel load 20 kN on 200 square or 225 diameter contact area

**Fig. 12.6** Standard fatigue vehicle

next section), and thus the standard fatigue vehicle would not give equivalence of fatigue damage for unwelded bars. However, it is appropriate for welded bars.

The standard fatigue vehicle specified in Part 10 of the Code has a gross weight of 320 kN, and consists of four wheels on each of four axles. The vehicle is similar to the shortest HB vehicle (see Fig. 3.6) but the transverse spacing of the wheels is different, as shown in Fig. 12.6.

Part 10 of the Code also gives a simplified load spectrum for use with the standard fatigue vehicle. The simplified spectrum gives the proportions of total commercial vehicles for various multiples of the standard fatigue vehicle gross weight of 320 kN.

## Compliance clauses for reinforcing bars

### Unwelded bars

Part 10 of the Code does not give a compliance clause for unwelded bars: instead maximum allowable stress ranges under the normal 'static' design loads at the serviceability limit state are specified in Part 4 of the Code. These ranges are 325 N/mm² for high yield bars and 265 N/mm² for mild steel bars, and are taken from BE 1/73.

The author is unaware of the origin of these stress ranges, but their implications are now considered in a simplistic manner.

Moss [269] has reported that the stress range ($\sigma_r$) – cycles to failure ($N$) relationship for a variety of high yield bars, when the bars are always in tension, is

$$N\sigma_r^{9.5} = 1.8 \times 10^{29} \qquad (12.5)$$

Thus, for a stress range of 325 N/mm², the fatigue life ($N$) is $0.25 \times 10^6$ cycles. However, a stress range as large as 325 N/mm² is only likely to occur when there is a reversal of stress during the stress cycle (i.e. the stress changes from compression to tension). In such circumstances the stress range for a particular number of cycles may be up to one-third greater than that predicted by equation (12.5). Hence, the upper limit to equation (12.5) can be expressed as

$$N\sigma_r^{9.5} = 2.8 \times 10^{30} \qquad (12.6)$$

Thus, for a stress range of 325 N/mm², the fatigue life from equation (12.6) is $3.8 \times 10^6$ cycles. Hence, the fatigue life of a bar, designed to have the maximum stress range of Part 4 of the Code, is likely to be of the order of $0.25 \times 10^6$ to $3.8 \times 10^6$ cycles. Thus, the specified stress ranges seem reasonable. However, it should be remembered that bending a bar can reduce its fatigue strength by up to 50% [265]. This fact should not affect shear links, since shear cracks are unlikely to occur under service load

conditions, and hence the stress range in a link is always very small.

### Welded bars

*General* Part 4 of the Code permits the use of welded bars, provided that the following four requirements are met:

1. Welding must be carried out in accordance with Parts 7 and 8 of the Code. Tack welding is not permitted since it can reduce fatigue strength considerably; for example, tack welding of stirrups to main bars can reduce the fatigue strength of the latter by about 35% (or fatigue life by 75%) [270].
2. The welded bar is not part of a top slab. This is, presumably, because such bars are subjected to the local effects of concentrated wheel loads in addition to the global effects of the standard fatigue vehicle.
3. The stress range is within that permitted by Part 10 of the Code. This point is discussed in the following sections of this chapter.
4. Lap welding should not be used, because adequate control cannot be exercised over the profile of the root beads.

*$\sigma_r$ – N relationships* Part 10 of the Code gives $\sigma_r$ – $N$ relationships for the following cases:

1. Welded intersections in fabric, or between hot rolled bars.
2. Butt weld between the ends of hot rolled bars.
3. Butt weld between the end of a hot rolled bar and the surface of a plate.
4. Fillet weld between the end of a hot rolled bar and the surface of a plate.

The appropriate design $\sigma_r$ – $N$ relationship, which corresponds to a 2.3% probability of failure is:

$$N\sigma_r^3 = K \times 10^{12} \qquad (12.7)$$

where $K$ is 1.52 for cases 1 and 2, 0.63 for case 3 and 0.43 for case 4. These relationships were derived, principally, for structural steelwork connections and their use is thus restricted in the Code to hot rolled bars. However, Moss [269] has tested butt welded connections in both hot rolled and cold worked bars, and he found that the Code relationship is a satisfactory lower bound fit to his test data.

The Code gives three methods of using equation (12.7) to assess fatigue life. The methods are now described briefly in ascending order of accuracy and complexity. Full descriptions are given in Part 10 of the Code.

*Assessment without damage calculation* Equation (12.7) has been used in conjunction with the simplified load spectrum, referred to earlier in this chapter, to produce graphs of limiting stress range ($\sigma_H$) against loaded length ($L$) for various categories of road. The loaded length is defined as the base length of the loop of the point load influence line which contains the greatest ordinate, as shown in Fig. 12.7. A full description of the derivation of the

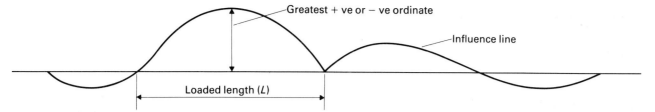

Fig. 12.7 Loaded length for fatigue calculation

graphs is given by Johnson and Buckby [24], and the graphs are given in Part 10 of the Code.

The graph of $\sigma_H$ against $L$ for a butt weld between the ends of two bars is shown in Fig. 12.8 for a dual three-lane motorway.

The stresses due to the standard (320 kN) fatigue vehicle as it crosses the bridge in each slow lane and each adjacent lane in turn should be calculated. At a particular design point on the bridge, the stress range is the greatest algebraic difference between these calculated stresses, irrespective of whether the maximum and minimum stresses result from the vehicle being in the same lane. The stress range, so calculated, should not exceed the appropriate limiting stress range ($\sigma_H$).

In some situations (e.g. at an expansion joint) the stresses calculated due to the passage of the standard fatigue vehicle should be increased by an impact factor, which can be up to 25%.

It is emphasised that a load spectrum is not used with this simplified method of compliance.

*Damage calculation, single vehicle method*   As with the previous method, a load spectrum is not used: instead, the standard fatigue vehicle is applied to each slow and each adjacent lane in turn. However, instead of just considering the greatest stress range (as is done in the previous method), the fatigue damage due to each stress cycle is assessed by means of a 'damage chart' given in Part 10 of the Code. The total damage is assumed to be the sum of the damages contributed by each individual stress cycle. The derivation of the damage charts and an example of

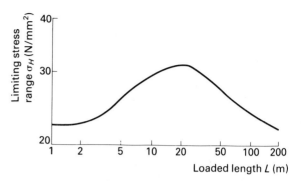

Fig. 12.8 Limiting stress range for butt welded bars for dual three-lane motorway

their use are given in Appendices C and D, respectively, of Part 10 of the Code.

*Damage calculation, vehicle spectrum method.* This method requires each vehicle in the load spectrum to be traversed along each lane, and consideration should be given to vehicles occurring simultaneously in one or more lanes. From the resulting stress histories, a stress spectrum, which gives the number of occurrences of various stress ranges, can be produced by the methods described in Part 10 of the Code. The fatigue damage due to each stress range can then be calculated from the $\sigma_r - N$ curves given in the Code. The cumulative damage is assessed by using Miner's rule, which states that the cumulative damage is equal to the sum of the individual damages [271]. This summation should not exceed unity if the fatigue life is to be considered acceptable.

## Chapter 13

# Temperature loading

## Introduction

As explained in Chapter 3, it is necessary to consider two aspects of temperature loading: overall temperature movements, and differential temperature effects. The overall movements are discussed first.

Having located the point on the structure which does not move (the stagnant point), it is simple to calculate the overall movement at any other point of the structure. If the articulation of the bridge is not complicated, it is possible to locate the stagnant point by inspection (e.g. at a fixed bearing). However, in general, it is necessary to consider the relative stiffnesses of the deck, piers and foundations in order to calculate the stagnant point. This calculation is discussed in detail by Zederbaum [272]. If any of the overall movements are restrained then stress resultants are induced in the structure. These stress resultants can be simply calculated from a knowledge of the restrained movements.

Since these overall movement calculations are well known and are not contentious, they are not discussed further in this chapter.

In contrast, many bridge engineers are uncertain as to how to include differential temperature effects into the design procedure. In particular, it is not clear how to deal with cracked sections, and there are differing views on the method of calculation of differential temperature effects at the ultimate limit state. The Code gives no guidance whatsoever on these matters. In view of this, the remainder of this chapter is concerned with differential temperature effects, and design procedures are suggested. These procedures are, to a certain extent, based upon current uncompleted research at the University of Birmingham and, thus, they should be considered as interim measures until' the research is completed.

## Serviceability limit state

### Component effects

#### Introduction

In the following a general non-linear differential temperature distribution is considered to be applied to a general section, as shown in Fig. 13.1.

It is important to realise that, when a temperature distribution is applied to a structure, temperature-induced strains occur, but temperature-induced stresses result only if such strains are restrained, either externally or internally.

There are two basic approaches to the determination of the effects of a non-linear temperature distribution, such as that shown in Fig. 13.1: the strain method or the stress method. The two methods are based upon the same assumptions of structural behaviour, and their end results are identical. In view of the fact that temperature loading is, in structural terms, an applied deformation, the author prefers the strain method; however, many bridge engineers prefer the stress method because they are used to working in terms of stresses, and because it is computationally more convenient for statically indeterminate structures. Each method is now presented.

In the following all stresses, strains and stress resultants are positive when tensile.

### Strain method

Consider the general section and temperature distribution of Fig. 13.1. At a distance $z$ above the bottom of the section, the temperature is $t_z$, the section breadth is $b_z$, the elastic modulus is $E_z$ and the coefficient of expansion is $\alpha_z$; each is a function of $z$. The coefficient of expansion may vary with depth due to different materials existing in the section. The elastic modulus may vary for two reasons:

1. Different materials existing in the section.
2. Stresses, due to co-existing force loading, varying through the depth of the section and, thus, being on different points of the material stress–strain curve.

The potential thermal strain ($\varepsilon$) at $z$, in the absence of any restraint, is

$$\varepsilon = \alpha_z t_z \qquad (13.1)$$

If it is assumed that plane sections remain plane, then the section must take up the strains indicated by the dashed line of Fig. 13.2(a). The latter strains can be defined in terms of a strain ($\varepsilon_o$) at $z = 0$, and a curvature ($\psi$). At any level the difference between the potential thermal strain and the strain that actually occurs is a strain which induces a thermal stress. Since no external force is applied to the sections, these thermal stresses must be self-equilibrating.

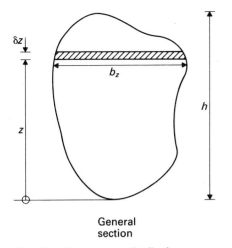 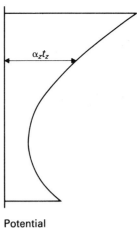

General section      Temperature distribution      Potential thermal strains

**Fig. 13.1** Temperature distribution

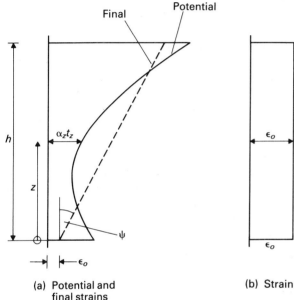 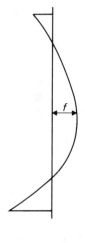

(a) Potential and final strains     (b) Strain     (c) Curvature     (d) Self-equilibrating stresses

**Fig. 13.2(a)–(d)** Thermal strain method

Hence, in the absence of any external restraint, the section experiences an axial strain, a curvature and a set of self-equilibrating stresses, as shown in Fig. 13.2.

At distance $z$ above the bottom of the section, the stress-inducing strain is

$$\varepsilon_o + \psi z - \alpha_z t_z$$

Thus, the stress at $z$ is

$$f = E_z (\varepsilon_o + \psi z - \alpha_z t_z) \tag{13.2}$$

For force and moment eqilibrium respectively:

$$\int_o^h f b_z dz = 0 \tag{13.3}$$

$$\int_o^h f b_z z dz = 0 \tag{13.4}$$

If equation (13.2) is substituted into equations (13.3) and (13.4), $\varepsilon_o$ and $\psi$ can be obtained as

$$\varepsilon_o = (F_3 F_4 - F_2 F_5)/F_1 F_3 - F_2^2) \tag{13.5}$$

$$\psi = (F_1 F_5 - F_2 F_4)/(F_1 F_3 - F_2^2) \tag{13.6}$$

where

$$F_1 = \int_o^h E_z b_z dz$$

$$F_2 = \int_o^h E_z b_z z dz$$

$$F_3 = \int_o^h E_z b_z z^2 dz$$

$$F_4 = \int_o^h \alpha_z E_z b_z t_z dz$$

$$F_5 = \int_o^h \alpha_z E_z b_z t_z z dz$$

$$\left. \right\} \tag{13.7}$$

Having obtained $\varepsilon_o$ and $\psi$, the self-equilibrating stresses can be obtained from equation (13.2).

*Uncracked section* For the case of a section which is uncracked, and in which the concrete stresses are sufficiently low for the elastic modulus to be considered constant throughout the section, equations (13.5) and (13.6) reduce to

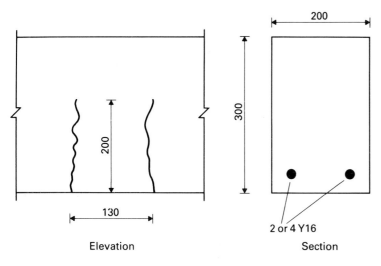

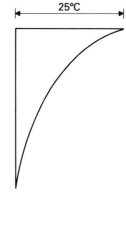

Elevation            Section            Temperatures
above ambient

**Fig. 13.3** Cracked section

$$\varepsilon_0 = \frac{\alpha_c}{AI}\left[ I\int_o^h b_z t_z dz - A\bar{z}\int_o^h b_z t_z(z - \bar{z})\ dz \right] \quad (13.8)$$

$$\psi = \frac{\alpha_c}{I}\int_o^h b_z t_z\ (z - \bar{z})\ dz \quad (13.9)$$

where $A$ is the cross-sectional area, $\bar{z}$ is the distance of the section centroid from the bottom of the section, $I$ is the second moment of area about the centroid and $\alpha_c$ is the coefficient of expansion of concrete. The strain at the centroid of the section $(\varepsilon)$ is given by

$$\bar{\varepsilon} = \varepsilon_o + \psi\ \bar{z}$$

$$\therefore\ \bar{\varepsilon} = \frac{\alpha_c}{A}\int_o^h b_z t_z dz \quad (13.10)$$

*Cracked section*  Tests carried out by Church [273] at the Transport and Road Research Laboratory, on both cracked and uncracked sections under the thermal loading shown in Fig. 13.3, have shown that the free thermal curvature of a cracked section is, typically, 20% greater than that of an uncracked section. Furthermore, the thermal curvature can be calculated from equation (13.6) by assuming the elastic modulus of the concrete, within the height of the crack, to be zero. If it is assumed that the crack extends to the neutral axis of the cracked transformed section, and the thermal stresses do not affect the position of the neutral axis, then equations (13.9) and (13.10) become

$$\psi = \frac{1}{E_c I}\left[ \alpha_s E_s A_s t_s\ (h - d - \bar{z}) + \right.$$

$$\left. \alpha_c E_c \int_{\bar{z}}^h b_z t_z\ (z - \bar{z})dz \right] \quad (13.11)$$

$$\bar{\varepsilon} = \frac{1}{E_c A}\left[ \alpha_s E_s A_s t_s + \alpha_c E_c \int_{\bar{z}}^h b_z t_z dz \right]$$

where

$\alpha_s$ = coefficient of expansion of reinforcement

$E_s$ = elastic modulus of reinforcement

$A_s$ = area of reinforcement

$t_s$ = temperature at level of reinforcement

$d$ = effective depth of reinforcement

$A$ = area of cracked transformed section

$I$ = second moment of area of cracked transformed section about its centroid

$\bar{z}$ = distance of centroid of cracked transformed section from the bottom of the section.

Tests have not yet been carried out under temperature distributions which cause the cracked part of the section to be hotter than the uncracked part. However, in such circumstances, the first of equations (13.11) predicts, for temperature differences of the shape shown in Fig. 13.3, that the free thermal curvature is, typically, 20% less than that of an uncracked section.

However, when the Code temperature distribution, with its non-zero temperature at the bottom of the section (see Fig. 3.2) is considered, it is found that the differences between the calculated free thermal curvatures for the uncracked and cracked section are much less than 20% (see Example 13.1). Thus the author would suggest that, for design purposes, the free thermal curvature could be calculated from equation (13.9) (which is for an uncracked section), irrespective of whether the section is cracked or uncracked.

### Stress method

In the stress method, the section is, at first, assumed to be fully restrained, so that no displacements take place, as shown in Fig. 13.4(a). The stress $f_o$, at $z$, due to the restraint is

$$f_o = - E_z\ \alpha_z t_z \quad (13.12)$$

Hence, the restraining force is

$$F = \int_o^h\ f_o b_z dz \quad (13.13)$$

If no longitudinal restraint is present, the restraining force must be released by application of a releasing force $F_r = -F$. The stress $(f_1)$ at $z$ due to the releasing force is

$$f_1 = - E_z\ \int_o^h f_o b_z dz \bigg/ \int_o^h E_z b_z dz \quad (13.14)$$

The net stress at $z$ is now $(f_o + f_1)$.

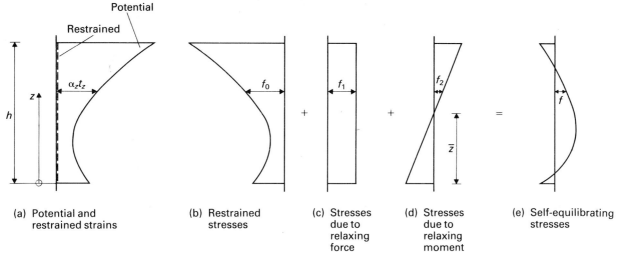

(a) Potential and restrained strains    (b) Restrained stresses    (c) Stresses due to relaxing force    (d) Stresses due to relaxing moment    (e) Self-equilibrating stresses

**Fig. 13.4(a)–(e)** Thermal stress method

The restraining moment ($M$) can be obtained by taking moments about $z = 0$; if a positive moment is sagging:

$$M = - \int_o^h (f_o + f_1) b_z z \, dz \qquad (13.15)$$

If no moment restraint is present, the restraining moment must be released by application of a releasing moment $M_r = -M$. The stress ($f_2$) at $z$ due to the releasing moment is

$$f_2 = -E_z(z - \bar{z}) (-M) \Big/ \int_o^h E_z b_z (z - \bar{z})^2 dz \qquad (13.16)$$

If the section is uncracked, equations (13.12) to (13.16) become

$$f_o = -E_c \, \alpha_c \, t_z$$

$$F = -\alpha_c E_c \int_o^h b_z t_z \, dz$$

$$f_1 = \alpha_c E_c \int_o^h b_z t_z \, dz / A \qquad (13.17)$$

$$M = \alpha_c E_c \int_o^h b_z t_z (z - \bar{z}) \, dz$$

$$f_2 = -(z - \bar{z})(-M)/I$$

If the section is cracked, the elastic modulus of the concrete, within the height of the crack, should be assumed to be zero. The equations for $F$ and $M$ then become:

$$F = -\alpha_s E_s A_s t_s - \alpha_c E_c \int_{\bar{z}}^h b_z t_z \, dz$$

$$M = \alpha_s E_s A_s t_s (h - d - \bar{z}) + \qquad (13.18)$$

$$\alpha_c E_c \int_{\bar{z}}^h b_z t_z (z - \bar{z}) \, dz$$

This approach is very similar to that adopted by Hambly [274] for a cracked section.

It can be seen from Fig. 13.4 that the self-equilibrating stresses can be calculated from

$$f = f_o + f_1 + f_2 \qquad (13.19)$$

The stresses calculated from equation (13.19) are identical to those calculated, by the strain method, from equation (13.2).

## Externally unrestrained structure

In a structure which is externally unrestrained, such as a simply supported beam, the strain $\varepsilon_o$ and the curvature $\psi$ can occur freely. Thus the only stresses in the structure are the self-equilibrating stresses calculated from equation (13.2) or (13.9).

However, these stresses do not occur at the free ends of the structure, where plane sections distort and do not remain plane. Hence the stresses build-up from zero to the values given by equations (13.2) and (13.19). Such a build-up of stress implies that, in order to maintain equilibrium, longitudinal shear stresses occur near to the ends of a member. The calculation of these shear stresses is demonstrated in Example 13.2.

## Externally restrained structure

If the strain $\varepsilon_o$ or the curvature $\psi$ is prevented from occurring by the presence of external restraints, then secondary stresses occur in addition to the self-equilibrating stresses. The secondary stresses can be calculated by using either a compatibility method or an equilibrium method. The author feels that the compatibility method explains the actual behaviour better, but the equilibrium method is generally preferred because it is computationally more convenient. The two methods are compared in the following section by considering a two-span beam.

It should be emphasised that bridge decks are two-dimensional in plan, and thus the transverse effects of temperature loading should also be considered. However, the same principles, which are illustrated for a one-dimensional structure, can also be applied to two-dimensional structures (see reference [99]).

### Compatibility method

The supports of the two-span beam shown in Fig. 13.5 are assumed to permit longitudinal movement, so that the strain $\varepsilon_o$ can occur. However, the free thermal curvature $\psi$ is prevented from occurring by the centre support. If the centre support were absent, the beam would take up the

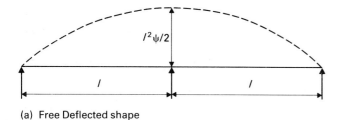

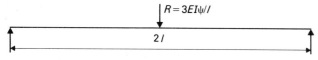

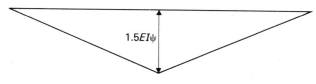

(a) Free Deflected shape

$R = 3EI\psi/l$

$2l$

(b) Force applied to give zero displacement at centre support

$1.5EI\psi$

(c) Thermal moments

**Fig. 13.5(a)–(c)** Compatibility method

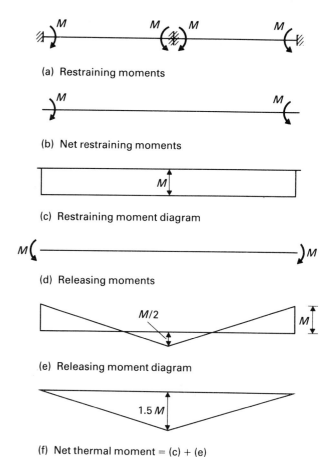

(a) Restraining moments

(b) Net restraining moments

(c) Restraining moment diagram

(d) Releasing moments

$M/2$

(e) Releasing moment diagram

$1.5\,M$

(f) Net thermal moment = (c) + (e)

**Fig. 13.6(a)–(f)** Equilibrium method

free thermal curvature, and the displacement at the position of the centre support would be $l^2\psi/2$, as shown in Fig. 13.5(a). It is now necessary to apply a vertical force $R$ at the centre support, as shown in Fig. 13.5(b), to restore the beam at this point to the level of the centre support. Hence, for the two-span beam,

$$R = 48\,EI\,(l^2\psi/2) / (2l)^3 = 3EI\psi/l$$

This force induces the thermal moments shown in Fig. 13.5(c); the maximum moment, at the support, is

$$M_s = R(2l)/4 = 1.5\,EI\psi \tag{13.20}$$

If the section is uncracked, the $EI$ value of the uncracked section should be used in equation (13.20). If the section is cracked, the author would suggest that the $EI$ value of the cracked transformed section should be used in equation (13.20). In addition, this value of $EI$ and the neutral axis depth appropriate to the cracked transformed section should be used to calculate the secondary stresses due to the thermal moment. However, although $\psi$ could be calculated from equations (13.6) and (13.11), it is probably sufficiently accurate to calculate $\psi$ from equation (13.9) using the second moment of area of the *uncracked* section, as discussed earlier in this chapter.

### Equilibrium method

At each support, the beam is first assumed to be fully restrained against rotation but not against longitudinal movement. Thus restraining moments ($M$), given by equation (13.15) and shown in Fig. 13.6(a), are set up. The net restraining moments are shown in Fig. 13.6(b), and the restraining moment diagram is shown in Fig. 13.6(c). Since no external moments are applied, it is necessary, for equilibrium, to cancel the end restraining moments by applying releasing moments which are equal and opposite to the restraining moments, as shown in Fig. 13.6(d). If the beam is analysed under the effects of the releasing

moments, the moment distribution shown in Fig. 13.6(e) is obtained. The final thermal moments, shown in Fig. 13.6(f), are obtained by summing the restraint moments and the distributed releasing moments. The maximum thermal moment is $1.5\,M$; by using equations (13.12), (13.14) and (13.15), this moment can be shown to be identical to that given by the compatibility method (equation (13.20)).

If the section is uncracked, the properties of the uncracked section should be used to calculate the secondary stresses due to the thermal moments.

It is mentioned earlier in this chapter that, if a section is cracked, its response to thermal loading can be calculated by assuming the elastic modulus of the concrete, within the crack height, to be zero. Thus, the restraint moment should be calculated from the second of equations (13.18) which was derived using this assumption. The properties of the cracked transformed section should be used to calculate the secondary stresses due to the thermal moments.

## Ultimate limit state

When considering thermal effects under ultimate load conditions it is essential to bear in mind that the thermal loading is a deformation rather than a force. The significance of this can be seen from Fig. 13.7, which compares the response of a material to an applied stress and an applied strain at both the serviceability and ultimate limit states.

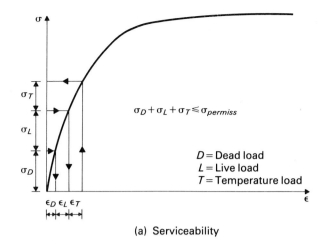

(a) Serviceability

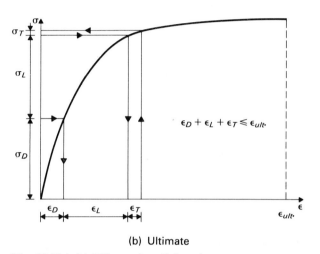

(b) Ultimate

**Fig. 13.7(a),(b)** Effects of applied strains

Although Fig. 13.7 is presented in terms of stress–strain curves, the following discussion is equally applicable to load-deflection or moment-rotation relationships.

It can be seen from Fig. 13.7 that, at the serviceability limit state, the applied thermal strain results in a relatively large thermal stress, but, at the ultimate limit state, only a small thermal stress arises.

In view of the comments made in the last paragraph, it is essential to consider carefully what the loads are, and what load effects result (see also Chapter 3). The author would suggest that the applied thermal strains or displacements should be interpreted as being *nominal loads*. These strains or displacements should be multiplied by the appropriate $\gamma_{fL}$ values to give the *design loads*. The *design load effects* are then the final strains or displacements, and the stresses or stress resultants which arise from any restraints. At the ultimate limit state, the stress or stress resultant design load effects are very small and, if full plasticity is assumed, are zero. However, due consideration should be taken at the ultimate limit state of the magnitudes of the thermal strains or displacements.

Thus, whereas at the serviceability limit state, it is necessary to limit the total (dead + live + thermal) stress so that it is less than the specified permissible stress; at the ultimate limit state, one is more concerned about strain capacity (i.e. ductility) and it is only necessary to limit the total (dead + live + thermal) strain so that it is less than the strain capacity of the material.

To summarise, the author would suggest that thermal stresses or stress resultants can be ignored at the ultimate limit state *provided that* it can be demonstrated that the structure is sufficiently ductile to absorb the thermal strains.

The strains associated with the self-equilibrating stresses are, typically, of the order of 0.0001. Such strains are very small compared with the strain capacity of concrete in compression, which the Code assumes to be 0.0035 (see Chapter 4). It thus seems reasonable to ignore, at the ultimate limit state, the strains associated with the self-equilibrating stresses.

One would expect structural concrete sections to possess adequate ductility, in terms of rotation capacity, so that thermal moments could be ignored at the ultimate limit state. However, it is not clear whether they are also sufficiently ductile in terms of shear behaviour. Tests, designed to examine these problems, are in progress at the University of Birmingham. To date, tests have been carried out on simply supported beams under various combinations of force and thermal loading. It has been found that temperature differences as large as 30°C, through the depth of a beam, with peak temperatures of up to 50°C do not affect the moment of resistance or rotation capacity [273]. Tests on statically indeterminate beams are about to commence to ascertain whether adequate ductility, in terms of bending and shear, is available to redistribute completely the thermal moments and shear forces, which arise from the continuity.

# Design procedure

## General

The logical way to allow for temperature effects in the design procedure is to check ductility at the ultimate limit state, and provide nominal reinforcement to control cracking, which may occur due to the temperature effects, at the serviceability limit state [275, 276]. However, the Code does not permit such an approach; thus, the following design procedure is suggested by the author.

## Ultimate limit state

1. Calculate the free thermal curvature and self-equilibrating stresses using the *uncracked* section.
2. Because of material plasticity, ignore the self-equilibrating stresses.
3. Calculate, from the free thermal curvature, the rotation required, assuming full plasticity of the section, such that no thermal continuity moments occur.
4. If the required rotation is less than the rotation capacity, ignore the thermal continuity moments.
5. If the required rotation exceeds the rotation capacity, add the thermal continuity moments to the moments due to the other loads and design accordingly.

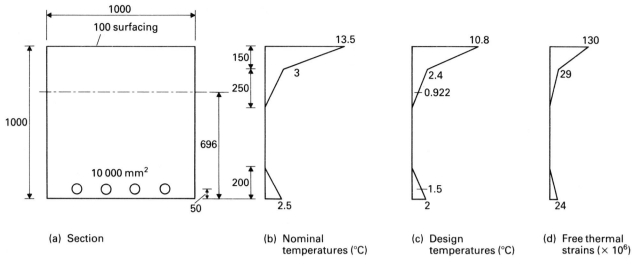

**Fig. 13.8(a)–(d)** Example 13.1

It should be noted that step 4 assumes adequate ductility in shear in addition to adequate rotation capacity. However, experimental evidence of adequate ductility in shear is not available at present.

## Serviceability limit state

1. Calculate the free thermal curvature (or the restraining moment) and the self-equilibrating stresses, using the uncracked or cracked section as appropriate.
2. Calculate the secondary stresses due to any external restraints.
3. Compare the total stresses with the allowable values. In the case of a prestressed member designed as Class 1 for imposed force loadings, it would seem reasonable to adopt the Class 2 allowable stresses when considering thermal loading in addition to the other loadings. Similarly, it would seem reasonable to adopt the Class 3 allowable stresses, under thermal loading, for a member designed to Class 2 for imposed force loadings.

It is worth mentioning that, for a reinforced concrete section, the effects of thermal loading at the serviceability limit state are less onerous when designing to the Code than to the present documents. This is because the Code does not require crack widths to be checked under thermal loading, and the Code allowable stresses ($0.5\,f_{cu}$ for concrete and $0.8\,f_y$ for reinforcement) are greater than those specified in BE 1/73 (see Chapter 4).

## Examples

### 13.1 Uncracked and cracked rectangular section

It is required to determine the response at the serviceability limit state of the section shown in Fig. 13.8(a) to the application of a differential temperature distribution. The

section is identical to that considered by Hambly [274]. The nominal positive temperature difference distribution, obtained from Figure 9 of Part 2 of the Code, is shown in Fig. 13.8(b). The concrete is assumed to be grade 30; thus, from Table 2 of Part 4 of the Code, the short-term elastic modulus of the concrete is 28 kN/mm². (It is not yet clear what value of elastic modulus to adopt, but the short-term value seems more appropriate than the long-term value.) The coefficients of thermal expansion of steel and concrete are each assumed to be $12 \times 10^{-6}/°C$.

The nominal temperature differences in Fig. 13.8(b) first have to be multiplied by a partial safety factor ($\gamma_{fL}$) of 0.8 (see Chapter 3) to give the design temperature differences of Fig. 13.8(c). The free thermal strains appropriate to the design temperature differences are shown in Fig. 13.8(d); these strains are the design loads.

In the following analysis, the section is considered to be both uncracked and cracked, and both the strain and stress methods are demonstrated.

*Uncracked*

Cross-sectional area $= A = 1000 \times 1000$

$$= 1 \times 10^6 \text{ mm}^2$$

Height of neutral axis $= \bar{z} = 500$ mm

Second moment of area $= I = 1000 \times 1000^3/12$

$$= 83.33 \times 10^9 \text{ mm}^4$$

$$\int_o^h b_z t_z dz = 1000\,[(6.6)\,(150) + (1.2)\,(250) + (1)\,(200)] = 1.49 \times 10^6$$

$$\int_o^h b_z t_z\,(z - \bar{z})dz$$

$$= 1000\,[(6.6)\,(150)\,(440.9) + (1.2)\,(250)\,(266.7) - (1)\,(200)\,(433.3)] = 0.43 \times 10^9$$

*Strain method*
From equation (13.10), the axial strain is

$$\bar{\varepsilon} = (12 \times 10^{-6})\,(1.49 \times 10^6)/1 \times 10^6$$

$$= 17.9 \times 10^{-6}$$

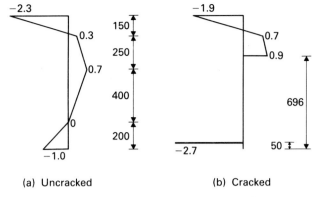

(a) Uncracked       (b) Cracked

**Fig. 13.9(a),(b)** Self-equilibrating stresses (N/mm²)

From equation (13.9), the curvature is

$$\psi = (12 \times 10^{-6}) \, (0.43 \times 10^9)/83.33 \times 10^9$$

$$= 61.9 \times 10^{-9} \text{ mm}^{-1}$$

Bottom fibre strain $= \varepsilon_o = \varepsilon - \psi\bar{z}$

$$= 17.9 \times 10^{-6} -$$

$$(61.9 \times 10^{-9}) \, (500)$$

$$= -13.1 \times 10^{-6}$$

The self-equilibrating stresses, shown in Fig. 13.9(a), can now be obtained from equation (13.2).

*Stress method*
From equation (13.17), the restraining force and moment are:

$$F = -(12 \times 10^{-6}) \, (28 \times 10^3) \, (1.49 \times 10^6)$$

$$= -0.501 \times 10^6 \text{N}$$

$$M = (12 \times 10^{-6})(28 \times 10^3) \, (0.43 \times 10^9)$$

$$= 0.144 \times 10^9 \text{ Nmm}$$

The self-equilibrating stresses can now be obtained from equation (13.19): they are identical to those, calculated by the strain method, in Fig. 13.9(a).

## Cracked

With an elastic modulus of the reinforcement of 200 kN/mm², the neutral axis depth is found to be 304 mm. Thus $\bar{z} = 696$ mm.
Area of transformed section $= A$

$$= (1000) \, (304) + (200/28) \, (10 \, 000)$$

$$= 0.3754 \times 10^6 \text{ mm}^2$$

Second moment of area of transformed section $= I$

$$= (1000) \, (304)^3/3 + (200/28) \, (10 \, 000) \, (646)^2$$

$$= 39.17 \times 10^9 \text{ mm}^4$$

$$\int_o^h E_z b_z t_z dz$$

$$= E_s A_s t_s + \int_{\bar{z}}^h E_c b t_z dz$$

$$= (200 \times 10^3) \, (10 \, 000) \, (1.5)$$

$$+ (28 \times 10^3) \, (1000) \, [(6.6) \, (150) + (1.661) \, (154)]$$

$$= 37.88 \times 10^9$$

$$\int_o^h E_z b_z t_z \, (z - \bar{z}) \, dz$$

$$= E_s A_s t_s \, (h - d - \bar{z}) + \int_{\bar{z}}^h E_c b t_z z dz$$

$$= (200 \times 10^3) \, (10 \, 000) \, (1.5) \, (-646) +$$

$$(28 \times 10^3) \, (1000) \, [(6.6) \, (150) \, (245) +$$

$$(1.661) \, (154) \, (88.4)]$$

$$= 5.487 \times 10^{12}$$

*Strain method*
From equations (13.11), the axial strain and curvature are

$$\bar{\varepsilon} = \frac{(12 \times 10^{-6}) \, (37.88 \times 10^9)}{(28 \times 10^3) \, (0.3754 \times 10^6)}$$

$$= 43.2 \times 10^{-6}$$

$$\psi = \frac{(12 \times 10^{-6})(5.487 \times 10^{12})}{(28 \times 10^3) \, (39.17 \times 10^9)}$$

$$= 60.0 \times 10^{-9} \text{ mm}^{-1}$$

Thus the cracked free thermal curvature is only 3% less than the uncracked free thermal curvature.

Bottom fibre strain $= \varepsilon_o = \bar{\varepsilon} - \psi\bar{z}$

$$= 43.2 \times 10^{-6} - (60 \times 10^{-9}) \, (696)$$

$$= 1.4 \times 10^{-6}$$

The self-equilibrating stresses, shown in Fig. 13.9(b), can now be obtained from equation (13.2). It can be seen that, after cracking, the extreme fibre compressive stress is reduced by 17% but the peak tensile stress is increased by 29%; however, the tensile stress is small.

*Stress method*
From equations (13.18), the restraining force and moment are:

$$F = -(12 \times 10^{-6}) \, (37.88 \times 10^9)$$

$$= -0.454 \times 10^6 \text{ N}$$

$$M = (12 \times 10^{-6})(5.487 \times 10^{12})$$

$$= 65.8 \times 10^6 \text{ Nmm}$$

The self-equilibrating stresses can now be obtained from equation (13.19): they are identical to those calculated by the strain method, in Fig. 13.9(b).

### Design load effects

If no external restraints are applied, the load effects are the axial strain, the curvature and the self-equilibrating stresses. These should be multiplied by the appropriate value of $\gamma_{f3}$ to give the design load effects: in fact, $\gamma_{f3}$ is unity at the serviceability limit state (see Chapter 4). Thus the stresses shown in Fig. 13.9 are the stresses which should be added to the dead and live load stresses, and the net stresses compared with the allowable values.

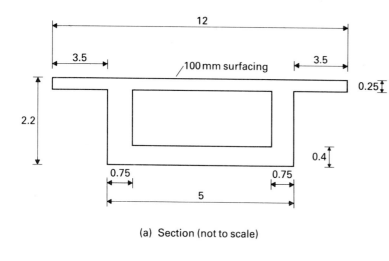

(a) Section (not to scale)

(b) Nominal temperatures (°C)

(c) Design temperatures (°C)

**Fig. 13.10(a)–(c)** Example 13.2

## 13.2 Box girder

A prestressed concrete continuous viaduct, with spans of 45 m, has the cross-section shown in Fig. 13.10(a). The concrete is of grade 50. It is required to determine the longitudinal effects of a positive temperature distribution at both the serviceability and ultimate limit states.

### General

Since the section is prestressed, it is assumed to be uncracked.

Cross-sectional area $= A = (12) (0.25) + (5) (0.4) +$
$$(2) (0.75) (1.55)$$
$$= 7.325 \text{ m}^2$$

First moments of area about soffit to determine $\bar{z}$:

$\bar{z} = [(12) (0.25) (2.075) + (5) (0.4) (0.2) +$
$$(2) (0.75) (1.55) (1.175)]/7.325$$
$$= 1.277 \text{ m}$$

Second moment of area $= I$

$= (12) (0.25)^3/12 + (12) (0.25) (0.798)^2 +$
$$(5) (0.4)^3/12 + (5) (0.4) (1.077)^2 +$$
$$(2) (0.75) (1.55)^3/12 +$$
$$(2) (0.75) (1.55) (0.102)^2$$
$$= 4.762 \text{ m}^4$$

From Table 2 of Part 4 of the Code, elastic modulus of concrete $= E = 34 \times 10^3 \text{ N/mm}^2$. The coefficient of expansion is $12 \times 10^{-6}/°C$.

### Serviceability limit state

The nominal temperature difference distribution, from Figure 9 of Part 2 of the Code, is shown in Fig. 13.10(b). These temperatures have to be multiplied by 0.8 (see Example 13.1) to obtain the design temperature differences of Fig. 13.10(c).

$$\int_o^h b_z t_z dz$$

$= (12) (0.15) (6.6) + (12) (0.1) (1.92) +$
$$(2) (0.75) (0.15) (0.72) + (5) (0.2) (1.0)$$
$$= 15.35$$

$$\int_o^h b_z t_z z dz$$

$= (12) (0.15) (6.6) (0.8639) +$
$$(12) (0.1) (1.92) (0.7276) +$$
$$(2) (0.75) (0.15) (0.72) (0.623) +$$
$$(5) (0.2) (1.0) (-1.21)$$
$$= 10.83$$

Using equations (13.9) and (13.10) (for the strain method), the curvature and axial strain are:

$\psi = (12 \times 10^{-6}) (10.83)/4.762$
$$= 27.3 \times 10^{-6} \text{ m}^{-1}$$
$\bar{\varepsilon} = (12 \times 10^{-6}) (15.35)/7.325$
$$= 25.1 \times 10^{-6}$$

The soffit strain is

$\varepsilon_o = \bar{\varepsilon} - \psi\bar{z} = 25.1 \times 10^{-6} -$
$$(27.3 \times 10^{-6}) (1.277) = -9.7 \times 10^{-6}$$

The self-equilibrating stresses, shown in Fig. 13.11(a), can now be obtained from equation (13.2). If it is assumed that the articulation is such that the axial strain can occur freely, then the only secondary stresses to occur are those due to the restraint to the curvature. This restraint produces the thermal bending moment diagram shown in Fig. 13.12. The maximum thermal moment occurs at the first interior support and is

$M_s = 1.27EI \ \psi = (1.27)(34 \times 10^6) (4.762) \times$
$$(27.3 \times 10^{-6})$$
$$= 5614 \text{ kNm}$$

The secondary stresses due to this moment are shown in Fig. 13.11(b). The net stresses, which are obtained by adding the secondary stresses to the self-equilibrating stresses, are shown in Fig. 13.11(c).

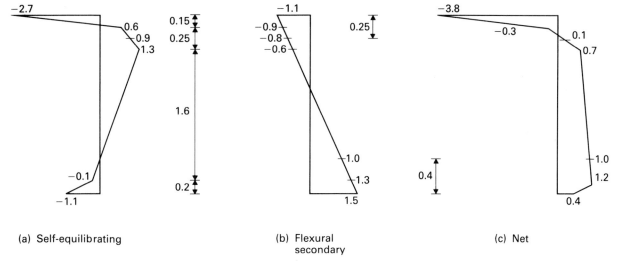

(a) Self-equilibrating

(b) Flexural secondary

(c) Net

**Fig. 13.11(a),(c)** Thermal stresses (N/mm²)

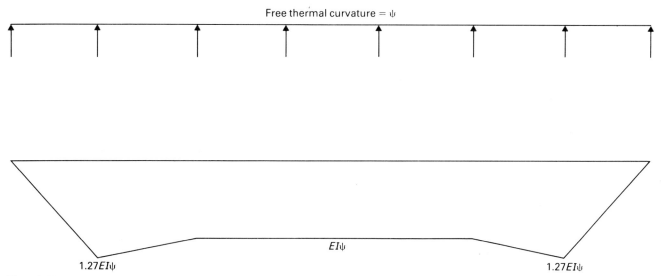

Free thermal curvature = $\psi$

$EI\psi$

$1.27EI\psi$

$1.27EI\psi$

**Fig. 13.12** Thermal moments in multi-span viaduct

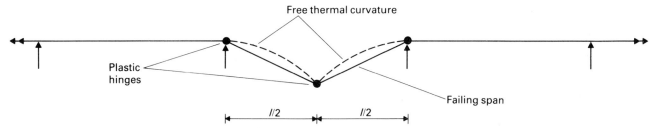

Free thermal curvature

Plastic hinges

Failing span

$l/2$

$l/2$

**Fig. 13.13** Thermal rotations in multi-span viaduct

*Longitudinal shear stresses* At the free end of the viaduct, the self-equilibrating stresses do not occur, but longitudinal shear stresses occur in the zone within which the self-equilibrating stresses build-up. The shear stress is greatest at the web–cantilever junction. The average compressive stress in the cantilever, away from the free end, is, from Fig. 13.11(a), (2.7 − 0.6)/2 = 1.1 N/mm². In accordance with St. Venant's principle this stress will be assumed to build-up over a length, along the span, equal to the breadth of the cantilever (i.e. 3.5 m). Then the average longitudinal shear stress at the web–cantilever junction is equal to the average compressive stress in the cantilever (i.e. 1.1 N/mm²).

*Transverse effects* In practice, the transverse effects of the temperature distribution should also be investigated.

### Ultimate limit state

The partial safety factor $\gamma_{fL}$ at the ultimate limit state is 1.0 (see Chapter 3). Hence the free thermal curvature at the ultimate limit state is $(27.3 \times 10^{-6})$ $(1.0/0.8)$ = $34.1 \times 10^{-6}$ m⁻¹. The worst effect that this positive curvature can have on the structure at the ultimate limit state is to cause rotation in a plastic hinge at the centre of a failing span, as shown in Fig. 13.13. This is because such a thermal rotation is of the same sign as the rotation due to force loading. (If a negative temperature difference distribution were being considered, the thermal rotation at a support hinge would be calculated.)

The total thermal rotation ($\theta_t$) is given by

$$\theta_t = \psi l/2$$

where $l$ is the distance between hinges. Thus $\theta_t = (34.1 \times 10^{-6})(22.5) = 0.77 \times 10^{-6}$ radians. To obtain the design load effect this rotation must be multiplied by $\gamma_{f3}$, which is 1.15 at the ultimate limit state. Thus the design rotation is $1.15 \times 0.77 \times 10^{-3} = 0.88 \times 10^{-3}$ radians.

The Code does not give permissible rotations, but the CEB Model Code [110] gives a relationship between permissible rotation and the ratio of neutral axis depth to effective depth. It is unlikely that the permissible rotation would be less than $5 \times 10^{-3}$ radians. This value is much greater than the design rotation. Thus, unless the moment due to the applied forces is considerably redistributed away from mid-span, the section has sufficient ductility to enable the thermal moments at the ultimate limit state to be ignored.

# Appendix A

# Equations for plate design

## Sign conventions

The positive directions of the applied stress resultants per unit length and the reinforcement direction $\alpha$ are shown in Fig. A1.

Stress resultants with the superscript* are the required resistive stress resultants per unit length in the reinforcement directions $x$, $y$ for orthogonal reinforcement and $x$, $\alpha$ for skew reinforcement.

The principal concrete force per unit length ($F_c$) which appears in equations (A19), (A21), (A23), (A26), (A28) and (A30) is tensile when positive.

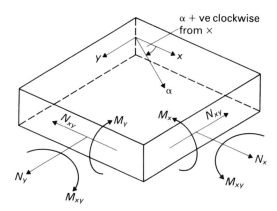

**Fig. A.1** Slab element

## Bending

### Orthogonal $x$, $y$ reinforcement

#### Bottom

Generally

$$M_x^* = M_x + |M_{xy}| \tag{A1}$$
$$M_y^* = M_y + |M_{xy}| \tag{A2}$$

If $M_x^* < 0$

$$M_x^* = 0$$

$$M_y^* = M_y + \left|\frac{M_{xy}^2}{M_x}\right| \tag{A3}$$

If $M_y^* < 0$

$$M_y^* = 0$$

$$M_x^* = M_x + \left|\frac{M_{xy}^2}{M_y}\right| \tag{A4}$$

#### Top

Generally

$$M_x^* = M_x - |M_{xy}| \tag{A5}$$

$$M_y^* = M_y - |M_{xy}| \tag{A6}$$

If $M_x^* > 0$

$$M_x^* = 0$$

$$M_y^* = M_y - \left|\frac{M_{xy}^2}{M_x}\right| \tag{A7}$$

If $M_y^* > 0$

$$M_y^* = 0$$

$$M_x^* = M_x - \left|\frac{M_{xy}^2}{M_y}\right| \tag{A8}$$

### Skew $x$, $\alpha$ reinforcement

#### Bottom

Generally

$$M_x^* = M_x + 2M_{xy} \cot \alpha + M_y \cot^2 \alpha + \left|\frac{M_{xy} + M_y \cot \alpha}{\sin \alpha}\right| \tag{A9}$$

$$M^*\alpha = \frac{M_y}{\sin^2 \alpha} + \left|\frac{M_{xy} + M_y \cot \alpha}{\sin \alpha}\right| \tag{A10}$$

If $M_x^* < 0$

$$M_x^* = 0$$

$$M_\alpha^* = \frac{1}{\sin^2 \alpha} \left( M_y + \left| \frac{(M_{xy} + M_y \cot \alpha)^2}{(M_x + 2M_{xy} \cot \alpha + M_y \cot^2 \alpha)} \right| \right) \quad \text{(A11)}$$

If $M_\alpha^* < 0$

$$M_\alpha^* = 0$$

$$M_x^* = M_x + 2M_{xy} \cot \alpha + M_y \cot^2 \alpha + \left| \frac{(M_{xy} + M_y \cot\alpha)^2}{M_y} \right| \quad \text{(A12)}$$

### Top

Generally

$$M_x^* = M_x + 2M_{xy} \cot \alpha + M_y \cot \alpha - \left| \frac{M_{xy} + M_y \cot \alpha}{\sin \alpha} \right| \quad \text{(A13)}$$

$$M_\alpha^* = \frac{M_y}{\sin^2 \alpha} - \left| \frac{M_{xy} + M_y \cot \alpha}{\sin \alpha} \right| \quad \text{(A14)}$$

If $M_x^* > 0$

$$M_x^* = 0$$

$$M_\alpha^* = \frac{1}{\sin^2 \alpha} \left( M_y - \left| \frac{(M_{xy} + M_y \cot \alpha)^2}{(M_x + 2M_{xy} \cot \alpha + M_y \cot^2 \alpha)} \right| \right) \quad \text{(A15)}$$

If $M^*_\alpha > 0$

$$M_\alpha^* = 0$$

$$M_x^* = M_x + 2M_{xy} \cot \alpha + M_y \cot^2 \alpha - \left| \frac{(M_{xy} + M_y \cot \alpha)^2}{M_y} \right| \quad \text{(A16)}$$

# In-plane forces

## Orthogonal x, y reinforcement

Generally

$$N_x^* = N_x + |N_{xy}| \quad \text{(A17)}$$
$$N_y^* = N_y + |N_{xy}| \quad \text{(A18)}$$
$$F_c = -2|N_{xy}| \quad \text{(A19)}$$

If $N_x < 0$

$$N_x^* = 0$$

$$N_y^* = N_y + \left| \frac{N_{xy}^2}{N_x} \right| \quad \text{(A20)}$$

$$F_c = N_x + \frac{N_{xy}^2}{N_x} \quad \text{(A21)}$$

If $N_y^* < 0$

$$N_y^* = 0$$

$$N_x^* = N_x + \left| \frac{N_{xy}^2}{N_y} \right| \quad \text{(A22)}$$

$$F_c = N_y + \frac{N_{xy}^2}{N_y} \quad \text{(A23)}$$

## Skew x, α reinforcement

Generally

$$N_x^* = N_x + 2N_{xy} \cot \alpha + N_y \cot^2 \alpha + \left| \frac{N_{xy} + N_y \cot \alpha}{\sin \alpha} \right| \quad \text{(A24)}$$

$$N_\alpha^* = \frac{N_y}{\sin^2 \alpha} + \left| \frac{N_{xy} + N_y \cot \alpha}{\sin \alpha} \right| \quad \text{(A25)}$$

$$F_c = -2 (N_{xy} + N_y \cot \alpha) (\cot\alpha \pm \operatorname{cosec} \alpha) \quad \text{(A26)}$$

In equation (A26), the sign in the last bracket is the same as the sign of $(N_{xy} + N_y \cot \alpha)$.

If $N_x^* < 0$

$$N_x^* = 0$$

$$N_\alpha^* = \frac{1}{\sin^2 \alpha} \left( N_y + \left| \frac{(N_{xy} + N_y \cot \alpha)^2}{(N_x + 2N_{xy} \cot \alpha + N_y \cot^2 \alpha)} \right| \right) \quad \text{(A27)}$$

$$F_c = \frac{(N_x + N_{xy} \cot \alpha)^2 + (N_{xy} + N_y \cot \alpha)^2}{N_x + 2N_{xy} \cot \alpha + N_y \cot^2 \alpha} \quad \text{(A28)}$$

If $N_\alpha^* < 0$

$$N_\alpha^* = 0$$

$$N_x^* = N_x + 2N_{xy} \cot \alpha + N_y \cot^2 \alpha + \left| \frac{(N_{xy} + N_y \cot \alpha)^2}{N_y} \right| \quad \text{(A29)}$$

$$F_c = N_y + \frac{N_{xy}^2}{N_y} \quad \text{(A30)}$$

# Appendix B

# Transverse shear in cellular and voided slabs

## Introduction

It is mentioned in Chapter 6 that no rules are given in the Code for the design of cellular or voided slabs to resist transverse shear. In this Appendix, the author suggests design approaches at the ultimate limit state.

All stress resultants in the following are per unit length.

## Cellular slabs

### General

The effect of a transverse shear force ($Q_y$) is to deform the webs and flanges of a cellular slab, as shown in Fig. B.1(b). Such deformation is generally referred to as Vierendeel truss action. The suggested design procedure is initially to consider the Vierendeel effects separately from those of global transverse bending, and then to combine the global and Vierendeel effects.

### Analysis of Vierendeel truss

Points of contraflexure may be assumed at the mid-points of the flanges and webs. Assuming the point of contraflexure in the web to be always at its mid-point implies that the stiffnesses of the two flanges are always equal, irrespective of their thicknesses and amounts of reinforcement. However, a more precise idealisation is probably not justified. The shear forces are assumed to be divided equally between the two flanges to give the loading, bending moment and shear force diagrams of Fig. B.1.

### Design

#### Webs

A web can be designed as a slab, in accordance with the methods described in Chapters 5 and 6, to resist the bending moments and shear forces of Figs. B.1(c) and (d), respectively.

#### Flanges

The flanges can be designed as slabs, in accordance with the method described in Chapter 6, to resist the shear forces of Fig. B.1(d).

In addition to the Vierendeel bending moments, the global transverse bending moment ($M_y$) induces a force of $M_y/h_e$, where $h_e$ is the lever arm shown in Fig. B.1(a), in both the compression and tension flanges. Thus, each flange should be designed as a slab eccentrically loaded by a moment $Q_y s/4$ (from Fig. B.1(c)) and either a compressive or tensile force, as appropriate, of $M_y/z$.

## Voided slabs

### General

The effect of a transverse shear force is to deform the webs and flanges of a voided slab in a similar manner to those of a cellular slab. However, since the web and flange thicknesses of a voided slab vary throughout their lengths, analysis of the Vierendeel effects is not readily carried out. In view of this, a method of design is suggested which is based on considerations of elastic analyses of voided slabs and the actual behaviour of transverse strips of reinforced concrete voided slabs subjected to shear [277]. The suggested ultimate limit state method is virtually identical to an unpublished working stress method proposed by Elliott [278] which, in turn, is based upon the test data and design recommendations of Aster [277]. Although Aster's tests to failure were conducted on transverse strips of voided slabs, a similar failure mode has been observed in a test on a model voided slab bridge deck by Elliott, Clark and Symmons [71].

The design procedure considers, independently, possible cracks initiating on the outside and inside of a void due to the Vierendeel effects of the transverse shear. The latter

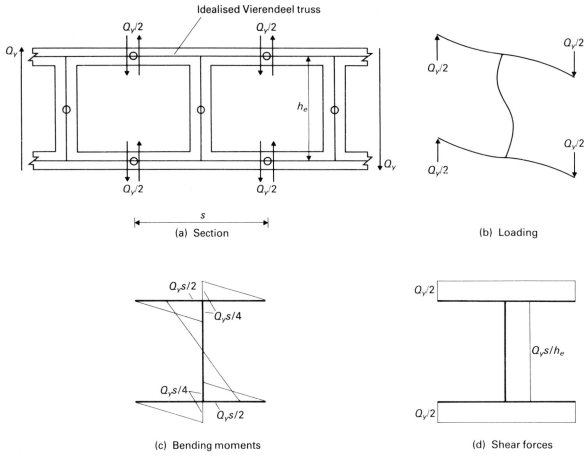

Fig. B.1(a)–(d) Cellular slab

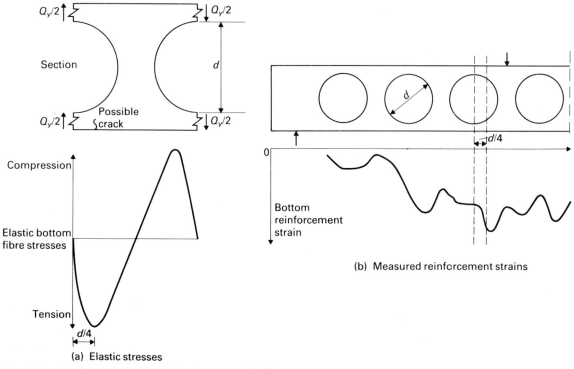

Fig. B.2(a),(b) Vierendeel stresses in voided slab [277]

effects and the global transverse bending effects are then combined.

In the following, the global transverse moment ($M_y$), co-existing with the transverse shear force ($Q_y$), is assumed to be sagging.

## Bottom flange design

Elastic analysis of the uncracked section shows that the distribution of extreme fibre stress, due to Vierendeel action, is as shown in Fig. B.2(a) [277]: the peak stress

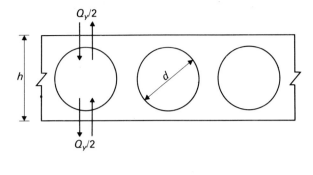

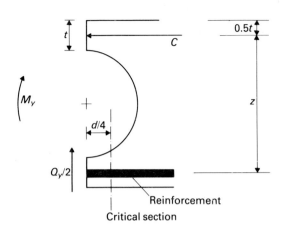

**Fig. B.3** Bottom flange of voided slab

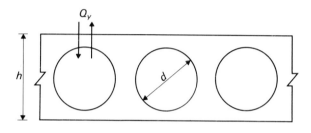

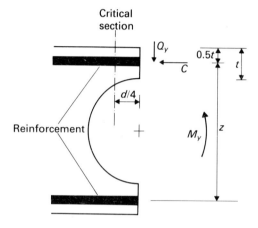

**Fig. B.4** Top flange of voided slab

occurs at about the quarter-point of the void (i.e., at $d/4$ from the void centre line, where $d$ is the void diameter). Thus a crack may initiate, from the bottom face of the slab, at this critical section.

It has also been observed that peak bottom flange reinforcement strains, in cracked concrete slab strips, occur at about $d/4$ from the void centre line. This is illustrated by Fig. B.2(b), which shows some of Aster's measured bottom reinforcement strains in a reinforced concrete transverse strip.

Fig. B.2(a) shows that the Vierendeel bending stress at the centre line of the void is zero; hence, only a shear force acts at this section, as shown in Fig. B.3. It is conservative, with regard to the design of the reinforcement in the bottom flange, to assume that the shear force ($Q_y$) is shared equally between the two flanges. In fact, less than $Q_y/2$ is carried by the bottom flange because it is cracked. Thus the Vierendeel bending moment at the critical section, $d/4$ from the void centre line, is:

$$M_V = (Q_y/2)\,(d/4) = Q_y d/8 \qquad \text{(B.1)}$$

The bottom flange reinforcement is also subjected to a tensile force of ($M_y/z$), where $M_y$ in this case is the maximum global transverse moment and $z$ is the lever arm for global bending shown in Fig. B.3. The resultant compressive concrete force ($C$) in the top flange is considered to act at mid-depth of the minimum flange thickness ($t$), because the design is being carried out at the ultimate limit

state and the concrete can be considered to be in a plastic condition.

The bottom flange reinforcement should be designed for the combined effects of the force $M_y/z$ and the Vierendeel moment $M_v$. The section depth should be that at the critical section.

## Top flange design

The extreme top fibre stress distribution, due to Vierendeel action, is similar in form to that, shown in Fig. B.2(a), for the bottom fibre. Thus, due to Vierendeel action, a crack may initiate, from the top face of the slab, at the critical section (distance $d/4$ from the void centre line). The Vierendeel bending stress is again zero at the centre line of the void, but it is now conservative, with regard to the design of the reinforcement in the top flange, to assume that all of the shear force is carried by the top flange. This assumption implies that the bottom flange is severely cracked due to global transverse flexure and cannot transmit any shear by aggregate interlock or dowel action.

The Vierendeel bending moment at the critical section is (see Fig. B.4):

$$M_V = Q_y d/4 \qquad \text{(B.2)}$$

The top flange is also subjected to a compressive force of ($M_y/z$) which counteracts the tension induced in the top

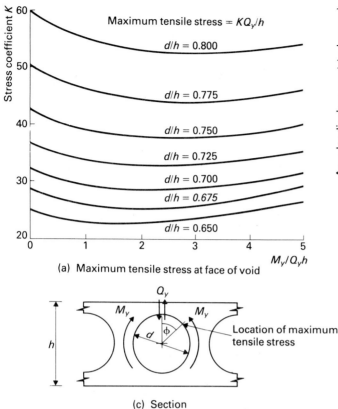

(a) Maximum tensile stress at face of void

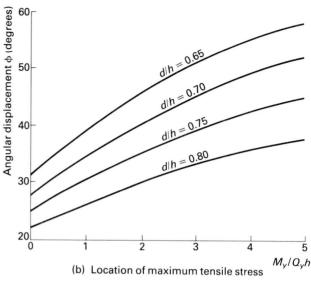

(b) Location of maximum tensile stress

(c) Section

**Fig. B.5(a)–(c)** Maximum tensile stress at face of void

flange reinforcement by the Vierendeel moment $M_V$. Hence, the greatest tension in the reinforcement is obtained when $M_y$ is a minimum.

The top flange should be designed as an eccentrically loaded column (see Chapter 9) to resist the compressive force ($M_y/z$), which acts at $t/2$ from the top face, and the moment $M_V$. The depth of the column should be taken as the flange thickness at the critical section.

## Detailing of flange reinforcement

The areas of flange reinforcement provided should exceed the Code minimum values discussed in Chapter 10, and the bar spacings should be less than the Code maximum values discussed in Chapters 7 and 10.

## Web design

It is desirable to design the section so that the occurrence of cracks initiating from the inside of a void is prevented, because it is difficult to detail reinforcement to control such cracks.

Elliott [278] has produced graphs which give the maximum tensile stress on the inside of a void due to combined transverse bending and shear: it is conservatively assumed that all of the shear is carried by the top flange. Elliott's graphs are reproduced in Fig. B.5.

The maximum tensile stress obtained from Fig. B.5(a) should be compared with an allowable tensile stress. The author would suggest that the latter stress should be taken as $0.45 \sqrt{f_{cu}}$: the derivation of this value, which is the

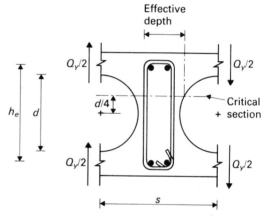

**Fig. B.6** Vertical web reinforcement in voided slab

Code allowable flexural tensile stress for a Class 2 pretensioned member, is given in Chapter 4.

Tensile stresses less than and greater than the allowable stress now have to be considered.

### Tensile stress less than allowable

Cracking at the inside of a void would not occur in this situation, and vertical reinforcement in the webs should be provided.

The work of Aster [277] indicates that the design can be carried out by considering the Vierendeel truss of Fig. B.1(b), for which the horizontal shear force at the point of contraflexure in the web is $Q_y s/h_e$. The critical section for Vierendeel bending of a web is considered to be at $d/4$ above the centre line of the void, as shown in Fig. B.6. The Vierendeel bending moment at this critical section is:

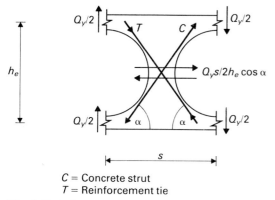

C = Concrete strut
T = Reinforcement tie

**Fig. B.7** Inclined web reinforcement in voided slab

$$M_V = (Q_y s/h_e)(d/4) = Q_y sd/4h_e \qquad (B.3)$$

Reinforcement at the critical section, with the effective depth shown in Fig. B.6, can be designed to resist the moment $M_V$.

The vertical reinforcement in the web is most conveniently provided in the form of vertical links, as shown in Fig. B.6; however, only one leg of such a link may be considered to contribute to the required area of reinforcement. This area should be added to that required to resist the *longitudinal* shear to give the total required area of link reinforcement.

### Tensile stress greater than allowable

If the tensile stress obtained from Fig. B.5(a) is greater than the allowable stress, cracking will occur on the inside of the void. In this situation, it is preferable to reduce the size of the voids, so as to reduce the tensile stress, or to alter the positions of the voids in the deck, so that they are not in areas of high transverse shear. If cracking is not precluded by either of these means, it is necessary to design the voided slab so that reinforcement crosses the crack, which initiates on the inside of the void. This can be done either by providing inclined reinforcement in the webs, or by providing a second layer of horizontal reinforcement in the flange, close to the void.

*Inclined reinforcement*  The forces acting in a web are shown in Fig. B.7. The horizontal shear force at the point of contraflexure of the web is $Q_y s/h_e$ (see discussion of vertical web reinforcement). For horizontal equilibrium

$$(T + C)\cos\alpha = Q_y s/h_e$$

But $T = C$, from vertical equilibrium; thus

$$T = Q_y s/2h_e \cos\alpha \qquad (B.4)$$

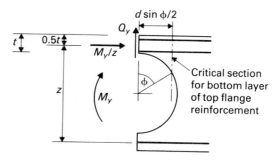

**Fig. B.8** Additional horizontal reinforcement in voided slab

Inclined reinforcement should be designed to resist this force. The reinforcement could take the form of, for example, inclined links or bars: the latter should be anchored by lapping with the top and bottom flange reinforcement.

*Additional horizontal reinforcement*  As an alternative to inclined reinforcement an additional layer of horizontal reinforcement may be provided as shown in Fig. B.8. The critical section for designing this reinforcement should be taken as the position of maximum tensile stress, obtained from Fig. B.5(b). The latter figure gives the position in terms of the angular displacement ($\phi$): its horizontal distance from the void centre line is thus $d \sin \phi/2$. It is conservative to assume that all of the transverse shear force is carried by the top flange and thus, from Fig. B.8, the Vierendeel bending moment at the critical section is:

$$M_V = Q_y d \sin \phi/2 \qquad (B.5)$$

The top flange is also subjected to a compressive force of $(M_y/z)$, which counteracts the tension induced in the reinforcement by the Vierendeel moment $M_V$. Hence, the greatest tension in the reinforcement is obtained when $M_y$ is a minimum.

The critical section should be designed as an eccentrically loaded column (see Chapter 9) to resist the compressive force $(M_y/z)$, which acts at $t/2$ from the top face, and the moment $M_V$. The depth of the column should be taken as the flange thickness at the critical section.

## Effect of global twisting moment

A global twisting moment induces forces in the flanges; these forces can be taken into account in the suggested design methods by replacing $M_y$ throughout by $M_y^*$ (obtained from the appropriate equation of Appendix A).

# References

1. British Standards Institution, *BS 153 Specification for steel girder bridges*, Parts 1 & 2, pp. 16; Part 3A, pp. 40; Parts 3B & 4, pp. 64; 1972.
2. Kerensky, O. A., Major British Standard covers design and construction of bridges', *BSI News*, July 1978, pp. 10–11.
3. Department of Transport, *Technical Memorandum (Bridges) BE 1/77 (1st amendment). Standard highway loadings*, May 1979, p. 47.
4. British Standards Institution, *BS 153. Specification for steel girder bridges. Part 3A*, 1972, p. 39.
5. Department of Transport, *Technical Memorandum (Bridges) BE 1/73 (1st revision, 1st amendment). Reinforced concrete for highway structures*, July 1979, p. 19.
6. Department of Transport, *Technical Memorandum (Bridges) BE 2/73 (3rd amendment). Prestressed concrete for highway structures*, June 1975, p. 22.
7. British Standards Institution, *The structural use of reinforced concrete in buildings. CP 114: Part 2: 1969*, p. 94.
8. British Standards Institution, *The structural use of prestressed concrete in buildings. CP 115: Part 2: 1969*, p. 44.
9. British Standards Institution, *The structural use of precast concrete. CP 116: Part 2: 1969*, p. 153.
10. Department of Transport, *Specification for road and bridge works*, 1976, p. 194.
11. Department of Transport, *Technical Memorandum (Bridges) BE 1/76 (3rd amendment). Design requirements for elastomeric bridge bearings*, September 1978, p. 14.
12. Department of Transport, *Interim Memorandum (Bridges) IM11 (1st addendum). PTFE in bridge bearings*, January 1972, p. 3.
13. Department of Transport, *Departmental Standard BD 1/78. Implementation of BS 5400 – The design and specification of steel, concrete and composite bridges*, September 1978, p. 3.
14. Ferguson, H., 'New code introduces limit-state approach to bridge design', *New Civil Engineer*, 6 July 1978, pp. 13–14.
15. British Standards Institution, *CP 110: Part 1: 1972. The structural use of concrete. Part 1. Design, materials and workmanship*, p. 155.
16. International Organization for Standardization, *General principles for the verification of the safety of structures. International Standard ISO 2394*, 1973, p. 5.
17. Beeby, A. W. and Taylor, H. P. J., *Accuracy in analysis – influence of $\gamma_{f3}$*, unpublished note submitted to a meeting of Comite Europeén du Beton Commission XI, Pavia, June 1977, pp. 12.
18. Henderson, W., Burt, M. E. and Goodearl, K. A., 'Bridge loading', *Proceedings of an International Conference on Steel Box Girder Bridges*, London, Institution of Civil Engineers, 1973, pp. 193–201.
19. Cope, R. J. and Rao, P. V., 'Non-linear finite element analysis of concrete slab structures', *Proceedings of the Institution of Civil Engineers*, Part 2, Vol. 63, March 1977, pp. 159–179.
20. Teychenné, D. C., Parrot, L. J. and Pomeroy, C. D., 'The estimation of the elastic modulus of concrete for the design of structures', *Building Research Establishment Current Paper CP 23/78*, March 1978, p. 11.
21. British Standards Institution, *Specification for hot rolled and hot rolled and processed high tensile alloy steel bars for the prestressing of concrete. BS 4486: 1980*, p. 4.
22. British Standards Institution, *Specification for nineteen-wire steel strand for prestressed concrete. BS 4757: 1971*, p. 14.
23. Roik, K. H. and Sedlacek, G., 'Extension of engineers' bending and torsion theory considering shear deformation', *Die Bautechnik*, January 1970, pp. 20–32.
24. Johnson, R. P. and Buckby, R. J., *Composite structures of steel and concrete. Vol. 2. Bridges, with a commentary on BS 5400: Part 5*, Crosby Lockwood Staples, London, 1979, p. 524.
25. Moffatt, K. R. and Dowling, P. J., 'Shear lag in steel box girder bridges', *The Structural Engineer*, Vol. 53, October 1975, pp. 439–448.
26. Moffatt, K. R. and Dowling, P. J., 'British shear lag rules for composite girders', *Proceedings of the American Society of Civil Engineers (Structural Division)*, Vol. 104, No. ST7, July 1978, pp. 1123–1130.
27. Clark, L. A., 'Elastic analysis of concrete bridges and the ultimate limit state', *The Highway Engineer*, October 1977, pp. 22–24.
28. Pucher, A., *Influence Surfaces of Elastic Plates*, Springer Verlag, Wien and New York, 1964.
29. Westergaard, H. M., 'Computations of stresses in bridge slabs due to wheel loads', *Public Roads*, Vol. 2, No. 1, March 1930, pp. 1–23.
30. Holland, A. D., and Deuce, T. L. G., 'A review of small span highway bridge design and standardisation', *Journal of the Institution of Highway Engineers*, August 1970, pp. 3–29.
31. Bergg, J. A., 'Eighty highway bridges in Kent', *Proceedings of the Institution of Civil Engineers*, Part 1, Vol. 54, November 1973, pp. 571–603.
32. Woolley, M. V., 'Economic road bridge design in concrete for the medium span range 15 m–45 m', *The Structural Engineer*, Vol. 52, No. 4, April 1974, pp. 119–128.
33. Pennells, E., *Concrete Bridge Designer's Manual*, Viewpoint Publications, 1978, p. 164.
34. Somerville, G. and Tiller, R. M., *Standard bridge beams for spans from 7 m to 36 m*, Cement and Concrete Association, Publication 32.005, November 1970, p. 32.
35. Dow-Mac Concrete Ltd., *Prestressed concrete bridge beams*.
36. Manton, B. H. and Wilson, C. B. *MoT/C & CA standard bridge beams*, Cement and Concrete Association, Publication 32.012, March 1971, p. 36.
37. Somerville, G., 'Three years' experience with M-beams', *Surveyor*, 27 July 1973, pp. 58–60.
38. Hook, D. M. A. and Richmond, B., 'Western Avenue Extension – precast concrete box beams in cellular bridge decks', *The Structural Engineer*, Vol. 48, No. 3, March 1970, pp. 120–128.
39. Chaplin, E. C., Garrett, R. J., Gordon, J. A. and Sharpe, D. J., 'The development of a design for a precast concrete bridge beam of U-section', *The Structural Engineer*, Vol. 51, No. 10, October 1971, pp. 383–388.

40. Swann, R. A., 'A feature survey of concrete box spine-beam bridges', *Cement and Concrete Association, Technical Report 42.469*, June 1972, p. 76.

41. Maisel, B. I. and Roll, F., 'Methods of analysis and design of concrete box beams with side cantilevers', *Cement and Concrete Association, Technical Report 42.494*, November 1974, p. 176.

42. Timoshenko, S. P. and Woinowsky-Krieger, S., *Theory of Plates and Shells*, McGraw-Hill, 1959, p. 580.

43. Plantema, F. J. *Sandwich construction*, John Wiley and Sons, 1966, p. 246.

44. Libove, C. and Batdorf, S. B., 'A general small deflection theory for flat sandwich plates', *U.S. National Advisory Committee for Aeronautics, Report No. 899*, 1948, p. 17.

45. Guyon, Y., 'Calcul des ponts dalles', *Annales des Ponts et Chaussées*, Vol. 119, No. 29, pp. 555–589, No. 36, pp. 683–718, 1949.

46. Massonnet, C., *Méthode de calcul des ponts à poutres multiple tenant compte de leur résistance à la torsion*, IABSE Publications, Vol. 10, 1950, pp. 147–182.

47. Morice, P. B. and Little, G., 'The analysis of right bridge decks subjected to abnormal loading', *Cement and Concrete Association, Publication 46.002*, 1956, p. 43.

48. Rowe, R. E., *Concrete Bridge Design*, C. R. Books, 1962, p. 336.

49. Cusens, A. R. and Pama, R. P., *Bridge Deck Analysis*, John Wiley and Sons, 1975, p. 278.

50. Department of Transport, *Technical Memorandum (Bridges) BE 6/74. Suite of bridge design and analysis programs. Program HECB/B/15 (ORTHOP) and Design Charts HECB/B5*, October 1974, p. 3.

51. Morley, C. T., 'Allowing for shear deformation in orthotropic plate theory', *Proceedings of the Institution of Civil Engineers*, Supplement paper 7367, 1971, p. 20.

52. Elliott, G., 'Partial loading on orthotropic plates', *Cement and Concrete Association, Technical Report 42.519*, October 1978, p. 60.

53. Goldberg, J. E. and Leve, H. L., 'Theory of prismatic folded plate structures', *IABSE Publication*, Vol. 17, 1957, pp. 59–86.

54. De Fries-Skene, A. and Scordelis, A. C., 'Direct stiffness solution for folded plates', *Proceedings of the American Society of Civil Engineers (Structural Division)*, Vol. 90, No. ST 4, August 1964, pp. 15–48

55. Department of Transport, *Technical Memorandum (Bridges) BE 3/74. Suite of bridge design and analysis programs. Program HECB/B/11 (MUPDI)*, August 1974, p. 2.

56. Department of Transport, *Technical Memorandum (Bridges) BE 4/76. Suite of bridge design and analysis programs. Program HECB/B/13 (STRAND 2)*, March 1976, p. 5.

57. Department of Transport, *Technical Memorandum (Bridges) BE 3/75. Suite of bridge design and analysis programs. Program HECB/B/14 (QUEST)*, February 1975, p. 2.

58. Department of Transport, *Technical Memorandum (Bridges) BE 4/75. Suite of bridge design and analysis programs. Program HECB/B/16 (CASKET)*, February 1975, p. 3.

59. Cheung, Y. K., 'The finite strip method in the analysis of elastic plates with two opposite simply supported ends', *Proceedings of the Institution of Civil Engineers*, Vol. 40. May 1968, pp. 1–7.

60. Loo, Y. C. and Cusens, A. R., 'A refined finite strip method for the analysis of orthotropic plates', *Proceedings of the Institution of Civil Engineers*, Vol. 48, January 1971, pp. 85–91.

61. Hambly, E. C., *Bridge Deck Behaviour*, Chapman and Hall Ltd, 1976, p. 272.

62. West, R., 'C & CA/CIRIA recommendations on the use of grillage analysis for slab and pseudo-slab bridge decks', *Cement and Concrete Association, Publication 46.017*, 1973, p. 24.

63. Robinson, K. E., 'The behaviour of simply supported skew bridge slabs under concentrated loads', *Cement and Concrete Association, Research Report 8*, 1959, p. 184.

64. Rusch, H. and Hergenroder, A., *Influence surfaces for moments in skew slabs*, Technological University of Munich, 1961, available as Cement and Concrete Association translation, p. 21 + 174 charts.

65. Balas, J. and Hanuska, A., *Influence Surfaces of Skew Plates*, Vydaratelstro Slovenskej Akademic, Czechoslavakia, 1964.

66. Naruoka, M. and Ohmura, H., 'On the analysis of a skew girder bridge by the theory of orthotropic plates', *IABSE Proceedings*, Vol. 19, 1959, pp. 231–256.

67. Yeginobali, A., 'Continuous skew slabs', *Ohio State University Engineering Experimental Station, Bulletin 178*, November 1959.

68. Yeginobali, A., 'Analysis of continuous skewed slab bridge decks', *Ohio State University Engineering Experimental Station, Report EES 170–2*, August 1962.

69. Schleicher, C. and Wegener, B., *Continuous Skew Slabs. Tables for Statical Analysis*, Verlag fur Bauwesen, Berlin, 1968.

70. Elliott, G. and Clark, L. A., *Circular voided concrete slab stiffnesses* in preparation.

71. Elliott, G., Clark, L. A. and Symmons, R. M., 'Test of a quarter-scale reinforced concrete voided slab bridge', *Cement and Concrete Association, Technical Report 42.527*, December 1979, p. 40.

72. Elliott, G. (Private communication).

73. Timoshenko, S. P. and Gere, J. M., *Mechanics of Materials*, Van Nostrand Reinhold, 1973, p. 608.

74. Holmberg, A., 'Shear-weak beams on elastic foundation', *IABSE Publication*, Vol. 10, 1960, pp. 69–85.

75. Clark, L. A., 'Comparison of various methods of calculating the torsional inertia of right voided slab bridges', *Cement and Concrete Association, Technical Report 42.508*, June 1975, p. 34.

76. Timoshenko, S. and Goodier, J. N., *Theory of Elasticity*, McGraw-Hill, 1951, p. 506.

77. Jackson, N., 'The torsional rigidity of concrete bridge decks', *Concrete*, Vol. 2, No. 11, November 1968, pp. 469–471.

78. Hillerborg, A., 'Strip method of design', *Viewpoint*, 1975, p. 256.

79. Wood, R. H. and Armer, G. S. T., 'The theory of the strip method for design of slabs', *Proceedings of the Institution of Civil Engineers*, Vol. 41, October 1968, pp. 285–311.

80. Kemp, K. O., 'A strip method of slab design with concentrated loads or supports', *The Structural Engineer*, Vol. 49, No. 12, December 1971, pp. 543–548.

81. Fernando, J. S. and Kemp, K. O., 'A generalized strip deflexion method of reinforced concrete slab design', *Proceedings of the Institution of Civil Engineers*, Part 2, Vol. 65, March 1978, pp. 163–174.

82. Spence, R. J. S. and Morley, C. T., 'The strength of single-cell concrete box girders of deformable cross-section', *Proceedings of the Institution of Civil Engineers*, Part 2, Vol. 59, December 1975, pp. 743–761.

83. Beeby, A. W., 'The analysis of beams in plane frames according to CP 110', *Cement and Concrete Association, Development Report 44.001*, October 1978, p. 34.

84. Jones, L. L. and Wood, R. H., *Yield-line Analysis of Slabs*, Thames and Hudson, Chatto and Windus, 1967, p. 401.

85. Jones, L. L., *Ultimate Load Analysis of Concrete Structures*, Chatto and Windus, 1962, p. 248.

86. Granholm, C. A. and Rowe, R. E., 'The ultimate load of simply supported skew slab bridges', *Cement and Concrete Association, Research Report 12*, June 1961, p. 16.

87. Clark, L. A., 'The service load response of short-span skew slab bridges designed by yield line theory', *Cement and Concrete Association, Technical Report 42.464*, May 1972, p. 40.

88. Nagaraja, R. and Lash, S. D., 'Ultimate load capacity of reinforced concrete beam-and-slab highway bridges', *Proceedings of the American Concrete Institute*, Vol. 67, December 1970, pp. 1003–1009.

89. Morley, C. T. and Spence, R. J. S., 'Ultimate strength of cellular concrete box-beams', *Proceedings of a Symposium on structural analysis, non-linear behaviour and techniques, Transport and Road Research Laboratory Supplementary Report 164 UC*, 1975, pp. 193–197.

90. Cookson, P. J., 'Collapse of concrete box girders involving distortion of the cross-section', PhD Thesis, University of Cambridge, 1976, p. 227.

91. Best, B. C. and Rowe, R. E., 'Abnormal loading on composite slab bridges', *Cement and Concrete Association, Research Report 7*, October 1959, p. 28.

92. British Standards Institution, *Code of Practice for foundations. CP 2004: 1972*, p. 158.

93. Hay, J. S., 'Wind forces on bridges – drag and lift coefficients', *Transport and Road Research Laboratory Report 613*, 1974, p. 8.

94. Emerson, M., 'Extreme values of bridge temperatures for design purposes', *Transport and Road Research Laboratory Report 744*, 1976, p. 25.

95. Emerson, M., 'Temperature differences in bridges: basis of design

requirements', *Transport and Road Research Laboratory Report 765*, 1977, p. 39.

96. Spratt, B. H., 'The structural use of lightweight aggregate concrete', *Cement and Concrete Association, Publication 45.023*, December 1974, p. 68.

97. Black, W., 'Notes on bridge bearings', *Transport and Road Research Laboratory Report 382*, 1971, p. 10.

98. Emerson, M., 'The calculation of the distribution of temperature in bridges, *Transport and Road Research Laboratory Report 561*, 1973, p. 20.

99. Blythe, D. W. R. and Lunniss, R. C., 'Temperature difference effects in concrete bridges', *Proceedings of a Symposium on bridge temperature, Transport and Road Research Laboratory Supplementary Report 442*, 1978, pp. 68–80.

100. Jones, M. R., 'Bridge temperatures calculated by a computer program', *Transport and Road Research Laboratory Report 702*, 1976, p. 21.

101. Mortlock, J. D., 'The instrumentation of bridges for the measurement of temperature and movement', *Transport and Road Research Laboratory Report 641*, 1974, p. 14.

102. Henderson, W., 'British highway bridge loading', *Proceedings of the Institution of Civil Engineers*, Part 2, June 1954, pp. 325–373.

103. Flint, A. R. and Edwards, L. S., 'Limit state design of highway bridges', *The Structural Engineer*, Vol. 48, No. 3, March 1970, pp. 93–108.

104. Rutley, K. S., 'Drivers' and Passengers' maximum tolerance to lateral acceleration', *Transport and Road Research Laboratory Technical Note 455*, 1970.

105. Burt, M. E., 'Forces on bridge due to braking vehicles', *Transport and Road Research Laboratory Technical Note 401*, 1969, p. 5.

106. Department of Transport, *Technical Memorandum (Bridges) BE5 (1st amendment). The design of highway bridge parapets*, August 1978, p. 15.

107. Blanchard, J., Davies, B. L. and Smith, J. W., 'Design criteria and analysis for dynamic loading of footbridges', *Proceedings of a Symposium on dynamic behaviour of bridges, Transport and Road Research Laboratory Supplementary Report 275*, 1977, pp. 90–100.

108. Neville, A. M., *Properties of Concrete*, Sir Isaac Pitman & Sons Ltd., London, 1965, p. 532.

109. Rowe, R. E., 'Current European views on structural safety', *Proceedings of the American Society of Civil Engineers (Structural Division)*, Vol. 96, No. ST3, March 1970, pp. 461–467.

110. Comité Euro-International du Béton, *CEB-FIP model code for concrete structures*, 1978.

111. Comité Europeén du Béton, *Recommendations for an international code of practice for reinforced concrete*, 1964, p. 156.

112. Bate, S. C. C., *et al.*, *Handbook on the unified code for structural concrete*, Cement and Concrete Association, 1972, p. 153.

113. Kajfasz, S., Somerville, G. and Rowe, R. E., *An investigation of the behaviour of composite beams*, Cement and Concrete Association, Research Report 15, November 1963, p. 44.

114. Mikhailov, O. V., 'Recent research on the action of unstressed concrete in composite structures (precast monolithic structures)', *Proceedings of the Third Congress of the Fédération Internationale de la Précontrainte, Berlin*, 1958, pp. 51–65.

115. American Concrete Institute – American Society of Civil Engineers Committee 323, 'Tentative recommendations for prestressed concrete', *Proceedings of the American Concrete Institute*, Vol. 54, No. 7, January 1958, pp. 545–578.

116. British Standards Institution, *Composite construction in structural steel and concrete. CP 117: Part 2: beams for bridges: 1967*, p. 32.

117. Hanson, N. W., 'Precast-prestressed concrete bridges. 2. Horizontal shear connections', *Journal of the Research and Development Laboratories of the Portland Cement Association*, Vol. 2, No. 2, May 1960, pp. 38–58.

118. Seamann, J. C. and Washa, G. W., 'Horizontal shear connections between precast beams and cast-in-place slabs', *Proceedings of the American Concrete Institute*, Vol. 61, No. 11, November 1964, pp. 1383–1410.

119. Beeby, A. W., 'Short term deformations of reinforced concrete members'. *Cement and Concrete Association, Technical Report 42.408*, March 1968, p. 32.

120. Bate, S. C. C., 'Some experimental data relating to the design of

prestressed concrete', *Civil Engineering and Public Works Review*, Vol. 53, 1958, pp. 1010–1012, 1158–1161, 1280–1284.

121. Concrete Society, 'Flat slabs in post-tensioned concrete with particular regard to the use of unbonded tendons – design recommendations', *Concrete Society Technical Report 17*, July 1979, p. 16.

122. Abeles, P. W., 'Design of partially prestressed concrete beams', *Proceedings of the American Concrete Institute*, Vol. 64, October 1967, pp. 669–677.

123. Beeby A. W. and Taylor, H. P. J., 'Cracking in partially prestressed members', *Sixth Congress of the Fédération Internationale de Précontrainte*, Prague, 1970, p. 12.

124. Schiessl, P., 'Admissible crack width in reinforced concrete structures', *Contribution II 3–17, Inter-Association Colloquium on the behaviour in service of concrete structures, Preliminary Report*, Vol. 2, Liege, 1975.

125. Beeby, A. W., 'Corrosion of reinforcing steel in concrete and its relation to cracking', *The Structural Engineer*, Vol. 56A, March 1978, pp. 77–81.

126. Clark, L. A., 'Tests on slab elements and skew slab bridges designed in accordance with the factored elastic moment field', *Cement and Concrete Association, Technical Report 42.474*, September 1972, p. 47.

127. Beeby, A. W., 'The design of sections for flexure and axial load according to CP 110', *Cement and Concrete Association, Development Report 44.002*, December 1978, p. 31.

128. British Standards Institution, *CP 110: Part 2: 1972. The structural use of concrete. Part 2. Design charts for singly reinforced beams, doubly reinforced beams and rectangular columns*, p. 90.

129. Pannell, F. N., 'The ultimate moment of resistance of unbonded prestressed concrete beams', *Magazine of Concrete Research*, Vol. 21, No. 66, March 1969, pp. 43–54.

130. British Standards Institution, *CP 110: Part 3: 1972. The Structural use of concrete. Part 3. Design charts for circular columns and prestressed beams*, p. 174.

131. Taylor, H. P. J. and Clarke, J. L., 'Design charts for prestressed concrete T-sections', *Cement and Concrete Association, Departmental Note DN/3011*, August 1973, p. 19.

132. Kemp, K. O., 'The yield criterion for orthotropically reinforced concrete slabs', *International Journal of Mechanical Sciences*, Vol. 7, 1965, pp. 737–746.

133. Wood, R. H., 'The reinforcement of slabs in accordance with a pre-determined field of moments', *Concrete*, Vol. 2, No. 2, February 1968, pp. 69–76.

134. Hillerborg, A., 'Reinforcement of slabs and shells designed according to the theory of elasticity', *Betong*, Vol. 38, No. 2, 1953, pp. 101–109.

135. Armer, G. S. T., 'Discussion of reference 133', *Concrete*, Vol. 2, No. 8, August 1968, pp. 319–320.

136. Uppenberg, M., *Skew concrete bridge slabs on line supports*, Institutionen for Brobyggnad Kungl, Tekniska Hogskolan, Stockholm, 1968.

137. Hallbjorn, L., *'Reinforced concrete skew bridge slabs supported on columns'*, Division of Structural Engineering and Bridge Building, Royal Institute of Technology, Stockholm, 1968.

138. Clements, S. W., Cranston, W. B. and Symmons, M. G., 'The influence of section breadth on rotation capacity', *Cement and Concrete Association, Technical Report 42.533*, September 1980, p. 27.

139. Morley, C. T., 'Skew reinforcement of concrete slabs against bending and torsional moments', *Proceedings of the Institution of Civil Engineers*, Vol. 42, January 1969, pp. 57–74.

140. Kemp, K. O., *Optimum reinforcement in a concrete slab subjected to multiple loadings*, International Association for Bridge and Structural Engineering, 1971, pp. 93–105.

141. Nielson, M. P., 'On the strength of reinforced concrete discs', *Acta Polytechnica Scandinavica*, Civil Engineering and Building Construction Series No. 70, 1969, p. 254.

142. Clark, L. A., 'The provision of tension and compression reinforcement to resist in-plane forces', *Magazine of Concrete Research*, Vol. 28, No. 94, March 1976, pp. 3–12.

143. Peter, J., 'Zur Bewehrung von Schieben und Schalen fur Hauptspannungen schiefwinklig zur Bewehrungsrichtung', Doctoral Thesis, Technical University of Stuttgart, 1964, p. 213.

144. Morley, C. T. and Gulvanessian, H., 'Optimum reinforcement of

concrete slab elements', *Proceedings of the Institution of Civil Engineers*, Part 2, Vol. 63, June 1977, pp. 441–454.

145.  Morley, C. T., 'Optimum reinforcement of concrete slab elements against combinations of moments and membrane forces', *Magazine of Concrete Research*, Vol. 22, No. 72, September 1970, pp. 155–162.

146.  Clark, L. A. and West, R., 'The behaviour of solid skew slab bridges under longitudinal prestress', *Cement and Concrete Association, Technical Report 42.501*, December 1974, p. 40.

147.  Clark, L. A. and West, R., 'The behaviour of skew solid and voided slab bridges under longitudinal prestress', *Seventh Congress of the Fédération Internationale de la Précontrainte*, New York, May 1974, p. 8.

148.  Shear Study Group. *The shear strength of reinforced concrete beams*, Institution of Structural Engineers, London, 1969, p. 170.

149.  Baker, A. L. L., Yu, C. W. and Regan, P. E., 'Explanatory note on the proposed Unified Code Clause on shear in reinforced concrete beams with special reference to the Report of the Shear Study Group', *The Structural Engineer*, Vol. 47, No. 7, July 1969, pp. 285–293.

150.  Regan, P. E., 'Safety in shear: CP 114 and CP 110', *Concrete*, Vol. 10, No. 10, October 1976, pp. 31–33.

151.  Taylor, H. P. J., 'Investigation of the dowel shear forces carried by the tensile steel in reinforced concrete beams', *Cement and Concrete Association, Technical Report 42.431*, February 1968, p. 23.

152.  Taylor, H. P. J., 'Investigation of the forces carried across cracks in reinforced concrete beams in shear by interlock of aggregate', *Cement and Concrete Association, Technical Report 42.447*, November 1970, p. 22.

153.  Pederson, C., 'Shear in beams with bent-up bars', *Final Report of the International Association for Bridge and Structural Engineering Colloquium on Plasticity in Reinforced Concrete*, Copenhagen, 1979, pp. 79–86.

154.  Clarke, J. L. and Taylor, H. P. J., 'Web crushing – a review of research', *Cement and Concrete Association, Technical Report 42.509*, August 1975, p. 16.

155.  Taylor, H. P. J., 'Shear strength of large beams', *Proceedings of the American Society of Civil Engineers (Structural Division)* Vol. 98, No. ST11, November 1972, pp. 2473–2490.

156.  Regan, P. E., 'Design for punching shear', *The Structural Engineer*, Vol. 52, No. 6, June 1974, pp. 197–207.

157.  Nylander, H. and Sundquist, H., 'Punching of bridge slabs', *Proceedings of a Conference on Developments in Bridge Design and Construction*, Cardiff, 1971, pp. 94–99.

158.  Moe, J., 'Shearing strength of reinforced concrete slabs and footings under concentrated loads', *Portland Cement Association, Development Department Bulletin D47*, April 1961, p. 135.

159.  Hanson, J. M., 'Influence of embedded service ducts on strengths of flat plate structures', *Portland Cement Association, Research and Development Bulletin RD 005.01D*, 1970.

160.  Hawkins, N. M., 'The shear provision of AS CA35–SAA Code for prestressed concrete', *Civil Engineering Transactions, Institute of Engineers, Australia*, Vol. CE6, September 1964, pp. 103–116.

161.  Reynolds, G. C., Clarke, J. L. and Taylor, H. P. J. 'Shear provisions for prestressed concrete in the Unified Code, CP 110: 1972', *Cement and Concrete Association, Technical Report 42.500*, October 1974, p. 16.

162.  Sozen, M. A. and Hawkins, N. M. 'Discussion of a paper by ACI–ASCE Committee 326. Shear and diagonal tension', *Proceedings of the American Concrete Institute*, Vol. 59, No. 9, September 1962, pp. 1341–1347.

163.  MacGregor, J. G. and Hanson, J. M. 'Proposed changes in shear provisions for reinforced and prestressed concrete beams', *Proceedings of the American Concrete Institute*, Vol. 66, No. 4, April 1969, pp. 276–288.

164.  MacGregor, J. G., Sozen, M. A. and Siess, C. P., 'Effect of draped reinforcement on behaviour of prestressed concrete beams', *Proceedings of the American Concrete Institute*, Vol. 57, No. 6, December 1960, pp. 649–677.

165.  Bennett, E. W. and Balasooriya, B. M. A., 'Shear strength of prestressed beams with thin webs failing in inclined compression', *Proceedings of the American Concrete Institute*, Vol. 68, No. 3, March 1971, pp. 204–212.

166.  Edwards, A. D., 'The structural behaviour of a prestressed box beam with thin webs under combined bending and shear', *Proceedings*

of the Institution of Civil Engineers, Part 2, Vol. 63, March 1977, pp. 123–135.

167.  Leonhardt, F., 'Abminderung der Tragfähigkeit des Betons infolge stabformiger, rechtwinklig zur Druckrichtung angedrahte Einlagen', Included in: Knittel, G. amd Kupfer, M. (Eds), *Stahlbetonbau: Berichte aus Forschung and Praxis. Festschrift Rusch*, Wilhelm Ernst & Sohn, Berlin, 1969, pp. 71–78.

168.  American Concrete Institute, *Building code requirements for reinforced concrete (ACI 318–77)*, 1977, p. 103.

169.  Hawkins, N. M., Crisswell, M. E. and Roll, F., 'Shear strength of slabs without shear reinforcement', *American Concrete Institute, Special Publication SP–42*, Vol. 2, 1974, pp. 677–720.

170.  Clark, L. A. and West, R., 'The torsional stiffness of support diaphragms in beam and slab bridges', *Cement and Concrete Association, Technical Report 42.510*, August 1975, pp. 20.

171.  Nadai, A., *Theory of Flow and Fracture of Solids*, McGraw-Hill, New York, 1950.

172.  American Concrete Institute Committee 438, 'Tentative recommendations for the design of reinforced concrete members to resist torsion', *Proceedings of the American Concrete Institute*, Vol. 66, No. 1, January 1969, pp. 1–8.

173.  Cowan, H. J. and Armstrong, S., 'Experiments on the strength of reinforced and prestressed concrete beams and of concrete encased steel joists in combined bending and torsion', *Magazine of Concrete Research*, Vol. 7, No. 19, March 1955, pp. 3–20.

174.  Brown, E. L., 'Strength of reinforced concrete T-beams under combined direct shear and torsion', *Proceedings of the American Concrete Institute*, Vol. 51, May 1955, pp. 889–902.

175.  Hsu, T. T. C., 'Torsion of structural concrete – plain concrete rectangular sections', *American Concrete Institute Special Publication SP–18*, 1968, pp. 203–238.

176.  Mitchell, D. and Collins, M. P., 'Detailing for torsion', *Proceedings of the American Concrete Institute*, Vol. 7, September 1976, pp. 506–511.

177.  Swann, R. A., 'Experimental basis for a design method for rectangular reinforced concrete beams in torsion', *Cement and Concrete Association, Technical Report 42.452*, December 1970, p. 38.

178.  Swann, R. A., 'The effect of size on the torsional strength of rectangular reinforced concrete beams', *Cement and Concrete Association, Technical Report 42.453*, March 1971, p. 8.

179.  Swann, R. A. and Williams, A., 'Combined loading tests on model prestressed concrete box beams with steel mesh', *Cement and Concrete Association, Technical Report 42.485*, October 1973, p. 43.

180.  Lampert, P., 'Torsion and bending in reinforced and prestressed concrete members', *Proceedings of the Institution of Civil Engineers*, Vol. 50, December 1971, pp. 487–505.

181.  Maisel, B. I. and Swann, R. A., 'The design of concrete box spine-beam bridges', *Construction Industry Research and Information Association, Report 52*, November 1974, p. 52.

182.  Beeby, A. W., 'The prediction and control of flexural cracking in reinforced concrete members', *American Concrete Institute Special Publication 30 (Cracking, deflection and ultimate load of concrete slab systems)*, 1971, p. 55.

183.  Clark, L. A. and Speirs, D. M., 'Tension stiffening in reinforced concrete beams and slabs under short-term load', *Cement and Concrete Association, Technical Report 42.521*, July 1978, p. 19.

184.  Jofriet, J. C. and McNeice, G. M., 'Finite element analysis of reinforced concrete slabs', *Proceedings of the American Society of Civil Engineers (Structural Division)*, Vol. 97, No. ST3, March 1971, pp. 785–806.

185.  Hughes, B. P., 'Early thermal movement and cracking of concrete', *Concrete*, Vol. 7, No. 5, May 1973, pp. 43, 44.

186.  Rao, P. S. and Subrahmanyan, B. V. 'Trisegmental moment-curvature relations for reinforced concrete members', *Proceedings of the American Concrete Institute*, Vol. 70, No. 6, May 1973, pp. 346–351.

187.  Clark, L. A. and Cranston, W. B. 'The influence of bar spacing on tension stiffening in reinforced concrete slabs', *Proceedings of International Conference on Concrete Slabs*, Dundee, 1979, pp. 118–128.

188.  Stevens, R. F., 'Deflections of reinforced concrete beams', *Proceedings of the Institution of Civil Engineers*, Part 2, Vol. 53, September 1972, pp. 207–224.

189. Beeby, A. W., 'Cracking and corrosion', *Concrete in the Oceans*, Technical Report No. 1, 1978, p. 77.

190. Base, G. D., Read, J. B., Beeby, A. W. and Taylor, H. P. J., 'An investigation of the crack control characteristics of various types of bar in reinforced concrete beams', *Cement and Concrete Association, Research Report 18*, December 1966; Part 1, p. 44; Part 2, p. 31.

191. Clark, L. A., 'Crack control in slab bridges', *Cement and Concrete Association, Departmental Note 3009*, October 1972, p. 24.

192. Clark, L. A., 'Flexural cracking in slab bridges', *Cement and Concrete Association, Technical Report 42.479*, May 1973, p. 12.

193. Clark, L. A., 'Strength and serviceability considerations in the arrangement of the reinforcement in concrete skew slab bridges', *Construction Industry Research and Information Association, Report 51*, October 1974, p. 15.

194. Clark, L. A. and Elliott, G., 'Crack control in concrete bridges', *The Structural Engineer*, Vol. 58A, No. 5, May 1980, pp. 157–162.

195. Holnsberg, A., 'Crack width prediction and minimum reinforcement for crack control', *Soertryk af Bygningsstatiske Meddelelser*, Vol. 44, No. 2, 1973.

196. Beeby, A. W., 'An investigation of cracking in slabs spanning one way', *Cement and Concrete Association, Technical Report 42.433*, April 1970, p. 31.

197. Walley, F. and Bate, S. C. C., *A guide to the B.S. Code of Practice for prestressed concrete*, Concrete Publications Ltd., 1961, p. 95.

198. Neville, A. M., *Creep of Concrete: Plain, reinforced and prestressed*, North-Holland Publishing Co., 1970, p. 622.

199. Cooley, E. H., 'Friction in post-tensioned prestressing systems', *Cement and Concrete Association, Research Report No. 1*, October 1953, p. 37.

200. Construction Industry Research and Information Association, *Prestressed concrete – friction losses during stressing, Report 74*, February 1978, p. 52.

201. Longbottom, K. W. and Mallett, G. P., 'Prestressing steels', *The Structural Engineer*, Vol. 51, No. 12, December 1973, pp. 455–471.

202. Creasy, L. R., 'Prestressed concrete cylindrical tanks', *Proceedings of the Institution of Civil Engineers*, Vol. 9, January 1958, pp. 87–114.

203. Allen, A. H., 'Reinforced concrete design to CP 110 – simply explained', *Cement and Concrete Association, Publication 12.062*, 1974, p. 227.

204. Parrott, L. J., 'Simplified methods of predicting the deformation of structural concrete', *Cement and Concrete Association, Development Report 3*, October 1979, p. 11.

205. Branson, D. E., *Deformation of Concrete Structures*, McGraw-Hill International Book Company, New York, 1977, p. 546.

206. Hobbs, D. W., 'Shrinkage-induced curvature of reinforced concrete members', *Cement and Concrete Association, Development Report 4*, November 1979, p. 19.

207. Somerville, G., 'The behaviour and design of reinforced concrete corbels', *Cement and Concrete Association, Technical Report 42.472*, August 1972, p. 12.

208. Clarke, J. L., 'Behaviour and design of small nibs', *Cement and Concrete Association, Technical Report 42.512*, March 1976, p. 8.

209. Williams, A., 'The bearing capacity of concrete loaded over a limited area', *Cement and Concrete Association, Technical Report 42.526*, August 1979, p. 70.

210. Base, G. D., 'Tests on four prototype reinforced concrete hinges', *Cement and Concrete Association, Research Report 17*, May 1965, p. 28.

211. Department of Transport, *'Technical Memorandum (Bridges) BE 5/75. Rules for the design and use of Freyssinet concrete hinges in highway structures'*, March 1975, p. 11.

212. Reynolds, G. C., 'The strength of half-joints in reinforced concrete beams', *Cement and Concrete Association, Technical Report 42.415*, June 1969, p. 9.

213. Fédération Internationale de la Précontrainte, 'Shear at the interface of precast and in situ concrete', *Technical Report FIP/9/4*, August 1978, p. 15.

214. Pritchard, B. P., 'The use of continuous precast beam decks for the M11 Woodford Interchange viaducts', *The Structural Engineer*, Vol. 54, No. 10, October 1976, pp. 377–382.

215. Kaar, P. H., King, L. B. and Hognestad, E., 'Precast-prestressed concrete bridges. 1. Pilot tests of continuous girders', *Journal of the Research and Development Laboratories of the Portland Cement Association*, Vol. 2, No. 2, May 1960, pp. 21–37.

216. Mattock, A. H. and Kaar, P. H., 'Precast-prestressed concrete bridges. 3. Further tests of continuous girders', *Journal of the Research and Development Laboratories of the Portland Cement Association*, Vol. 2, No. 2, May 1960, pp. 38–58.

217. Beckett, D., *An Introduction to Structural Design (1) Concrete bridges*, Surrey University Press, Henley-on-Thames, 1973, pp. 93–96.

218. Mattock, A. H. and Kaar, P. H., 'Precast-prestressed concrete bridges. 4. Shear tests of continuous girders', *Journal of the Research and Development Laboratories of the Portland Cement Association*, Vol. 3, No. 1, January 1961, pp. 19–46.

219. Sturrock, R. D., 'Tests on model bridge beams in precast to in situ concrete construction', *Cement and Concrete Association, Technical Report 42.488*, January 1974, p. 17.

220. Mattock, A. H., 'Precast-prestressed concrete bridges. 5. Creep and shrinkage studies', *Journal of the Research and Development Laboratories of the Portland Cement Association*, Vol. 3, No. 3, September 1961, pp. 30–70.

221. Building Research Establishment, *Bridge Foundations and Substructures*, HMSO, 1979.

222. Lee, D. J., 'The theory and practice of bearings and expansion joints for bridges', *Cement and Concrete Association, Publication 12.036*, 1971, p. 65.

223. Cranston, W. B., 'Analysis and design of reinforced concrete columns', *Cement and Concrete Association, Research Report 41.020*, 1972, p. 54.

224. Marshall, W. T., 'A survey of the problem of lateral instability in reinforced concrete beams', *Proceedings of the Institution of Civil Engineers*, Vol. 43, July 1969, pp. 397–406.

225. Timoshenko, S. P., *Theory of Elastic Stability*, McGraw-Hill Book Co., New York and London, 1936, p. 518.

226. Comité Europeén du Béton, *'International recommendations for the design and construction of concrete structures'*, 1970, p. 80.

227. Larsson, L. E., 'Bearing capacity of plain and reinforced concrete walls', Doctoral Thesis, Chalmers Technical University, Goteborg, Sweden, 1959, p. 248.

228. British Standards Institution, *Structural recommendations for loadbearing walls. CP 110: Part 2: 1970*, p. 40.

229. Seddon, A. E., 'The strength of concrete walls under axial and eccentric loads', *Proceedings of a Symposium on The Strength of Concrete Structures*, London, May 1956, pp. 445–473.

230. British Standards Institution, *Design and construction of reinforced and prestressed concrete structures for the storage of water and other aqueous liquids. CP 2007: Part 2: 1970*, p. 50.

231. British Standards Institution, *The structural use of concrete for retaining aqueous liquids. BS 5337: 1976*, p. 16.

232. Lindsell, P., 'Model analysis of a bridge abutment', *The Structural Engineer*, Vol. 57A, No. 6, June 1979, pp. 183–191.

233. Whittle, R. T. and Beattie, D., 'Standard pile caps', *Concrete*, Vol. 6, No. 1, January 1972, pp. 34–36 and No. 2, February 1972, pp. 29–31.

234. Clarke, J. L., 'Behaviour and design of pile caps with four piles', *Cement and Concrete Association, Technical Report 42.489*, November 1973, p. 19.

235. Yan, H. T., 'Bloom base allowable in the design of pile caps', *Civil Engineering and Public Works Review*, Vol. 49, No. 575, May 1954, pp. 493–495 and No. 576, June 1954, pp. 622–623.

236. American Concrete Institute Committee 105, 'Reinforced concrete column investigation – tentative final report of Committee 105', *Proceedings of the American Concrete Institute*, Vol. 29, No. 5, February 1933, pp. 275–282.

237. Snowdon, L. C., 'Classifying reinforcing bars for bond strength', *Building Research Station, Current Paper 36/70*, November 1970.

238. Ferguson, P. M. and Breen, J. E., 'Lapped splices for high-strength reinforcing bars', *Proceedings of the American Concrete Institute*, Vol. 62, September 1965, pp. 1063–1078.

239. Nielsen, M. P. and Braestrup, M. W. 'Plastic shear strength of reinforced concrete beams', *Bygningsstatiske Meddelelser*, Vol. 46, No. 3, 1975, pp. 61–99.

240. Ferguson, P. M. and Matloob, F. N., 'Effect of bar cut-off on

bond and shear strength of reinforced concrete beams', *Proceedings of the American Concrete Institute*, Vol. 56, No. 1, July 1959, pp. 5–24.

241. Green, J. K., 'Detailing for standard prestressed concrete bridge beams', *Cement and Concrete Association, Publication 48.018*, December 1973, p. 21.

242. Base, G. D., 'An investigation of transmission length in pretensioned concrete', *Cement and Concrete Association, Research Report 5*, August 1958, p. 29.

243. Base, G. D., 'An investigation of the use of strand in pretensioned prestressed concrete beams', *Cement and Concrete Association, Research Report 11*, January 1961, p. 12.

244. Zielinski, J. and Rowe, R. E., 'An investigation of the stress distribution in the anchorage zones of post-tensioned concrete members', *Cement and Concrete Association, Research Report 9*, September 1960, p. 32.

245. Zielinski, J. and Rowe, R. E., 'The stress distribution associated with groups of anchorages in post-tensioned concrete members', *Cement and Concrete Association, Research Report 13*, October 1962, p. 39.

246. Clarke, J. L., 'A guide to the design of anchor blocks for post-tensioned prestressed concrete members', *Construction Industry Research and Information Association, Guide 1*, June 1976, p. 34.

247. French Government, 'Conception et calcul du beton precontraint', *Publication 73–64 BIS*.

248. Lenschow, R. J. and Sozen, M. A., 'Practical analysis of the anchorage zone problem in prestressed beams', *Proceedings of the American Concrete Institute*, Vol. 62, No. 11, November 1965, pp. 1421–1439.

249. Howells, H. and Raithby, K. D., 'Static and repeated loading tests on lightweight prestressed concrete bridge beams', *Transport and Road Research Laboratory, Report 804*, 1977, p. 9.

250. Kerensky, O. A., Robinson, J. and Smith B. L., 'The design and construction of Friarton Bridge', *The Structural Engineer*, Vol. 58A, No. 12, December 1980, pp. 395–404.

251. Department of Transport, 'Lightweight aggregate concrete for use in highway structures', *Technical Memorandum (Bridges) No. BE 11*, April 1969, p. 4.

252. Grimer, F. J., 'The durability of steel embedded in lightweight concrete', *Building Research Station, Current Paper (Engineering Series) 49*, 1967, p. 17.

253. Swamy, R. N., Sittampalm, K., Theodorakopoulos, D., Ajibade, A. O. and Winata, R., 'Use of lightweight aggregate concrete for structural applications', in *Advances in Concrete Slab Technology*, Dhir, R. K. and Munday, J. G. L. (Eds), Pergamon Press, 1979, pp. 40–48.

254. Teychenne, D. C., 'Structural concrete made with lightweight aggregate', *Concrete*, Vol. 1, No. 4, April 1967, pp. 111–122.

255. Hanson, J. A., 'Tensile strength and diagonal tension resistance of structural lightweight concrete', *Proceedings of the American Concrete Institute*, Vol. 58, No. 1, July 1961, pp. 1–40.

256. Ivey, D. L. and Buth, E., 'Shear capacity of lightweight concrete beams', *Proceedings of the American Concrete Institute*, Vol. 64, No. 10, October 1967, pp. 634–643.

257. Shideler, J. J., 'Lightweight-aggregate concrete for structural use', *Proceedings of the American Concrete Institute*, Vol. 54, No. 4, October 1957, pp. 299–328.

258. Short, A. and Kinniburgh, W., *Lightweight Concrete*, Applied Science Publishers (third ed), 1978, p. 464.

259. Swamy, R. N., 'Prestressed lightweight concrete', in *Developments in Prestressed Concrete – 1*, Sawko, F. (Ed.), Applied Science Publishers, 1978, pp. 149–191.

260. Evans, R. H. and Paterson, W. S., 'Long-term deformation characteristics of Lytag lightweight-aggregate concrete', *The Structural Engineer*, Vol. 45, No. 1, January 1967, pp. 13–21.

261. Wills, J., 'A simplified method to calculate dynamic behaviour of footbridges', *Transport and Road Research Laboratory, Supplementary Report 432*, 1978, p. 10.

262. Gorman, D. J., *Free Vibration of Beams and Shafts*, John Wiley and Sons, 1975, p. 386.

263. Wills, J., 'Correlation of calculated and measured dynamic behaviour of bridges', *Proceedings of a Symposium on dynamic behaviour of bridges, Transport and Road Research Laboratory, Supplementary Report 275*, 1977, pp. 70–89.

264. Tilly, G. P., 'Damping of highway bridges: a review', *Proceedings of a Symposium on dynamic behaviour of bridges, Transport and Road Research Laboratory, Supplementary Report 275*, 1977, pp. 1–9.

265. American Concrete Institute Committee 215, 'Considerations for design of concrete structures subjected to fatigue loading', *Proceedings of the American Concrete Institute*, Vol. 71, No. 3, March 1974, pp. 97–121.

266. Bennett, E. W., 'Fatigue in concrete', *Concrete*, Vol. 8, No. 5, May 1974, pp. 43–45.

267. Price, K. M. and Edwards, A. D., 'Fatigue strength of prestressed concrete flexural members', *Proceedings of the Institution of Civil Engineers*, Vol. 47, October 1970, pp. 205–226.

268. Edwards, A. D. and Picard, A., 'Fatigue characteristics of prestressing strand', *Proceedings of the Institution of Civil Engineers*, Vol. 53, Part 2, September 1972, pp. 323–336.

269. Moss, D. S., 'Axial fatigue of high-yield reinforcing bars in air', *Transport and Road Research Laboratory, Supplementary Report 622*, 1980, p. 29.

270. Burton, K. T. and Hognestad, E., 'Fatigue tests of reinforcing bars – tack welding of stirrups', *Proceedings of the American Concrete Institute*, Vol. 64, No. 5, May 1967, pp. 244–252.

271. Miner, M. A., 'Cumulative damage in fatigue', *Transactions of the American Society of Mechanical Engineers*, Vol. 67, Series E, September 1945, pp. A159–A164.

272. Zederbaum, J., 'Factors influencing the longitudinal movement of a concrete bridge system with special reference to deck contraction', *American Concrete Institute Special Publication 23 (First International Symposium on Concrete Bridge Design)*, 1967, pp. 75–95.

273. Church, J. G. 'The effects of diurnal thermal loading on cracked, concrete bridges', *MSc Thesis, University of Birmingham*, 1981.

274. Hambly, E. C., 'Temperature distributions and stresses in concrete bridges', *The Structural Engineer*, Vol. 56A, No. 5, May 1978, pp. 143–148.

275. Clark, L. A., 'Discussion of reference 274', *The Structural Engineer*, Vol. 56A, No. 9, September 1978, p. 244.

276. Hambly, E. C., 'Reply to reference 275', *The Structural Engineer*, Vol. 57A, No. 1, January 1979, pp. 28–29.

277. Aster, H., 'The analysis of rectangular hollow reinforced concrete slabs supported on four sides', *PhD Thesis, Technological University of Stuttgart*, April 1968, p. 111.

278. Elliott, G. E., 'Designing for transverse shear in voided slabs', *Cement and Concrete Association*, unpublished, p. 4.

279. Badoux, J. C. and Hulsbos, C. L., 'Horizontal shear connection in composite concrete beams under repeated loads', *Proceedings of the American Concrete Institute*, Vol. 64, No. 12, December 1967, pp. 811–819.

280. The Concrete Society, 'A review of the international use of lightweight concrete in highway bridges', *Concrete Society Technical Report No. 20*, August 1981, p. 15.

# Index